An

Identification Guide

to the

Larval Marine Invertebrates

of the

Pacific Northwest

An
Identification Guide
to the
Larval Marine Invertebrates
of the
Pacific Northwest

Edited by Alan L. Shanks

OREGON INSTITUTE OF MARINE BIOLOGY

Oregon State University Press
Corvallis

Dedication

This volume is dedicated to the memory of my mother, Beverly Rowe Shanks. When I was a little boy in kindergarten, I began a shell collection. My mother bought me my first identification guide, *The Little Golden Book of Seashells*. A few years later, as my collection grew in size, she bought me a copy of *Light's Manual* and then a copy of Ricketts and Calvin, *Between Pacific Tides*. By this time, the fifth grade, my future career was determined. Over the years I have collected quite a number of identification guides and, in memory of my mother, this is my contribution to this literature.

The paper in this book meets the guidelines for permanence and durability of the Committee on Production Guidelines for Book Longevity of the Council on Library Resources and the minimum requirements of the American National Standard for Permanence of Paper for Printed Library Materials Z39.48-1984.

Library of Congress Cataloging-in-Publication Data
Library of Congress Cataloging-in-Publication Data
An identification guide to the larval marine invertebrates of the pacific northwest / edited by Alan L. Shanks.— 1st ed.
    p. cm.
Includes bibliographical references (p.   ).
 ISBN 0-87071-531-3 (alk. paper)
 1. Marine invertebrates—Larvae—Northwest Coast of North America—Identification. I. Shanks, Alan L. II. Title.
 QL365.4.N66 I33 2002
 592.177'432—dc21

                                2001002936

Oregon State University Press
101 Waldo Hall
Corvallis OR 97331-6407
541-737-3166 • fax 541-737-3170
http://osu.orst.edu/dept/press

**OREGON STATE
UNIVERSITY**

Contents

Introduction 1

1 Porifera: The Sponges 5
   by Steven Sadro

2 Cnidaria (Coelenterata) 13
   by Steven Sadro

3. Platyhelminthes: The Flatworms
   with an Emphasis on Marine Turbellaria 24
   by Alan L. Shanks

4 Nemertea: The Ribbon Worms 28
   by Kevin B. Johnson

5 Entoprocta 39
   by Alan L. Shanks

6 Polychaeta 41
   by Lana Crumrine

7 Sipuncula: The Peanut Worms 80
   by Kevin B. Johnson

8 Echiura and Pogonophora: The Coelomate Worms 85
   by Alan L. Shanks

9 Mollusca: Gastropoda 88
   by Jeffrey H. R. Goddard

10 Mollusca: Bivalvia 131
   by Laura A. Brink

11 Mollusca: The Smaller Groups
   Polyplacophora, Scaphopoda, and Cephalopoda 152
   by Alan L. Shanks

12 Arthropoda, Cirripedia: The Barnacles 157
   by Andrew J. Arnsberg

13 Arthropoda: Isopoda 178
   by Steven Sadro

14 Arthropoda: Decapoda 181
   by Amy L. Puls

15 Phoronida 253
   by Kevin B. Johnson

16 Bryozoa **260**
by Katherine Rafferty

17 Brachiopoda **269**
by Alan L. Shanks

18 Echinodermata **272**
by Bruce A. Miller

19 Hemichordata, Class Enteropneusta:
The Acorn Worms **293**
by Alan L. Shanks

20 Urochordata: Ascidiacea **296**
by Steven Sadro

Index **304**

Introduction

Many marine invertebrates produce larvae that spend time in the plankton. Many species are broadcast spawners; eggs and sperm are shed into the water, where fertilization occurs, and the zygote then goes through its development in the water column. Other species have internal fertilization and brood the larvae to a more advanced stage of development before they are released into the plankton. Finally, some species brood their young to such an advanced stage of development that they are released as crawl-away juveniles. There are many reasons one might study larval invertebrates: they have fascinating and complex patterns of development, they are uniquely adapted to a temporary life as zooplankters, they play a major role in establishing adult population size, and many larval types are quite beautiful.

The geographic range covered by this manual is roughly from British Columbia to northern California, though many of our species are also found along the coast of California and even Baja California. For example, of the 46 species of crabs listed in the *Intertidal Invertebrates of California* (Morris et al., 1980), 54% have ranges that extend into southern Oregon. Of the 33 crabs listed in *Light's Manual: Intertidal Invertebrates of the Central California Coast* (Smith and Carlton, 1975), 85% are found in the Pacific Northwest. Thus, researchers and students along the entire West Coast should find this guide to larval invertebrates useful.

Populations of benthic marine invertebrates are often characterized by large fluctuations in the size of adult populations (McEdward, 1995). Most of these benthic invertebrates produce larvae that spend, depending on the species, from minutes to months developing in the plankton. Current research suggests that the fluctuations in adult population size maybe caused by variations in the relative recruitment of their larvae; they have to survive their pelagic phase. Larval mortality can be from predation, starvation, or transport of the larvae away from suitable habitats in which to settle (Rumrill, 1990; Thorson 1946).

The importance of the larval period in determining adult population size has been recognized for decades (Thorson, 1946) and has inspired a great deal of research. An important limiting factor in this area of research is the problem of larval taxonomy; it is difficult to study any aspect of larval biology if one cannot identify the larvae. For example, the larvae of estuarine-dependent invertebrates (e.g., ghost shrimp)

I

can be exported from an estuary; their development would then take place over the continental shelf, after which they must reinvade the estuary. Alternately, the entire larval development can occur within the estuary. It is impossible to address even such a simple topic as the location of larval development if one cannot identify the larvae.

Because of the economic importance of fishes, their larvae have received a great deal of research, and excellent volumes describing them are available for both the east and west coasts of North America (e.g., Fahay 1983; Matarese et al., 1989). No similar volumes are available for any invertebrate groups from the West Coast. Papers, graduate dissertations, and reports that describe the larvae of West Coast invertebrates have been published, but because these descriptions are scattered throughout a large literature (more than 100 years' publications in at least several dozen journals) their utility is diminished. Further, the larvae of many common coastal invertebrates have not been described. The purpose of this volume is to begin to fill this void in our knowledge.

This identification guide is aimed at individuals who are attempting to identify larval invertebrates from zooplankton samples. We assume that the user has a basic knowledge of invertebrate zoology. To assist even the novice user of this volume, however, each chapter includes a brief description of the typical life cycle(s) of the organisms in the group; the key characteristics of their larvae; a list of the shallow-water invertebrate species found in the Pacific Northwest and whether there are descriptions of their larvae; a key to the larvae for which we have published descriptions; and a reference section. In the keys, we attempt to use characteristics that are readily apparent in preserved samples, under a dissecting (or compound) microscope and without the need for dissection.

There are several limitations to this manual. The first, and most obvious, is that this guide has been constructed from published descriptions of larvae. For a few taxonomic groups the larvae of most of the local species have been described (e.g., barnacles and phoronids), but for the vast majority of groups only a portion, often a small portion, of the larvae have been described. Depending on the taxonomic group, this limitation means that one's identification of a larva may range from being pretty accurate to quite problematic. Second, this guide was to a large extant written by people who have not been trained as taxonomists. Much of it was written by undergraduate and graduate students. Because of our inexperience, errors will have crept into the guide. Third, this guide is untried. We expect that as it is used the users will find errors; this is inevitable. We would love to get your feedback so that we can improve and update the volume.

We hope that a byproduct of this guide will be new descriptions of previously undescribed larvae. There are two methods to identify unknown larvae. Living larvae can be removed from a plankton sample and reared until they have metamorphosed into their adult morphology, at which point one of the numerous guides to coastal invertebrates can be used to identify the species. To a large extent this is what Thorson did for his classic volume on Danish marine invertebrate larvae (Thorson, 1946). The alternate is to spawn the adults and rear the larvae. An excellent guide to this procedure for a large number of species can be found in Strathmann (1987). We will be happy to receive new larval descriptions and incorporate them into future guides.

During the preparation of this book we found many references in which it was obvious that researchers had raised a particular larval type through its development, but, because the purpose of the paper was something other than the publication of a larval description, none was provided. This was particularly frustrating during the writing of the chapter on bivalves. To study their potential for mariculture, many shallow-water West Coast bivalves have been raised in the lab. But few of the associated publications provide any description of the larvae. We strongly encourage researchers to include descriptions and pictures of larvae in their papers. Ideally, this description should extend across the ontogenetic stages of the larvae, presenting photographs or line drawing of the different larval stages and key characteristics that would help one to identify the larvae when found in the plankton. This information should add no more than a page or two to an article. Over time, these additions to the published descriptions will greatly expand our current knowledge.

References

Fahay, M. P. (1983). Guide to the early stages of marine fishes occurring in the western North Atlantic Ocean, Cape Hatteras to the southern Scotian Shelf. J. Northwest Atlantic Fish. Sci. 4:1-423.

Matarese, A., A. W. Kendell, Jr., D. M. Blood and B. M. Vinter. (1989). Laboratory guide to early life history stages of Northeast Pacific fishes. In: NOAA.

McEdward, L. R. (1995). Ecology of Marine Invertebrate Larvae. CRC Press, Inc., Boca Raton, Florida. 480 p.

Morris, R. H., D. P. Abbott and E. C. Haderlie (1980). Intertidal
    Invetebrates of California. Stanford University Press, Stanford,
    California. 690 p.
Rumrill, S. S. (1990). Natural mortality of marine invertebrate
    larvae. Ophelia 21:163-98.
Smith, R. I. and J. T. Carlton (1975). Light's Manual: Intertidal
    Invetebrates of the Central California Coast. University of
    California Press, Berkeley. 717 p.
Strathmann, M. F. (1987). Reproduction and Development of
    Marine Invertebrates of the Northern Pacific Coast. University
    of Washington Press, Seattle. 670 p.
Thorson, G. (1946). Reproduction and larval development of
    Danish marine bottom invertebrates, with special reference to
    the planktonic larvae in the sound (Oresund). Medd. Dan. Fisk.
    Havunders., Ser Plank. 4:1-523.

1

Porifera:
The Sponges

Steven Sadro

Found at all depths and latitudes, sponges are among the most ubiquitous of marine organisms. With the advent of modern laboratory and field techniques (e.g., electron microscopy, histochemistry, the development of molecular biology and biochemistry, and the use of SCUBA for collection purposes) our knowledge of sponge biology has mushroomed (see reviews by Fell, 1974; Bergquist et al., 1979). The phylum Porifera is divided into four classes, three of which occur off the coast of the Pacific Northwest. Of the three classes present locally, this chapter deals only with the classes Calcarea (Table 1) and Demospongiae (Table 2). It does not present the Hexactinellida, which are found only in deep water. Of the two classes of sponges presented here, the class Demospongiae is the larger; indeed, containing 80% of the known sponges, it is the largest of all the classes (Fell, 1974).

Reproduction and Development
Sponges differ from other invertebrates in the maintenance of an almost protozoan like independence of their constituent cells (Bergquist et al., 1979). For this reason they are considered the most primitive of the multicellular animals. They lack organs but have well-developed connective tissue in which differentiated cells perform a variety of functions.

Sponges reproduce both asexually and sexually. Asexual reproduction takes a variety of forms (e.g. fragmentation, budding, formation of direct developing gemmules, formation of pseudolarvae) (Wilson, 1902; Fell, 1974) and serves both as a dispersal mechanism and a method of survival during periods of extremely unfavorable conditions. Many features of sexual reproduction in sponges have been described in detail (Fry, 1970; Brien, 1973; Fell, 1974; Bergquist et al., 1979). Consistent with their lack of differentiated organs, sponges do not posses true gonads. Rather, a major portion of the sponge body is involved in reproduction. Generally sponges are hermaphrodites. Sexual dimorphism does not exist in sponges. Most sponges are viviparous and, consequently, the eggs are retained and fertilized internally. Some sponges are, however, oviparous; either fertilization takes place internally with the zygote eventually being released into the sea or oocytes are released and fertilization occurs externally. Within the sponge connective tissue, eggs develop from ameobocytes and sperm develops from either ameobocytes or transformed collar cells. Spermatozoa are shed into the excurrent canals and released into the sea.

Sponge larvae are relatively uniform in their morphology. They are always ciliated, but there can be regions of longer cilia or areas that lack cilia completely. There are two general types of sponge larvae, solid parenchymella larvae and hollow amphiblastula larvae. Sizes range from 50 µm to 5 mm in length. Differential pigmentation of the posterior or anterior pole is not unusual and commonly coincides with areas lacking cilia.

The solid parenchymella larvae (also known as stereo-gastrula larvae) (Fig. 1A) lack an internal cavity and bear flagellated cells over their entire surface, except (often) at the posterior pole. These larvae are similar in appearance to the cnidarian planulae (see Chapter 2). Many parenchymella larvae contain spicules that are frequently arranged in a bundle near the posterior pole (calcareous spicules would be apparent under crossed polarized light; see Chapter 10, Bivalves, for technique). Larvae may lack spicules at release, but these may develop later during the free-swimming phase. The amphi-blastula larvae (Fig. 1B) are hollow blastulae with one hemisphere composed of small flagellated cells and the other composed of large, nonflagellated macromeres (i.e. cruciform, macrogranular, and agranular cells) (Minchin, 1896; Lévi, 1963; Fell, 1974; Franzen, 1988). In terms of cell differentiation, parenchymella larvae can be very simple or quite complex, whereas amphiblastula larvae are typically simple. Neither has any organization, however, beyond the cellular level of differentiation. Both amphiblastula and parenchymella larvae exhibit pronounced phototaxis and geotaxis and often reverse their response to light and gravity as metamorphosis approaches (Bergquist et al., 1979).

Fig. 1. Transverse sections through generalized sponge larvae. (A) Solid parenchymella larva.(B) Hollow amphiblastula larva. (A from Ruppert and Barnes, 1994; B from Minchin, 1896, Fig 3.)

A
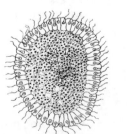

B
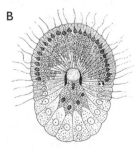

Most sponge larvae spend only a brief time in the plankton, usually less than three days, before exhibiting settlement behavior. As settlement approaches, the larvae enter a short creeping stage (2–3 hours) that may be interrupted by additional periods of swimming before settlement and metamorphosis finally occur. The larvae of some Demospongiae species have no swimming period at all; they sink to the substrate after expulsion, where they creep until settlement. Some species do not release any larvae; instead, propagules are incubated within the endosome of the parent (Bergquist et al., 1979).

Identification of Local Taxa

Lacking morphological information on sponge larvae, this chapter does not attempt to serve as a taxonomic key. Rather, it compiles useful diagnostic characteristics based on information gathered from a limited number of sponge species within a limited number of orders. It is important to note that inferences are made regarding similar orders where no information is available. Identification to species is not possible from the information provided here, although identification to class and, in some cases, to order is sometimes possible. As is the case with many poorly studied larval groups, the only way to make an identification to species is to collect ripe adults and raise the larvae to the adult stage.

Relatively few characteristics are available to differentiate sponge larvae. Amphiblastula larvae are easily distinguished from parenchymella larvae based on their size (amphiblastula larvae are typically much smaller than parenchymella larvae) and their general morphology. Differentiating among amphiblastula larvae is difficult. The parenchymella larvae are morphologically more diverse and are consequently easier to distinguish. Important morphological characters include body length, distribution of cilia, cilia length, presence or absence of discrete bands of cilia (usually near the anterior or posterior pole), absence of cilia at the posterior or anterior pole, and pigmentation. Additional useful characteristics include larval phototactic and geotactic behavior and direction of rotation during swimming.

Class Calcarea

All species of the class Calcarea produce relatively small (usually <100 μm long) amphiblastula larvae. Most are simple flagellated blastulae with some nonflagellated cells at the posterior pole, but a few examples of more complex forms exist (e.g., *Granita compressa*). It is not possible to differentiate further

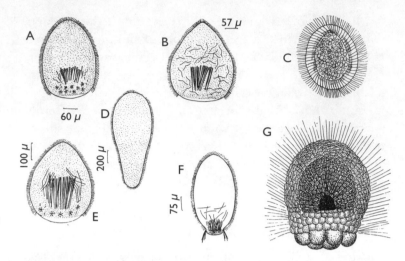

Fig. 2. Larvae from the classes Demospongiae and Calcarea. 1, 2, and 4–6 illustrate Demospongiae sponge larvae with internal spicules present. (A) **Microciona** *coccinea*. (B) **Ophlitaspongia** *seriata*. (C) Parenchymella larva of **Clathrina**, a genera in the class Calcarea. (D) **Halichondria** *moorei*. (E) **Mycale** *macilenta*. (F) **Haliclona** sp. (G) Amphiblastula larva of **Leucosolenia** *variabilis*, a species in the class Calcarea. Genera in bold have local representatives. (A, B, D–F from Bergquist and Sinclair ,1968; C from Barnes, 1968, Fig. 4-12E; G from Minchin, 1896, Fig. 1)

among the amphiblastula larvae without specialized histological and biochemical techniques. Illustrations of typical calcareous sponge larvae are presented in Fig. 2. Table 1 lists local taxa.

Class Demospongiae

The majority of the species in the class Demospongiae produce parenchymella larvae, although some species in Clionidae and Plakinidae produce amphiblastula (Fell, 1974; Bergquist et al., 1979). Typical Demospongiae larvae are depicted in Figs. 2 and 3. Table 2 lists local taxa.

Information compiled by Bergquist et al. (1979) and others on the morphological and behavioral characteristics of the Demospongiae has been used to generate the following generalized larval descriptions, by order. Genera on which the descriptions are based are given in parentheses. Asterisks (*) indicate taxa represented locally.

Order Homoscleromorpha. *(Oscarella, Plakina*)* Larvae from two families have been described. Both possess small, uniformly ciliated and pigmented amphiblastula larvae.

Order Hadromerida. *(Polymastia*, Cliona*, Tethya*)* Larvae from three families have been described. All *Polymastia* and some *Cliona* species produce small amphiblastula larvae; all *Tethya* and most *Cliona* species produce small parenchymella larvae.

Order Dendoroceratida. *(Aplysilla*, Halisarca*)* Larvae from two families have been described, within which there is considerable variability in morphology and larval behavior. Both produce parenchymella larvae. *Aplysilla* species are uniformly ciliated, but the anterior pole can be bare of cilia and the posterior pole typically has a ring of longer cilia.

Table 1. Species in the class Calcarea from the Pacific Northwest (from Kozloff, 1996)

Subclass Calcinea	**Order Sycettida**	**Family Amphoriscidae**
Family Clathrinidae	**Family Sycettidae**	*Leucilla nuttingi*
Clathrina blanca	*Scypha compacta*	
Clathrina coriacea	*Scypha mundula*	
Clathrina sp.	*Scypha protecta*	
Subclass Calcaronea	*Scypha* spp.	
Order Leucosoleniida	*?Tenthrenodes* sp.	
Family Leucosoleniidae	**Family Grantiidae**	
Leucosolenia eleanor	*Grantia comoxensis*	
Leucosolenia nautilia	*Grantia ?compressa*	
Leucosolenia spp.	*Leucandra heathi*	
	Leucandra ?levis	
	Leucandra pyriformis	
	Leucandra taylori	
	Leucopsila stylifera	
	Sycandra ?utriculus	

Halisarca species are colorless, whereas *Aplysilla* species are pigmented. Both rotate clockwise when swimming.

Order Poecilosclerida. *(Paracornulum, Coelosphaera, Ophlita-spongia*, Microciona*, Holoplocamia, Phorbas, Lissodendoryx*, Mycale*, Anchinoe, Tedania*)* Larvae from seven families have been described, all of which produce parenchymella larvae. Larvae are all of small to medium size (300–800 μm length) and ovoid in shape. They are uniformly ciliated, but in all cases there is a bare unpigmented region at the posterior pole. Only the ciliated areas are pigmented. *Lissodendoryx* species have a ring of cilia around the posterior pole. Spicules are present in larvae from all genera. All larvae have counterclockwise rotation while swimming. Only *Lissodendoryx* and *Mycale* species respond to light.

Order Haplosclerida. *(Adocia*, Haliclona*. Chalinula, Cally-spongia, Reniera*, Sigmadocia*)* Larvae from three families have been described. Their larvae present a complicated assortment of morphologies and behaviors. All genera produce paren-chymella larvae, but larval size ranges over an order of magnitude (length, 100–1,000 μm). In all cases, ciliated cells are unpigmented while unciliated cells are pigmented and the pigmentation is more pronounced posteriorly. There are three distinct larval types. *Reniera* species have uniform ciliation over the entire larval body. Larvae from species in the genera *Haliclona, Sigmadocia,* and *Adocia* possess an unciliated, pigmented posterior cap fringed by longer cilia. Larvae of *Adocia* species have a bare unpigmented anterior pole. All larvae rotate clockwise while swimming. Larvae of *Haliclona*

Fig. 3. Larvae from the class Demospongiae. (A) **Halichondria** *melanadocia*. (B) Large, uniformly ciliated crawling larvae of **Halichondria** sp. (C) Completely ciliated oval larvae typical of Haplosclerida with spicules. (D) Poecilosclerid larvae (*Phorbas* sp.) with uniform cilia, oval to spherical shape, and bare posterior region. (E) Larva of *Spongia* sp. (Dictyoceratida), illustrating posterior ciliary tuft, posterior bare ring, and left wound metacronal spiral. Genus in bold has local representatives. (A from Woollacott, 1990, Fig. 2; B–E from Bergquist et al., 1979)

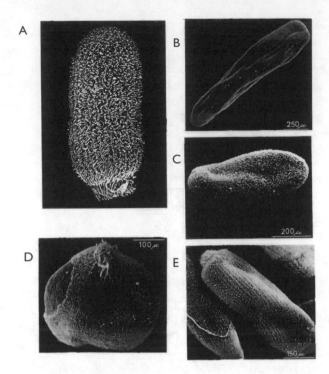

and *Adocia* species are photopositive, larvae of *Sigmadocia* species are photonegative, and larvae of *Reniera* species show no light response.

Order Halichondriida. *(Halichondria*, Ulosa, Hymeniacidon*)* Larvae from two families have been described, and all have parenchymella larvae. Larval size ranges from 200 to 1,500 μm. There are two distinct morphologies within the order. Some *Halichondria* species are non-swimming, crawling forms. These large, elongate, posteriorly tapering larvae are usually uniformly ciliated and generally rotate counterclockwise during their crawling motion. *Halichondria* and *Hymeniacidon* species produce small, oval swimming larvae. Ciliated areas are pigments. These swimming larvae either have uniform ciliation with or without a bare posterior region or (some *Halichondria* species) complete ciliary coverage with a ring of longer cilia at the posterior pole. Swimming larvae rotate clockwise. Positive phototaxis is seen in species of *Halichondria*.

Table 2. Species in the class Demospongiae from the Pacific Northwest (from Kozloff, 1996)

Order Homoscleromorphida
Family Plakinidae
Plakina ?brachylopha
Plakina ?trilopha
Plakina sp.

Order Choristida
Family Stellettidae
Penares cortius
Stelletta clarella

Family Geodiidae
Geodia mesotriaena
Geodinella robusta

Family Pachastrellidae
Poecillastra rickettsi

Order Spirophorida
Family Tetillidae
Craniella spinosa
Craniella villosa

Order Hadromerida
Family Suberitidae
Laxosuberites sp.
Prosuberites sp.
Pseudosuberites spp.
Suberites montiniger
Suberites simplex
Suberites ?suberea
Suberites sp.

Family Polymastiidae
Polymastia pacifica
Polymastia pachymastia
Weberella ?verrucosa

Family Clionidae
Cliona ?argus
Cliona ?celata
Cliona lobatta
Cliona ?warreni
Cliona sp.

Family Tethidae
Tethya californiana

Family Latrunculiidae
Latrunculia sp.

Order Axinellida
Family Axinellidae
Axinella sp.
Phakettia ?beringensis
Pseudaxinella ?rosacea
Stylissa stipitata
Syringella amphispicula

Family Desmoxyidae
?Higginsia sp.

Family Raspailiidae
Hemectyon hyle

Order Halichondriida
Family Halichondriidae
Ciocalyptus penicillus
Eumastia sitiens
Halichondria bowerbanki
Halichondria panicea
Halochondria spp.
Topsentia disparilis
Hymeniacidon ?perleve
Hymeniacidon sinapium
Hymeniacidon ungodon
?Hymeniacidon sp.
Prianos problematicus
Stylinos sp.

Order Poecilosclerida
Family Mycalidae
Mycale adhaerens
Mycale bamfieldense
Mycale bellabellensis
Mycale hispida
Mycale macginitiei
Mycale richardsoni
Mycale ?toporoki
Mycalecarmia lobata
Paresperella psila

Family Hamacanthidae
Zygherpe hyaloderma

Family Cladorhizidae
Asbestopluma occidentalis

Family Biemnidae
Biemna rhadia

Family Esperiopsidae
Neoesperiopsis digitata
Neoesperiopsis infundibula
Neoesperiopsis rigida
Neoesperiopsis vancouvrensis

Family Myxillidae
Acarnus erithacus
Ectyomyxilla parasitica
Forcepia ?japonica
Hymendecyon lyoni
Iophon chelifer
Iophon piceus
Jones amaknakensis
Lissodendoryx firma
Lissodendoryx sp.
Merriamum oxeota
Myxilla behringensis
Myxilla lacunosa
Stelodoryx alaskensis

Family Tedaniidae
Tedania fragilis
Tedania gurjanovae
?Tedanione obscurata

Family Hymedesmiidae
Anaata brepha
Anaata spongigartina
Arndtanchora sp.
Hymedesanisochela rayae
Hymenamphiastra cyanocrypta
Hymedesmio spp.
?Hymenanchora sp.
?Stylopus arndti

Family Anchinoidae
Podotuberculum hoffmanni
Hamigera ?lundbecki

Family Clathriidae
Axocielita originalis
?Dictyociona asodes
Microciona microjoanna
Microciona prolifera
Microciona primitiva
Ophlitaspongia pennata
Thalysias laevigata

Family Plocamiidae
Anthoarcuata graceae
Plocamia karykina
Plocamilla illgi
Plocamilla lambei
Stelotrochota hartmani

Order Haplosclerida
Family Haliclonidae
?Adocia spp.
Adocia gellindra
Haliclona ?ecbasis
Haliclona ?permollis
Orina sp.
Pachychalina spp.
Reniera mollis
Sigmadocia edaphus
Sigmadocia spp.
Toxidocia spp.

Order Petrosiida
Family Petrosiidae
Xestospongia trindanea
Xestospongia vanilla

Family Dysideidae
Dysidea fragilis
Spongionella sp.

table continues

Table 2. Species in the class Demospongiae from the Pacific Northwest (continued)

Order Dendroceratida
Family Aplysillidae
Aplysilla ?glacialis
Chelonaplysilla polygraphis
Pleraplysilla **sp**.

Family Halisarcidae
Halisarca sacra
Order Verongiida
Family Verongiidae
Hexadella **sp**.

Undetermined Family
Psammopemma **sp**.

References

Barnes, R. D. (1968). Invertebrate Zoology, 2nd Edition. W. B. Saunders Co., Philadelphia.

Bergquist, P. R. and M. E. Sinclair (1968). The morphology and behaviour of larvae of some intertidal sponges. N. Z. J. Mar. Freshwater Res. 2:426–37.

Bergquist, P. R., M. E. Sinclair, C. R. Green, et al. (1979). Comparative morphology and behavior of larvae of Demospongiae. In: Colloques Internationaux du C.N.R.S. N° 291—Biologie des spongiaires, C. Levi and N. Boury-Esnault (eds.), pp. 103–11.

Brien, P. (1973). Les demosponges. Morphologie et reproduction. In: Traite de Zoologie 3:P.-P, Grassé (ed.), pp 133–461. Masson et Cie, Paris.

Fell, P. E. (1974). Porifera. In: Reproduction of Marine Invertebrates, A. C. Giese and J. S. Pearse (eds.). Academic Press, New York.

Franzen, W. (1988). Oogenesis and larval development of *Scypha ciliata*. Zoomorph. 349–57.

Fry, W. G. (1970). The biology of the Porifera. In: The Biology of the Porifera, W. G. Fry (ed.). Symp. Zool. Soc. Lond. Academic Press, London, New York.

Kozloff, E. N. (1996). Marine Invertebrates of the Pacific Northwest. University of Washington Press, Seattle.

Lévi, C. (1963). Gastrulation and larval phylogeny in sponges. In: The Lower Metazoa, Comparative Biology and Phylogeny, E. C. Dougherty (ed.), pp. 375–82. University of California Press, Berkeley and Los Angeles.

Minchin, E. A. (1896). Note on the larva and the post larval development *Leucosolenia variabilis* n. sp. with remarks on the development of other Asconidae. Proc. Roy. Soc. London 60:42–52.

Ruppert, E. E. and R. D. Barnes (1994). Invertebrate Zoology. Saunders College Publishing, New York.

Wilson, H. V. (1902). On the asexual origin of the sponge larvae. Amer. Natur. 36:451–59.

Woollacott, R. M. (1990). Structure and swimming behavior of the larvae of *Halichondria melanadocia* (Porifera : Demospongiae). J. Morph. 205:135–45.

2

Cnidaria (Coelenterata)

Steven Sadro

The cnidarians (coelenterates), encompassing hydroids, sea anemones, corals, and jellyfish, are a large (ca 5,500 species), highly diverse group. They are ubiquitous, occurring at all latitudes and depths. The phylum is divided into four classes, all found in the waters of the Pacific Northwest. This chapter is restricted to the two classes with a dominant polyp form, the Hydrozoa (Table 1) and Anthozoa (Table 2), and excludes the Scyphozoa, Siphonophora, and Cubozoa, which have a dominant medusoid form. Keys to the local Scyphozoa and Siphonophora can be found in Kozloff (1996), and Wrobel and Mills (1998) present a beautiful pictorial guide to these groups.

Reproduction and Development

The relatively simple cnidarian structural organization contrasts with the complexity of their life cycles (Fig. 1). The ability to form colonies or clones through asexual reproduction and the life cycle mode known as "alteration of generations" are the two fundamental aspects of the cnidarian life cycle that contribute to the group's great diversity (Campbell, 1974; Brusca and Brusca, 1990). The life cycle of many cnidarians alternates between sexual and asexual reproducing forms. Although not all cnidarians display this type of life cycle, those that do not are thought to have derived from taxa that did. The free-swimming medusoid is the sexually reproducing stage. It is generated through asexual budding of the polyp form. Most polyp and some medusae forms are capable of reproducing themselves by budding, and when budding is not followed by complete separation of the new cloned individuals colonies are formed (e.g., *Anthopleura elegantissima*). In some groups, either the medusoid or the polyp stage may be missing; for example, the medusoid stage is absent or vestigial in the anthozoans. The majority of the following information on both hydrozoan and anthozoan reproduction is from Strathmann (1987) and Campbell (1974).

The hydroids (order Hydoida) and hydrocorals (order Stylasterina), class Hydrozoa, reproduce asexually (through fission and budding) and sexually. There is such diversity in hydrozoan sexual reproduction and development that it is difficult to generalize. In most species the sexes are separate.

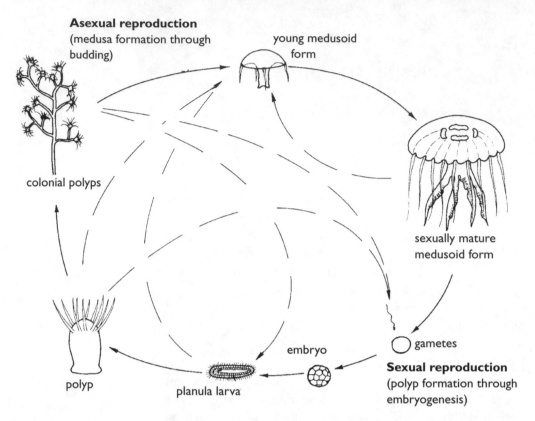

Asexual reproduction
(medusa formation through budding)

young medusoid form

colonial polyps

sexually mature medusoid form

polyp

planula larva

embryo

gametes

Sexual reproduction
(polyp formation through embryogenesis)

Fig. 1. Basic cnidarian life cycle showing alteration between sexual and asexual "generations." Alternative cycles found in some taxa are indicated by broken lines. (From Campbell, 1974)

Gonads form in specialized structures known as gonophores. In species with diphasic life cycles, gonophores develop into hydromedusae (free-swimming, bell-shaped medusae) that then develop gonads as they feed in the plankton. In some species the medusoid form is lacking and the mudusa is retained on the polyp in various reduced forms. Alternately, the reduced medusa is free in the water column for only a brief period of spawning. Either fertilization in Hydromedusae is internal or the gametes are shed freely in the water column. The propagules develop into planulae, which settle to the bottom and establish a new colony. Wrobel and Mills (1998) and Kozloff (1996) provide an excellent guide to the medusea of common local hydrozoans.

The corals, anemones, sea pens, and solitary cup corals, class Anthozoa, reproduce asexually (through fission, budding, or pedal laceration) and sexually. Some anthozoan species (not reported locally) asexually produce propagules morphologically similar to sexually produced planulae (Black and Johnson, 1979; Carter and Thorp, 1979; Strathmann, 1987). Sexual reproduction and development is more uniform in anthozoans than in hydrozoans. Most species are believed to produce separate sexes, but gonads may be diffuse and scattered throughout the mesoglea and, hence, difficult to detect. Cases of hermaphroditism in species that usually have separate sexes have been documented (Jennison, 1979). Some species (e.g., *Epiactis prolifera* and *Aulactinia incubans*) are simultaneous or sequential hermaphrodites. Although most anthozoans shed large, yolky eggs, some species brood. External brooders retain their eggs or embryos on the parental column, and internal brooders retain their eggs within the enteron in the column or in the tentacles. Internal brooding can continue to the planula (e.g., *Balanophyllia elegans* and *Cribrinopsis fernaldi*) or tentaculate stage (e.g., *Aulactinia incubans, Epiactis fernaldi,* and *E. ritteri*). Internal brooders release their young through the mouth or pores in the tentacle tips.

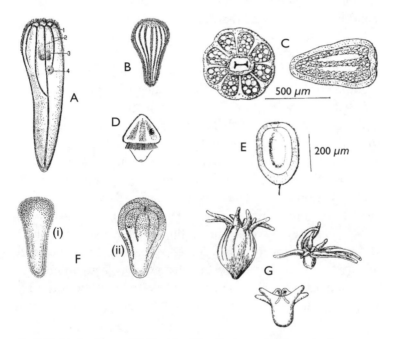

Fig. 2. Cnidarian larvae. (A) Planula of the sea pansy, *Renilla* sp. (order Pennatulacea). (B) Planula of the anemone *Lebrunia* (order Actiniaria). (C) Brooded planula of the anemone **Urticina** *filina* (*Tealia filina*, order Actiniaria). (D) Planula from the order Zoanthidea (note that only one species from this order, *Epizoanthus scotinus,* is found in Pacific Northwest waters). (E) Planktotrophic planula of the anemone **Metridium senile** (Order Actiniaria). (F) Early- (i) and late- (ii) stage planulae of the stone coral, *Siderastraea* sp. (order Scleractinia). (G) Actinula larvae from burrowing anemones in the order Ceriantharia (note that only one species from this order, *Pachycerianthus fimbriatus,* is found in Pacific Northwest waters). Genera and species with local representatives are bold. (A, B, D, F, G from Hyman, 1940; C, E from Thorson, 1946)

Identification of Local Taxa

A limited number of characteristics can be used to describe cnidarian larvae. There are two basic larval types, the planula and the actinula (Fig. 2). Further distinction among larval types is more challenging. Planulae can be characterized by size, color, ciliary pattern, and presence or absence of an apical tuft. They can be further differentiated into feeding and non-feeding forms, with the non-feeding or lecithitrophic forms typically being larger than their feeding counterparts. Little information is available on distinguishing characteristics of actinula larvae.

Lacking morphological information on cnidarian larvae, this chapter does not attempt to serve as a species identification key but instead compiles useful diagnostic information. The best way to identify a larval type is to either collect gametes from adults or raise larvae to metamorphosis and a recognizable adult stage. Strathmann (1987) describes methods for collecting gametes and rearing larvae.

Class Hydrozoa

In most hydrozoans (Table 1) the embryos develop into a non-feeding planula larvae that usually settles to the bottom within a few days. Planulae are generally club-shaped or ovoid and uniformly ciliated (Fig. 3). They possesses both nematoblasts and nematocytes. They never have an apical ciliary tuft nor do they develop septa, characteristics of some anthozoan planulae. Hydrozoan planulae are free swimming, but in some hydroids and hydrocorals (stylasterines) they may be demersal, drifting near the bottom or creeping along it until metamorphosis. Some types of planulae secret mucus threads that may alter their dispersal (Strathmann, 1987). Some species (e.g., *Hybodocon* and *Tubularia* species) have a post-settlement motile juvenile stage called an actinula. This stage creeps along the bottom until it eventually attaches and develops into a polyp.

Fig. 3. Longitudinal section of a representative hydrozoan planula, *Gonothyraea* (from Brusca and Brusca, 1990, Fig. 39)

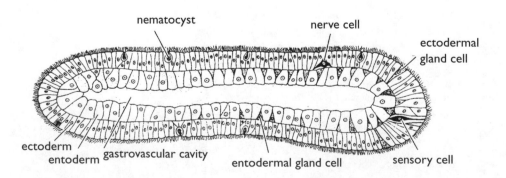

Table 1. Species in the class Hydrozoa from the Pacific Northwest (from Kozloff, 1996)

Order Hydroida
Suborder Athecata
  (Anthomedusae)
Family Corymorphidae
Corymorpha sp.

Family Euphysidae
Euphysa ruthae
Euphysa spp.

Family Tubulariidae
Hybocodon prolifer
Tubularia crocea
Tubularia harrimani
Tubularia indivisa
Tubularia marina
undescribed species

Family Corynidae
Coryne sp.
Sarsia japonica
Sarsia spp.

Family Boreohydridae
undescribed species

Family Cladonematidae
Cladonema californicum

Family Rhysiddae
Rhysia sp.

Family Clavidae
Cordylophora caspia
Hataia parva
Rhizogeton sp.

Family Hydractiniidae
Hydractinia aggregata
Hydractinia laevispina
Hydractinia milleri
Hydractinia sp.
undescribed species

Family Rathkeidae
Rathkea octopunctata

Family Bougainvilliidae
Bimeria spp.
Bougainvillia spp.
Garveia annulata
Garveia groenlandica

Family Pandeidae
Leucckartiara spp.
Neoturris spp.
Perigonimus spp.

Family Halimedusidae
Halmedusa typus

Family Calycopsidae
Bythotiara huntsmani

Family Trichydridae
Trichydra pudica

Family Eudendriidae
Eudendrium spp.

Family Protohydridae
Protohydra ?leuckarti

Suborder Thecata
  (Leptomedusae)
Family Tiarannidae
Stegopoma spp.

Family Laodiceidae
Staurophora spp.
Ptychogena spp.

Family Mitrocomidae
Foersteria spp.
Mitrocoma spp.
Mitrocomella spp.
Tiaropsidium spp.
Tiaropsis spp.

Family Haleciidae
Halecium spp.
Hydrodendron spp.

Family Campanulariidae
Campanularia spp.
Clytia spp.
Gonothyraea spp.
Obelia bidentata
Obelia dichotoma
Obelia geniculata
Orthopyxis spp.
Rhizocaulus verticillatus

Family Campanulinidae
Calycella spp.

Family Bonneviellidae
Bonneviella spp.

Family Lafoeidae
Cryptolaria spp.
Filellum spp.
Grammaria spp.
Hebella spp.
Lafoea spp.
Zygophylax spp.

Family Sertulariidae
Abietinaria spp.
Diphasia spp.
Dynamena spp.
Hydrallmania spp.
Sertularella spp.
Sertularia spp.
Symplectoscyphus spp.
Thuiaria spp.

Family Plumulariidae
Plumularia spp.

Family Aglaopheniidae
Algaophenia spp.
Cladocarpus spp.
Thecocarpus spp.

Family Eirenidae
Eutonina indicans

Family Aequoreidae
Aequorea victoria

Suborder Limnomedusae
Family Olindiasidae
Gonionemus vertens
Monobrachium parasiticum

Family Proboscidactylidae
Proboscidactyla flavicirrata

Order Stylasterina
  (Hydrocorals)
Family Stylasteridae
Allopora petrograpta
Allopora porphyra
Allopora venusta
Allopora verrilli
Errinopora pourtalesii

Suborder Physonectae
Family Agalmidae
Cordagalma cordiformis
Nanomia cara

Family Physophoridae
Physophora hydrostatica

Suborder Calycophorae
Family Diphyidae
Chelophyes appendiculata
Dimophyes arctica
Lensia baryi
Lensia conoidea
Muggiaea atlantica

Family Prayidae
Desmophyes annectens
Praya dubia
Praya reticulata

Family Sphaeronectidae
Sphaeronectes gracilis

Class Anthozoa

Most anthozoans (Table 2) produce planula larvae. The local exception is the species *Pachycerianthus fimbriatus* (family Cerianthidae), which produces an actinula larva (Fig. 2G). Planulae can be divided into three types: feeding, non-feeding with pelagic development, and non-feeding with demersal or benthic development.

Feeding planulae (Fig. 4A–D) develop from small (100–250 µm diam.) yolky eggs that are spawned into the water column (Strathmann, 1987). The planulae are ovoid and taper toward the anterior (aboral) end (Gemmill, 1920; Widersten, 1968, 1973; Chia and Koss, 1979). They swim actively, and at least the ones that have been studied are photopositive. A long apical tuft of cilia is swept side to side as the larvae swim. Later in development, mesenteries form and protrude into a spacious gastrocoel. In these more developed planulae, the mesenteries are quite obvious (Figs. 2B, 4B). Feeding methods of planulae are diverse. Some species filter particles out of the water directly (e.g., *Metridium* spp.; Fig. 2E), other species produce strands of mucus that are ingested and any adherent particles consumed (e.g., *Anthopleura xanthogrammica* and *Caryophyllia smithi*), and some species are endoparasites, feeding on particles in the gastrovascular cavities of hydromedusae (e.g., *Peachia quinquecapitata*).

Non-feeding planulae with pelagic development (Figs. 2C, 4E) develop from free-spawned, large yolky eggs (500–850 µm diam.) (Strathmann, 1987). These planulae commonly take the form of a ciliated ovoid that tapers toward the posterior (oral) end (Widersten, 1968; Stricker, 1985). The planulae lack apical tuft and apical organ. Larvae may be pelagic for a week or more. Later-stage planulae develop mesenteries.

Non-feeding planulae with demersal or benthic development (Fig. 4F–H) develop from large, yolky eggs (500–800 µm diam.) that are shed with a mucus coat. They are retained near the adult (Strathmann, 1987). These ciliated, lecithotrophic planulae lack both the apical tuft and apical organ (Nyholm, 1949; Widersten, 1968). Older larvae develop mesenteries. Larvae of this type (e.g., *Ptilosarcus gurneyi* and *Halcampa decementaculata*) are free swimming for approximately a week. They generally remain near the bottom or creep across the bottom.

Order Pennatulacea, *Ptilosarcus gurneyi*. Eggs are pink-orange in color, initially fusiform, and then rounding to 500–600 µm diameter. Eggs in the lab are at least initially bouyant (Chia and Crawford, 1973). Larvae are lecithotrophic and develop-

Table 2. Species in the class Anthozoa from the Pacific Northwest (from Kozloff, 1996)

Class Anthozoa
Subclass Alcyonaria
  (Ococorallia)
Order Alcyonacea
Suborder Stolonifera
Family Clavulariidae
Clavularia moresbii
Clavularia **spp.**
?Sarcodictyon **sp.**

Suborder Alcyoniina
Family Alcyoniddae
?Alcyonium **spp.**

Family Nephtheidae
Gersemia rubiformis

Suborder Holaxonia
Family Acanthogorgiidae
Calcigorgia spiculifera

Family Plexauridae
Swiftia kofoidi
Swiftia simplex
Swiftia spauldingi
Swiftia torreyi

Family Chrysogorgiidae
Radiceps **sp.**

Family Isididae
Acanella **sp.**

Family Primnoidae
Callogorgia kinoshitae
Parastennella **sp.**
Primnoa willeyi

Suborder Scleraxonia
Family Anthothelidae
Anthothela pacifica

Family Paragrogiidae
Paragorgia pacifica

Order Pennatulacea
Suborder Sessiliflorae
Family Kophobelemnidae
Kophobelemnon affine
Kophobelemnon biflorum
Kophobelemnon hispidum

Family Anthoptilidae
Anthoptilum grandiflorum

Family Funiculinidae
Funiculina parkeri

Family Protoptilidae
Helicoptilum rigidum

Family Scleroptilidae
Scleroptilum **sp.**

Family Umbellulidae
Umbellula lindahli

Suborder Subselliflorae
Family Virgulariidae
Balticina californica
Balticina septentrionalis
Stylatula elongata
Virgularia **spp.**

Family Pennatulidae
Pennatula phosphorea
Ptilosarcus gurneyi

Subclass Ceriantipatharia
Order Ceriantharia
Suborder Spirularina
Family Cerianthidae
Pachycerianthus fimbriatus

Order Antipatharia
Suborder Antipathina
Family Antipathidae
Antipathes **sp.**

Subclass Zoantharia
Order Scleractinia
  (Madreporaria)
Suborder Caryophylliina
Family Caryophylliidae
Caryophyllia alaskensis
Cyathoceras quaylei
Desmophyllum cristagalli
Paracyathus stearnsi
Lophelia californica
Solenosmilia variabilis

Suborder Dendrophylliina
Family Dendrophylliidae
Balanophyllia elegans

Order Actiniaria
Suborder Nynantheae
Family Edwardsiidae
Edwardsia sipunculoides
Nematostella vectensis

Family Halcampoididae
Halcampoides purpurea

Family Haloclavidae
Bicidium aequoreae
Peachia quinquecapitata

Family Halcampidae
Halcampa crypta
Halcampa decemtentaculata

Family Actiniidae
Anthopleura artemisia
Anthopleura elegantissima
Anthopleura xanthogrammica
Aulactinia incubans
Cribrinopsis fernaldi
Cribrinopsis williamsi
Urticina columbiana
Urticina coriacea
Urticia crassicornis
Urticia lofotensis
Urticia piscivora
Urticia **sp.**

Family Liponematidae
Liponema brevicornis

Family Actinostolidae
Paractinostola faeculenta
Stomphia coccinea
Stomphia didemon
Stomphia **sp.**

Family Hormathiidae
Stephanauge annularis

Family Metridiidae
Metridium senile
Metridium **sp.**

Family Haliplanellidae
Haliplanella lineata

Order Corallimorpharia
Family Corallimorphidae
Corallimorphus **sp.**
Corynactis californica

Order Zoanthidea
Family Epizoanthidae
Epizoanthus scotinus

ment is demersal. Developing larvae appear similar to those of *Renilla* (see Fig. 2A). At 18 days, planula are pear-shaped and uniformly ciliated. They swim with the narrow anterior end forward (Chia and Crawford, 1977). Therer is no apical organ. The posterior end consists of eight tentacular buds surrounding the pharynx. The cilia around the pharynx are longer then those on the remainder of the body. Planula settle on sand covered with organic film.

Order Ceriantharia, *Pachycerianthus fimbriatus.* The burrowing anemone, *P. fimbriatus,* is the only local repres-entative of this order. The development of Ceriantharia species from other regions has been described by Nyholm (1943). The larvae are planktontrophic and carnivorous. They are the only local taxa that produce a pelagic actinula larva.

Order Zoanthinaria, *Epizoanthus scotinus.* This is the only local representative of this order. Their larvae have not been described. The planula of a zoanthinarian in Fig. 2D is, compared to other Cnidaria larvae, unique in appearance. If the local zoanthinarian has a similar-shaped planula, then one should be able to differentiate it from other local planulae.

Order Scleractinia, *Balanophyllia elegans.* This cup-coral broods it large red-orange embryos internally until the planula stage. The non-feeding planula are demersal, large (3–5 mm long by 1–2 mm wide), completely covered with short (ca. 20 µm) flagella, and colored crimson red (Fadlallah and Pearse, 1982). If suspended in the water column the planulae return quickly to the bottom, where they adhere and resume crawling.

Order Scleractinia, *Caryophyllia alaskensis.* The development of this species has not been described, but that of a closely related species *(C. smithi)* has (Tranter et al., 1982). Eggs are fertilized externally. Ova are either brown or cream in color, spherical or slightly oval in shape, and 130–150 µm diameter. Planktotrophic planulae develop 48 hours after fertilization, at which time feeding commences. An apical tuft (30–60 µm long) is present during the first six weeks of development. After eight to ten weeks, lab-reared planulae are 800–1,000 µm in length.

Order Actiniaria, *Anthopleura elegantissima.* Eggs are freely spawned, brown, spherical, and ranging in size from 120 to 250 µm. The yolk in the eggs is evenly dispersed and contains no symbiotic algae. Three-week-old planulae are oval to cylindrical and uniformly ciliated (Chia and Koss, 1979). They are 170–190 µm long with an apical tuft 70 µm long. They swim

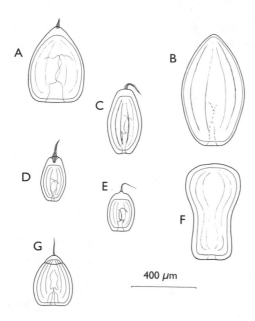

Fig. 4. Anthozoan planulae. (A–D) Planktotrophic planulae with pelagic development. (E) Lecithotrophic planula with pelagic development. (F–H) Lecithotrophic planulae with demersal development. Faint lines with the planulae represent the internal mesentaries. (modified from Widersten, 1973)

400 μm

actively with the apical tuft forward. The cilia of the apical tuft sweep the water as a unit. The planulae are feeding.

Order Actiniaria, *Anthopleura xanthogrammica*. Eggs are 175–225 μm diameter, spherical, and purple in color. The planulae are feeding and similar in appearance to those of *A. elegantissima*. The planulae are initially ca 150 μm long and 100 μm diameter. After several days of development the planulae are about 250 μm long and 150 μm in diameter. Within a week after fertilization the planulae posses numerous nematocysts. Primary septa and contractile elements begin to appear in the first and second week after fertilization (Siebert, 1974).

Order Actiniaria, *Halcampa decementaculata*. The eggs are yellowish gray-green, 310–330 μm diameter, and surrounded by a clear mucilaginous jelly coat. A planula emerges from the jelly coat after the third day. Planulae are 440 by 330 μm and are not active swimmers. They settle within 28 days.

Order Actiniaria, *Metridium* **sp.** The eggs are pink and 120–195 μm diameter. Early planulae are 180–190 μm long, excluding the apical tuft. A blastoporal indentation is present on the postereor end. The planulae are strong swimmers, positively phototactic, and planktotrophic. By five to seven days the planulae have grown to 225 μm long, with a 200 μm long apical tuft (Fig. 2E). By the eighth or ninth day the planulae are 320 μm long and 165 μm in diameter at the posterior end. The enteron is brownish and pointed at the anterior end. The enteron can be protruded as a button or collar with thin cilia.

Order Actiniaria, *Urticina ?crassicornis.* The eggs are yellow to greenish tan, 500–700 μm diameter; they float with the animal pole down. The non-feeding planulae are uniformly ciliated and lack both apical tuft and apical organ. After 18 days or fewer the planulae settle. At settlement the planulae are ca 600 μm long and the endoderm is divided by well-developed septa (see Fig. 2C3; Stricker, 1985).

References

Black, R. and M. S. Johnson (1979). Asexual viviparity and population genetics of *Actinia tenebrosa.* Mar. Biol. 53:27–31.

Brusca, R. C. and G. J. Brusca (1990). Invertebrates. Sunderland: Sinauer Associates, Inc.

Campbell, R. D. (1974). Cnidaria. In: Reproduction of Marine Invertebrates, A. C. Giese and J. S. Pearse (eds.), pp. 133–39. Academic Press, New York.

Carter, M. A. and C. H. Thorp (1979). The reproduction of *Actinia equina* L. var. *mesembryanthemum.* J. Mar. Biol. Assoc. (U.K.) 59:989–1001.

Chia, F.-S. and B. J. Crawford (1973). Some observations on gametogenesis, larval development and substratum selection of the Sea Pen *Ptilosarcus guerneyi.* Mar. Biol. 23:73–82.

——— (1977). Comparative fine structural studies of planulae and primary polyps of identical age of the Sea Pen, *Ptilosarcus gurneyi.* J. Morph. 151:131–58.

Chia, F.-S. and R. Koss (1979). Fine structural studies on the nervous system and the apical organ in the planula larva of the sea anemone *Anthopleura elegantissima.* J. Morph. 160:275–98.

Fadlallah, Y. H. and J. S. Pearse (1982). Sexual reproduction in solitary corals: overlapping oogenic and brooding cycles, and benthic planulas in *Balanophyllia elegans.* Mar. Biol. 71:223–31.

Gemmill, J. F. (1920). The development of the sea anemones, *Metridium dianthus* (Ellis) and *Adamsia palliata* (Bohad). Phil. Trans. Roy. Soc. Lond., ser. B, 290:351–75.

Hyman, L. H. (1940). The Invertebrates: Protozoa through Ctenophora. McGraw-Hill Book Co., New York.

Jennison, B. L. (1979). Gametogenesis and reproductive cycles in the sea anemone *Antheropleura elegantissima* (Brandt, 1835). Can. J. Zoo. 57:403–11.

Kozloff, E. N. (1996). Marine Invertebrates of the Pacific Northwest.Univeristy of Washington Press, Seattle. 539 p.

Nyholm, K. G. (1943). Zur entwicklung und entwiclungbiologie der Ceriantharien und aktinern. Zool. Bidr. Uppsala 22:87–248.

——— (1949). On the development and dispersal of Athernia actinia with special reference to *Halcampa duodecimcirrata* (M. Sars). Zool. Bidr. Uppsala 27:467–505.

Siebert, A. E. Jr. (1974). A description of the embryology, larval development, and feeding of the sea anemones *Anthopleura elegantissima* and *A. xanthogrammica.* Can. J. Zool. 52:1383–88.

Strathmann, M. F. (1987). Reproduction and Development of Marine Invertebrates of the Northern Pacific Coast. University of Washington Press, Seattle. 670 p.

Stricker, S. A. (1985). An ultrastructural study of larval settlement in the sea anemone *Urticina crassicornis* (Cnidaria, Actinaria). J. Morph. 186:237–53.

Thorson, G. (1946). Reproduction and larval development of Danish marine bottom invertebrates, with special reference to the planktonic larvae in the sound (Øresund). Meddelselser Fra Kommissionen for Denmarks Fiskere-Og HavundersØgelser. Serie:Plankton 4.

Tranter, P. R. G., D. N. Nicholson, and D. Kinchington (1982). A description of spawning and post-gastrula development of the cool temperate coral, *Caryophyllia smithi*. J. Mar. Biol. Ass. (U.K.) 62:845–54.

Widersten, B. (1968). On the morphology and development of some cnidarian larvae. Zool. Bidr. Uppsala 37:139–82.

———— (1973). On the morphology of actiniarian larvae. Zool. Scr. 2:119–24.

Wrobel, D. and C. Mills (1998). Pacific Coast Pelagic Invertebrates: A guide to the common gelatinous animals. Sea Challenges: Monterey Bay Aquarium, Monterey, California. 108 p.

3

Platyhelminthes: The Flatworms with an Emphasis on Marine Turbellaria

Alan L. Shanks

The phylum Platyhelminthes is made up of the free-living and parasitic flatworms. All free-living flatworms as well as some symbionts of invertebrates are in the class Turbellaria (Strathmann, 1987). Members of the other three classes (e.g., Monogtenea, Trematoda, and Cestoda) are parasitic and not reviewed here. For further information on these classes of flatworms, see references cited in Strathmann (1987).

Tubellarians are hermaphroditic. All fertilization is internal, with cross-mutual fertilization common (Brusca, 1975). Most species produce relatively few zygotes that are either brooded or encapsulated and go through direct development (Brusca and Brusca, 1990). In direct development, the gastrula flattens its oral surface against the substrate and matures. Direct development leads to crawl-away juveniles.

In the class Tubellaria, only some species in the order Polycladida (all species in the suborder Cotylea and some species in the suborder Acotylea; Table 1) produce larvae with indirect development (Hyman, 1951). In these species a free-swimming eight-lobed Müller's larva is formed (Fig. 1). The lobes are ciliated and provide propulsion. These larvae are pelagic for several days before they settle to the bottom with their oral surface downward and metamorphose into juvenile flatworms (Brusca and Brusca, 1990). Some parasitic polyclades (some species in the genus *Stylochus*) produce a free-swimming four-lobed larvae know as a Götte's larvae.

References

Brusca, G. J. (1975). General Patterns of Invertebrate Development. Mad River Press, Eureka, California.

Brusca, R. C. and G. J. Brusca (1990). Invertebrates. Sinauer Associates, Inc., Sunderland.

Hyman, L. H. (1951). The Invertebrates: *Platyhelminthes* and *Rhynchocoela* the *acoelomate Bilateria*. McGraw-Hill, London and New York.

Kozloff, E. N. (1996). Marine Invertebrates of the Pacific Northwest. University of Washington Press, Seattle.

Strathmann, M. F. (1987). Reproduction and Development of Marine Invertebrates of the Northern Pacific Coast. University of Washington Press, Seattle.

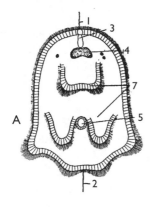

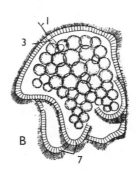

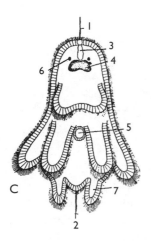

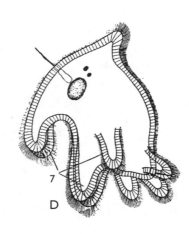

Fig. 1. Müller's larvae. (A, B) Front and side view of larvae with developing lobes. (C, D) Advanced Müller's larvae from the front and the side. 1, apical sensory tuft; 4, caudal tuft; 5, frontal gland; 6, brain; 7, mouth; 8, eyes; 9, ciliated lobes. (from Hyman, 1951, Fig. 67)

Table 1. Species in the order Polycladida from the Pacific Northwest (from Kozloff, 1996)

Suborder Acotylea
Kaburakia excelsa
Stylochus atentaculatus
Stylochus tripartitus

Family Leptoplanidae
Freemania litoricola
Leptoplana vesiculata
Notoplana atomata
Notoplana celeris
Notoplana inquieta
Notoplana inquilina
Notoplana longastyletta
Notoplana natans
Notoplana rupicola
Notoplana sanguinea
Notoplana sanjuania
Phylloplana viridis
Stylochoplana chloranota

Family Callioplanidae
Pseudostylochus burchami
Pseudostylochus ostreophagus

Family Cestoplanidae
Cestoplana sp.

Suborder Cotylea
Family Pseudocerotidae
Pseudoceros canadensis

Family Euryleptidae
Stylostomum album
Eurylepta aurantiaca
Eurylepta leoparda
Stylostomum sanjuania

4

Nemertea: The Ribbon Worms

Kevin B. Johnson

Nemerteans are an ancient group of worms derived from the flat worms. Worldwide, there are about 800 species in the phylum Nemertea. They are found at all depths in the ocean, as well as in some freshwater and damp terrestrial habits, and a few species are commensals. Locally there are 39 species (Table 1).

The nemerteans are characterized by soft, elongate, and non-segmented bodies that are covered with cilia. Their bodies are highly contractile, a characteristic for which they are famous; intertidal species that when contracted are around 8 cm long can be 45 cm long when fully extended (Haderlie, 1980). Nemerteans are predators, capturing prey with a unique eversible proboscis that can be shot out to capture a victim.

Reproduction and Development

The phylum Nemertea displays a wide variety of reproductive strategies. Many species are dioecious, though cases of hermaphroditism have been observed (Stricker, 1987). Although the ability to regenerate the body is common (e.g., Nusbaum and Oxner, 1910, 1911), asexual reproduction is not a normal part of the reproductive cycle for most nemerteans. Fertilization is often external. Eggs may either be spawned freely into the water column, protected in a gelatinous egg mass or other encasement, or, in a few instances, held internally by species that are ovoviviparous.

The development of most nemerteans (orders Palaeo-nemertea, Hoplonemertea, and Bdellonemertea) appears to be direct, with wormlike (vermiform) young developing in benthic egg cases before hatching. According to Stricker (1987), vermiform juvenile nemerteans typically remain on the benthos after hatching, but some may spend a brief period in the plankton. Cantell (1989), however, in listing known life histories of fourteen "direct" developing species, defines half as "pelagic." Also, there seems to be confusion about whether to call vermiform, adult-like young "larvae" or "juveniles" (this problem is not unique to nemerteans). Some experts use the term "larva" when referring to pelagic, but apparently direct-developing forms (e.g., Iwata, 1960a, 1968; Cantell, 1989). Others, however, consider the hoplonemerteans, palaeonemerteans, and bdellonemerteans to undergo more traditional direct development and to lack a larval stage altogether (e.g., Stricker, 1987; Brusca and Brusca, 1990). These different points of view may arise from conflicting reports about whether the epithelium of pre-settling nemerteans is shed during "metamorphosis" (Stricker and Reed, 1981). Most palaeonemerteans, hoplonemerteans, and bdellonemerteans, whether undergoing pelagic or benthic development, are thought to develop from the embryo directly into the oval, ciliated vermiform stage, which proceeds to the benthic juvenile *without* shedding the epithelium. Because the post-embryonic young of palaeonemerteans, hoplonemerteans, and bdellonemerteans are adult-like in form and undergo no dramatic metamorphosis, I refer to them as juveniles.

The fourth order of nemerteans, the Heteronemertea, often undergo indirect development with a pelagic stage known as a pilidium larva. The pilidium larva looks like a helmet or hat, with ciliated lobes for locomotion and capturing food particles. Others exhibit a Desor's larval stage and develop within a benthic egg case. *Micrura akkeshiensis* passes through the Iwata larval stage (Iwata, 1958), which is similar in certain respects to a pilidium larva but lacks the lobes and helmet shape. Additional details on the reproductive biology of nemerteans are available in Cantell (1989).

Identification of Regional Taxa

Table 1 summarizes species and developmental modes of nemerteans thought to be found in the Pacific Northwest (Coe, 1905; Austin, 1985; Stricker, 1996). The Nemertea are continuously undergoing extensive taxonomic revision. Consequently, Coe (1905) and more recent observations are most useful with knowledge of synonymies and recent taxonomic changes (see Gibson, 1995). It should be noted that developmental life history in the majority of species found in the Pacific Northwest is unknown. In these cases, information in this chapter provides the developmental modes of taxonomically related species with the caution that developmental life history cannot be confidently stated until directly observed for the species in question.

Orders Palaeonemertea, Hoplonemertea, and Bdellonemertea

Juveniles of the orders Palaeonemertea (class Anopla), Hoplonemertea (class Enopla), and Bdellonemertea (class Enopla) are generally wormlike and develop directly into the

Table 1. Species in the phylum Nemertea from the Pacific Northwest (taxonomic authority from Gibson, 1995)

Taxa	Development[a]
Class Anopla	
Order Heteronemertea	
Cerebratulus albifrons	Pilidium (genus)
Cerebratulus californiensis	**Pilidium (Coe, 1940)**
Cerebratulus herculeus	Pilidium (genus)
Cerebratulus marginatus	**Pilidium (Coe, 1905)**
Cerebratulus montgomeryi	Pilidium (genus)
Cerebratulus occidentalis	Pilidium (genus)
Lineus pictifrons	Pilidium-like or benthic direct (genus)
Lineus ruber	**Desor's (Gontcharoff, 1960)**
Lineus bilineatus	**Pilidium (Cantell, 1989)**
Lineus rubescens	Pilidium-like or benthic direct (genus)
Micrura alaskensis	**Pilidium (Stricker & Folsom, 1998)**
Micrura verrilli	**Pilidium (Coe, 1940)**
Micrura wilsoni	Pilidium (genus)
Myoisophagos sanguineus	Pilidium-like or benthic direct (order)
Order Palaeonemertea	
Carinoma mutabilis	**Direct, pelagic not specified (Coe, 1940)**
Carinomella lactea	Pelagic direct (genus)
Procephalothrix spiralis	Pelagic direct (family)
Tubulanus capistratus	Pelagic direct (genus)
Tubulanus polymorphus	**Pelagic direct (Stricker, 1987)**
Tubulanus sexlineatus	Pelagic direct (genus)
Class Enopla	
Order Hoplonemertea	
Amphiporus angulatus	Pelagic or benthic direct (order)
Amphiporus bimaculatus	Pelagic or benthic direct (order)
Amphiporus cruentatus	Pelagic or benthic direct (order)
Amphiporus formidabilis	Pelagic or benthic direct (order)
Amphiporus imparispinosus	Pelagic or benthic direct (order)
Carcinonemertes epialti	**Pelagic direct (Stricker & Reed, 1981)**
Carcinonemertes errans	Pelagic direct (genus)
Emplectonema buergeri	Pelagic direct (genus)
Emplectonema gracile	**Pelagic direct (Stricker & Cloney, 1982)**
Nemertopsis gracilis	Pelagic direct (family)
Oerstedia dorsalis	**Pelagic direct (Gontcharoff, 1961)**
Paranemertes peregrina	**Pelagic direct (Roe, 1976)**
Tetrastemma nigrifrons	Benthic direct (genus)
Tetrastemma candidum	Benthic direct (genus)
Tetrastemma phyllospadicola	**Benthic direct (Stricker, 1985)**
Zygonemertes virescens	Pelagic or benthic direct (order)
Order Bdellonemertea	
Malacobdella grossa	**Pelagic direct (Hammarsten, 1918)**
Malacobdella siliquae	Pelagic direct (genus)
Malacobdella macomae	Pelagic direct (genus)

[a]If development has not been studied, then the development of the closest relative is given (shared taxonomic level in parentheses). Species for which the development is known are in bold.

adult nemertean. These juveniles are typically elongate, covered with short cilia, and may bear an apical tuft of long cilia. Additional tufts of long cilia, but shorter than the apical tuft, may be present at the posterior or paired in anteriolateral positions. Direct development has been attributed to eight local species (see Table 1). Some general descriptions of juvenile appearance are published. For instance, the juvenile of *Oerstedia dorsalis* is milky white in color, has two pairs of ocelli, and may be 0.43 mm in length at hatching (Iwata, 1960a). Bear in mind, however, that attributes such as color, size, and ocelli number are probably only superficially descriptive, and undescribed juveniles may possess the same traits. Consequently, no information is currently available with which the many vermiform nemertean juveniles may be confidently distinguished. To provide an idea of their general appearance, Fig. 1 gives examples of pelagic hoplonemertean and palaeonemertean juvenile forms from around the world.

Some claim it is possible to distinguish between palaeonemertean juveniles and other vermiform juveniles on the basis of the presence or absence of a functional proboscis. According to Iwata (1960b), palaeonemerteans do not develop a functional proboscis until after "metamorphosis," whereas enoplans (especially hoplonemerteans) possess a proboscis in the pelagic stage and use it to capture prey. An example of a proboscis is given in Fig. 2. Cantell (1989) points out, however, that Jägersten (1972) observed an apparently functional proboscis in the palaeonemertean *Cephalothrix*. Therefore, caution should be used before assigning a juvenile to an order based solely on the presence or absence of a proboscis. Also, Iwata (1960a) states that the larve of many hoplonemertean and bdellonemertean

Fig. 1. Oval-shaped vermiform pelagic juveniles of palaeonemerteans (A–C) and hoplonemerteans (D–F). A, *Procephalothrix filiformis*; B, *Procephalothrix simulu;s* C, *Tubulanus* sp.; D, *Oerstedia dorsalis*; E, *Emplectonema gracile*; F, *Carcinonemertes* sp. Two species from the Pacific Northwest are shown: *Oerstedia dorsalis* and *Emplectonema gracile* (D, E). (A, E adapted from Iwata, 1960a; B, C adapted from Iwata, 1968; D adapted from Gontcharoff, 1961; F adapted from Humes, 1942)

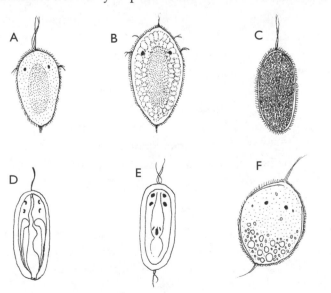

Fig. 2. Unidentified pelagic nemertean juvenile from the west coast of Sweden, bearing "larval" proboscis (PR). EY, eye; ST, statocyst; length = 0.6 mm. Statocysts are present in relatively few species. (after Jägersten, 1972)

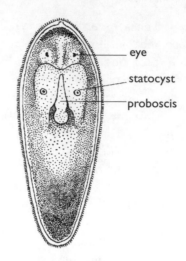

species are elliptical and have long anterior and short posterior tufts. As can be seen from Fig. 1, however, palaeonemerteans can also be elliptical in shape and possess similar ciliary patterns.

Order Heteronemertea

Nemerteans of the order Heteronemertea primarily display indirect development with a planktonic pilidium larva. Based on known development of congeners (see Table 1), local heteronemerteans without a pilidium larva probably either undergo benthic direct development or pass through a Desor's larval stage within an egg case.

Four genera of heteronemerteans are found along the coast of the Pacific Northwest: *Cerebratulus, Lineus, Micrura,* and *Myoisophagos.* The latter genus is represented in the region by the heteronemertean previously known as *Lineus vegetus,* which was recently synonymized with *Myoisophagos sanguineus* (Riser, 1994). Development has been studied for only six of the region's heteronemertean species, one of which possesses a Desor's larva and has no pelagic phase. Reviewed below are examples of pilidium larvae from the genera *Cerebratulus, Lineus,* and *Micrura.* Sources for these adapted illustrations do not always clearly indicate the larval size; in such cases, size is not indicated. On the basis of personal observations of Oregon coast pilidium larvae, it seems that pilidium length is commonly between 0.2 and 1.5 mm, depending on stage and species.

Local representatives of genera are used as examples below when possible, but most illustrations are of species from outside the Pacific Northwest. These diagrams are provided to show the range of pilidium form. As more is learned about the local pilidium larvae, it may be possible to create a key that uses a combination of size, pigment patterns, and lobe patterns to assign pilidium larvae to genus or species. This system has been suggested (Cantell, 1969) for Swedish coast pilidium larvae (see further discussion at end of chapter).

Genus *Cerebratulus*. It is thought that all members of the genus *Cerebratulus* produce a planktonic pilidium larva. This is indeed the case for the two local *Cerebratulus* species studied, *C. marginatus* (Fig. 3) and *C. californiensis*. *Cerebratulus* larvae tend to be large (>1 mm in height) and have a pyramidal shape (a pointed apex). Larvae depicted in Fig. 3 lack a pointed apex, possibly because of distortion due to preservation (Fig. 3A) or because they are too young to display the mature larval shape (Fig. 3B). *Cerebratulus* pilidium larvae sometimes possess subdivided lobes (e.g., Fig. 3A) that are distinct from the lobes of other heteronemertean genera.

Genus *Lineus*. *Lineus bilineatus* is found on the Pacific Northwest coast and is known to produce a pilidium larva. In contrast, *Lineus ruber*, also found locally, produces a benthic, encased Desor's larva. Members of this genus may also develop directly in egg cases, free-living on the benthos, or in the water column. Advanced *Lineus* pilidium larvae so far described tend to be smaller than the larvae of *Cerebratulus* species (i.e., <1 mm in height), bear a rounded apex and episphere, and have continuous lobes with no subdivision (T. Lacalli, pers. comm.). Examples of pilidium larvae from this genus, both collected along the coast of Japan, are given in Fig. 4.

Genus *Micrura*. *Micrura*, the least-studied of the pilidium-producing genera, is represented by three species in the Pacific Northwest. Two of these species are known to produce a pilidium larva, *M. alaskensis* (Stricker and Folsom, 1998) and *M. verrilli* (Coe, 1940). No illustrations of local *Micrura* larvae are available, and I am unable to locate any good descriptions of advanced pilidium larva from this genus. The illustration

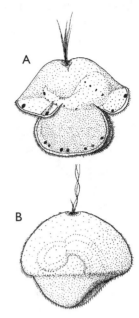

Fig. 3. (A) Advanced *Cerebratulus lacteus* pilidium larva; shape of apex possibly distorted by preservation. (B) Ten-day-old pilidium larva of *Cerebratulus marginatus*, a species found on the Oregon coast; this larva is probably too young and indistinct to differentiate it from the pilidium larvae of other heteronemertean species and genera. (A adapted from Verrill, 1892; B adapted from Coe, 1905)

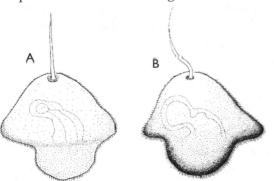

Fig. 4.(A) Pilidium larva of *Lineus torquatus*, body height without apical tuft = 0.24 mm. (B) Pilidium larva of *Lineus alborostratus*, body height without apical tuft = 0.3 mm. (A adapted from Iwata, 1957; B adapted from Iwata, 1960a)

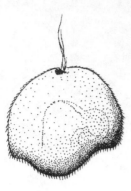

Fig. 5. Pilidium larva of *Micrura leidyi*; this larva is likely too young and indistinct to be differentiated from the pilidium larvae of other heteronemertean species and genera. (adapted from Coe, 1943)

of an early pilidium from the New England coast presented here (Fig. 5), an adaptation of one of the few published drawings of a *Micrura* pilidium larva, is too young and indistinct to be differentiated from other heteronemertean pilidium larvae.

Characteristics distinguishing the pilidium larvae in Figs. 3–5 may not be consistent characteristics of their genera. It does appear, however, that advanced *Cerebratulus* pilidium larvae tend to be relatively large, exceeding 1 mm in height, have a pointed apex, and sometimes have subdivided lobes, whereas advanced *Lineus* larvae tend to be smaller and more regular in shape (see Fig. 4 for examples). Until useful observations are made of the advanced larvae of unstudied heteronemerteans, however, this should be viewed only as a provocative pattern in need of validation. Observations are especially needed of *Micrura* species, for which little or no information is available on the advanced pilidium larva. Many detailed studies of nemertean development must be completed before the pilidium larvae found off the Pacific Northwest can be related confidently to species.

History of Pilidium Larva Classification

More than a century ago the first attempts were made to classify pilidium larvae (Bürger, 1895). For pilidium larvae in the Gulf of Naples, Bürger used traits such as shape of the helmet, shape of the lateral lobes, and length of the apical ciliary tuft. In addition to using these traits, Schmidt (1937) presented a method of identifying pilidium larvae that stressed the size ratio between lobes and placed pilidium larvae into descriptive groups. Dawydoff (1940) also classified pilidium larvae in descriptive groups, using names that indicate larval morphology such as *Pilidium elongatum*, *P. magnum*, *P. minutum*, *P. longivertex*, and *P. depressum*. Cantell (1969) points out that larval groupings such as these can embrace species of different families, are therefore not natural taxonomic units, and should probably be avoided. Cantell also provides the most recent attempt to link pilidium larvae with their corresponding adults and uses a more modern view of nemertean relationships. Characteristics used by Cantell to describe pilidium larvae found off the west coast of Sweden include helmet shape, lateral lobe shape, apical tuft length, body size, pigmentation, and planktonic season. He argues convincingly that, taken together, these traits can provide information on pilidium identity. Characteristics of imaginal juveniles, visible within some transparent pilidium larvae, and newly metamorphosed (hatched) heteronemerteans may also provide clues to larval taxonomy. These characteristics include eye placement and color, head shape, cephalic furrows, and presence or absence of the caudal cirrus (tail).

Development is undescribed for the majority of the local heteronemerteans. As more observations are made on nemertean development, it will be possible to relate pilidium larvae to their respective adults using morphology, pigmentation, and temporal distribution of the larvae. Too little information is currently available to apply Cantell's (1969) system even at the genus level. Riser's observation in 1974 that our knowledge was too incomplete to relate pilidium larvae to adult species is still very true for most pilidium larvae in the world, including those of the Pacific Northwest.

Making New Observations

Indeed, more information is needed about the development of many nemertean species. For those making new observations on pilidium larvae, whether on lab-reared or field-caught animals, the following thoughts and tips are provided to help make observations most useful. Common problems associated with describing pilidium larvae arise from descriptions of fixed specimens or very young larvae, morphologically indistinct from other species.

As with all zooplankton, specimens are best observed alive. This is especially important for the soft-bodied pilidium larvae, because the helmet-like form is easily distorted by muscle contractions during fixation. For instance, *Cerebratulus* pilidium larvae typically have a pointed apex, being rather pyramidal in shape. The pilidium larva of *C. lacteus*, adapted from Verrill (1892) and depicted in Fig. 3A, probably possesses a more pointed apex that has been distorted by muscle contraction in the fixation process. Contraction of muscles not only distorts shape but can make the larva appear smaller than its actual size. It is best, if possible, to collect a wide range of sizes and developmental stages. This helps the observer know how large the larvae can get before they are ready to metamorphose. Finally, two points regarding collection and preservation: First, when samples are fixed, it is wise to sort quickly and separate soft-bodied forms, such as pilidium larvae, from the rest of the sample; this can prevent further damage in a vial or jar of settled plankton. Second, plankton tows for soft-bodied forms should be gentle, preferably towed by hand and using a blind cod-end (Reeve, 1981). This guards against the distortion of delicate forms caused by water flowing forcefully through the plankton net.

Many available illustrations depict young pilidium larvae, which are morphologically indistinct from the pilidium larvae of other species (e.g., Figs. 3B, 5). These early, undifferentiated pilidium larvae are often raised from the fertilized eggs of adults in the laboratory. This yields the significant advantage of knowing the identity of the parents, but the pilidium larvae must be raised to a late, mature stage to be distinct. Attempts to raise pilidium larvae to later stages have been largely unsuccessful, possibly because of improper suspension in the culture vessel or ignorance of the appropriate diet. Pilidium larvae caught in the field are often more advanced developmentally, but the identity of the adult may not be known. It is possible to raise newly metamorphosed juvenile nemerteans, captured as advanced pilidium larvae in the plankton, to the point where the species can be identified. The diet of juvenile nemerteans is better understood than that of planktonic pilidium larvae; attempts to raise juvenile worms can meet with some success. The important point, however, is that descriptions of pilidium larvae should include the most advanced distinct stages possible.

Acknowledgments

The author expresses sincere appreciation to the following experts, who shared their knowledge and contributed immensely to this project: R. Gibson of the Liverpool John Moores University, S. A. Stricker of the University of New Mexico at Albuquerque, T. Lacalli of the University of Saskatchewan, and F. Iwata of the Akkeshi Marine Biological Station, Hokkaido.

References

Austin, W. C. (1985). An Annotated Checklist of Marine Invertebrates in the Cold Temperate Northeast Pacific. Cowachin Bay, B.C., Khoyatan Marine Laboratory.

Brusca, R. C. and G. J. Brusca (1990). Invertebrates. Sinauer Associates, Inc., Sunderland.

Bürger, O. (1895). Die Nemertinen des Golfes von Neapel. Fauna u. Fl. des Golfes von Neapel 22:743.

Cantell, C.-E. (1969). Morphology, development, and biology of the pilidium larvae (Nemertini) from the Swedish west coast. Zool. Bidr. Uppsala 38:61–111.

———— (1989). Nemertina. In: Reproductive Biology of Invertebrates, Volume IV, Part A: Fertilization, Development, and Parental Care, K. G. Adiyodi and R. G. Adiyodi (eds.). John Wiley & Sons, New York.

Coe, W. R. (1905). Nemerteans of the west and northwest coasts of America. Bull. Mus. Comp. Zool. Harv. 67.

────── (1940). Revision of the nemertean fauna of the Pacific coasts of north, central, and northern South America. Allan Hancock Pacific Expeditions 2(13):247–323.

────── (1943). Biology of the nemerteans of the Atlantic coast of North America. Trans. Conn. Acad. of Arts Sci. 35:129–328.

Dawydoff, C. (1940). Les formes larvaires de polyclades et de némertes du plancton indochinois. Bull. Biol. France et Belgique 74:1–54.

Gibson, R. (1995). Nemertean genera and species of the world: an annotated checklist of original names and description citations, synonyms, current taxonomic status, habitats and recorded zoogeographic distribution. J. Nat. Hist. 29:271–561.

Gontcharoff, M. (1960). Le développement post-embryonaire et la croissance chez *Lineus ruber* et *Lineus viridis* (Nemertes, Lineidae). Ann. Sci. Nat. (Zool.) Paris 12(2):225–79.

────── (1961). Embranchement des nemertiens (Nemertini G. Cuvier 1817; Rhynchocoela M. Schultz 1851). Traite de Zoologie, Anatomie, Systematique, Biologie 4.

Hammarsten, O. D. (1918). Beitrag zur embryonalentwicklung der Malacobdella grossa. Arb. Zool. Inst., Stockholm 11:1–96.

Humes, A. G. (1942). The morphology, taxonomy, and bionomics of the nemertean genus Carcinonemertes. Ill. Biol. Mono. 18(4):1–105.

Iwata, F. (1957). On the early development of the Nemertine, *Lineus torquatus* Coe. J. Fac. Sci., Hokkaido University 13(1-4):54–58.

────── (1958). On the development of the nemertean *Micrura akkeshiensis*. Embr. 4(2):103–31.

────── (1960a). Studies on the comparative embryology of nemerteans with special reference to their interrelationships. Pub. Akkeshi Mar. Biol. Station 10:1–51.

────── (1960b). The life history of the Nemertea. Bull. Mar. Biol. Stat. Asamuchi 10:95–97.

────── (1968). Nemertini. Invertebrate Embryology. M. Kume and K. Dan. Belgrade, NOLIT, Publishing House.

Jägersten, G. (1972). Evolution of the Metazoan Life Cycle. Academic Press, London.

Nusbaum, J. and M. Oxner (1910). Studien über die regeneration der Nemertinen. I. Regeneration bei *Lineus ruber* (Müll). Arch. f. Entwmech. 30:74–132.

────── (1911). Weiterer studien über die regeneration der Nemertinen. I. Regeneration bei Lineus ruber (Müll). Arch. f. Entwmech. 32:349–96.

Reeve, M. R. (1981). Large cod-end reservoirs as an aid to the live collection of delicate zooplankton. Limnol Oceanog. 26(3):577–80.

Riser, N. W. (1974). Nemertinea. In: Reproduction of Marine Invertebrates *VI Acoelomate* and *Pseudocoelomate Metazoans*, A. C. Giese and J. S. Pearse (eds.). Academic Press, New York.

────── (1994). The morphology and generic relationships of some fissiparous heteronemertines. Proc. Biol. Soc. Wash. 107:548–56.

Roe, P. (1976). Life history and predator-prey interactions of the nemertean *Paranemertes peregrina* Coe. Biol. Bull. 150:80–106.

Schmidt, G.A. (1937). Bau und Entwicklung der Pilidium von *Cerebratulus pantherinus* un *marginatus* und die Frage der morphologischen Merkmale der Hauptformen der Pilidien. Zool. Jahrb. Anat. Ontog. 62:423–48.

Stricker, S. A. (1985). A new species of *Tetrastemma* (Nemertea, Monostilifera) from San Juan Island, Washington, USA. Can. J. Zoo. 63:682–90.

——— (1987). Phylum Nemertea. In: Reproduction and Development of Marine Invertebrates of the Northern Pacific Coast, M. F. Strathmann (ed.). University of Washington Press, Seattle.

——— (1996). Phylum Nemertea. In: Marine Invertebrates of the Pacific Northwest, E.N. Kozloff (ed.). University of Washington Press, Seattle.

Stricker, S. A. and R. A. Cloney (1982). Stylet formation in nemerteans. Bio. Bull. 162:387–405.

Stricker, S. A. and M. W. Folsom (1998). A comparative ultrastructural analysis of spermatogenesis in nemertean worms. Hydrobiol., 365:55–72.

Stricker, S. A. and C. G. Reed (1981). Larval morphology of the nemertean *Carcinonemertes epialti* (Nemertea: Hoplonemertea). J. Morph. 169:61–70.

Verrill, A. E. (1892). The marine nemerteans of New England and adjacent waters. Trans. Conn. Acad. 8:382–456.

5

Entoprocta

Alan L. Shanks

The entoprocts are a small phylum with only around 120 species worldwide. They are tiny animals, most only a few millimeters in length. Adults form colonies of polyps or zooids, each composed of a bowl-shaped calyx mounted on a stalk. The rim of the bowl has ciliated tentacles that produce a feeding current. The mouth and anus are connected by a U-shaped gut and both open within the calyx. The largest grouping in the phylum is the family Loxosomatidae. Most of the species in this family are symbionts on other invertebrates; nearly half of them reside on polychaete worm tubes (Mariscal, 1975; Brusca and Brusca, 1990). Eleven species are found locally (Table 1).

Asexual reproduction is common and in some species may be the most important means of reproduction. In addition, a curious form of precocious asexual budding has been described in the larvae (sexually produced) of several loxosomatids. In these species, free-swimming larvae are produced. These larvae, while pelagic, asexually produce adult buds. The miniature adults are held within pockets in the larvae. Prior to settlement of the larvae, the miniature adults are released through a rupture of the larval body wall. After release, the pelagic larvae die. In one species from Florida, half of the miniature adults produced by the larvae were reproductively mature males. In another species (*Loxosomella vivipara*), the adult buds were forming in the larvae while the larvae were still developing in the ovary of the parent. The larvae are released and swim for one to three days before the buds are released through the ruptured larval body wall (Mariscal, 1975).

Fertilization is internal and the zygotes are brooded within the atrium of the parent. Brooding continues until a relatively large free-swimming larva is released. Nutrition is provided by yolky eggs or direct nutrient transfer from the parent. In the later case, small eggs (10 μm diam.) grow into relatively large larvae (200 μm diam.) during development in the atrium. Fully developed larvae have been observed "stealing" food from the parental food groove prior to their release into the plankton. In some species, large nutrient-rich cells are released into the atrium by the parent and fed upon by the brooded larvae (Mariscal, 1975; Nielsen, 1990).

Table 1. Species in the phylum Entoprocta from the Pacific Northwest (from Kozloff, 1996)

Order Solitaria
Family Loxosomatidae
Loxosoma davenporti
Loxosomella nordgaardi
Loxosomella **spp.**

Order Coloniales
Family Loxokalypodidae
Loxokalypus socialis

Family Pedicellinidae
Myosoma spinosa
Pedicellina cernua

Family Barentsiidae
Barentsia benedeni
Barentsia gracilis
Barentsia misakiensis
Barentsia ramosa
Barentsia robust

Fig. 1. Entoproct larvae.
(A) *Loxosomella*
elegans. (B) *Loxosoma*
pectinaricola.
(C) *Loxosoma*
jaegersteni. (D)
Barentsia gracilis. (E)
Pedicellina *nutans*.
Highlighted species and
genera are represented
locally. In B and C, the
numerous circles on the
larvae are stalked
vesicles. (From Nielsen
1990, Figs. 3 and 5)

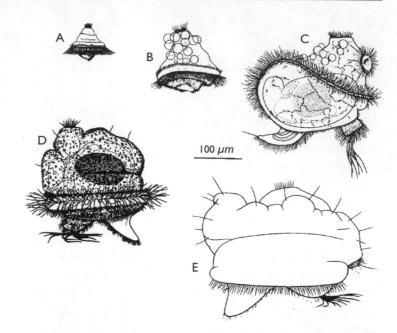

Some loxosomatid species produce feeding larvae that can
have long residence times in the plankton (up to 7 months,
Nielsen, 1990). Early stages in these feeding forms appear to
be typical trochophores (Fig. 1A); they would be difficult to
distinguish from the trochophores of polychaetes. Most species
produce non-feeding larvae. Many of the loxosomatid larvae
have numerous, conspicuous stalked vesicles (Fig. 1B, C). Larva
in the families Pedicellinidae and Barentsiidae are much like
miniature swimming adult calyxes (Fig. 1D, E). All that is
needed to turn these larvae into adults is an outgrowth of a
stalk and tentacles. The pelagic phase in the non-feeding larvae
is short (a few days). Some larval types (e.g., *Pedicellina nutans*,
Fig. 1E) are such poor swimmers that there is no pelagic phase;
they settle immediately and begin exploring the bottom.

References

Brusca, R. C. and G. J. Brusca (1990). Invertebrates. Sinauer
    Associates, Inc., Sunderland.
Kozloff, E. N. (1996). Marine Invertebrates of the Pacific Northwest.
    University of Washington Press, Seattle.
Mariscal, R. H. (1975). Entoprocta. In: Reproduction of Marine
    Invertebrates Vol. II, A. C. Giese and J. S. Pearse (eds.), pp. 1–42.
    Academic Press, New York.
Nielsen, C. (1990). Bryozoa Entoprocta. In: Reproductive Biology of
    Invertebrates. Vol. IV, Part B. Fertilization, Development, and
    Parental Care, K. G. Adiyodi and R. G. Adiyodi (eds.), pp. 201–
    10. John Wiley & Sons, New York.

6

Polychaeta

Lana Crumrine

Well over 200 species of the class Polychaeta are found in waters off the shores of the Pacific Northwest. Larval descriptions are not available for the majority of these species, though descriptions are available of the larvae for at least some species from most families. This chapter provides a dichotomous key to the polychaete larvae to the family level for those families with known or suspected pelagic larva. Descriptions have been gleaned from the literature from sites worldwide, and the keys are based on the assumption that developmental patterns are similar in different geographical locations. This is a large assumption; there are cases in which development varies with geography (e.g., Levin, 1984).

Identifying polychaetes at the trochophore stage can be difficult, and culturing larvae to advanced stages is advised by several experts in the field (Bhaud and Cazaux, 1987; Plate and Husemann, 1994).

Reproduction, Development, and Morphology

Within the polychaetes, the patterns of reproduction and larval development are quite variable. Sexes are separate in most species, though hermaphroditism is not uncommon. Some groups undergo a process called epitoky at sexual maturation; benthic adults develop swimming structures, internal organs degenerate, and mating occurs between adults swimming in the water column. Descriptions of reproductive pattern, gamete formation, and spawning can be found in Strathmann (1987). Larval polychaetes generally develop through three stages: the trochophore, metatrochophore, and nectochaete stages. Trochophores are ciliated larvae (see Fig. 1). A band of cilia, the prototroch, is used for locomotion and sometimes feeding. Trochophore larvae are generally broad anteriorly and taper posteriorly. The anterior and posterior sections of the larva are called the episphere and hyposphere, respectively. They are usually pelagic.

By the metatrochophore stage, two to three segments have usually formed. Parapodia (with or without setae) may by this developmental stage become apparent. Doral and ventral podia are know as neuropodia and notopodia, respectively. The anterior-most segment is called the prostomium. Just posterior

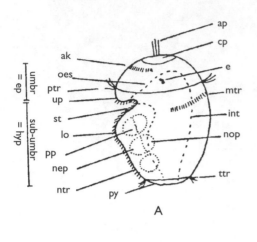

Fig. 1. Parts of a
generalized polychaete
trochophore (A) and
nectochaete (B). (From
Korn, 1960)
ac, anal cirrus; ak,
acrotroch; ap, apical tuft;
br, chaetae; cp, cerebral
plate; dc, dorsal cirrus; e,
eye; el, elytra; ep,
episphere; hyp,
hyposhere; int, intestine;
lo, lower lip; ltl, lateral
tentacle; mt, median
tentacle; mtr, metatroch;
nep, neuropod; nop,
notopod; ntr, neurotroch;
oes, oesophagus; p, palp;
pa, paratroch; pp,
parapod; pr, prostomium;
proc, proctodaeum; ptr,
prototroch; py, pygium; st,
stomodaeum; sub-umbr,
subumbrella; tlc,
tentacular cirrus; ttr,
telotroch; umbr, umbrella;
up, upper lip; vc, ventral
cirrus

to the prostomium is the peristomium. These two anterior
segments form the head. Eyes and antennae form in the
prostomium. The mouth, palps, or tentacular cirri form in the
peristomium. Larvae may be planktotrophic and develop a
mouth and gut while in the plankton. Others are lecithotrophic
and survive on yolky substances. Once a larva has developed
setae, the segments with setae are called setigers. The most
posterior segment is called the pygidium. New segments
develop from the pygidium in a stepwise manner.

Advanced larvae are called nectochaetes (see Fig. 1).
Nectochaetes typically have many more setigers than
metatrochophores. Depending on the species, the larval
ciliation may have been lost and the larvae may no longer be
pelagic. After settlement, larvae may go through a benthic stage
called the erpochaete stage, or they may metamorphose to
become juvenile worms (Strathmann, 1987).

Polychaete larvae are morphologically complex and diverse.
Because of this complexity, a large number of technical terms
have been coined to help describe larval anatomy. Fig. 1 and
the following glossary should help to make sense of this
"foreign language." The definitions in the glossary are from
Lacalli (1980) and Bhaud and Cazaux (1987).

achaetous: without setae
acicula (pl. aciculae): a stout chitinous rod embedded in one or both
  parapodial lobes
acrotroch: circlet of cilia in front of the prototroch
antenna: sensory appendage arising from the anterior or the dorsal surface
  of the prostomium

anal cirrus: elongated appendage arising from the pygidium

apical tuft: bundle or group of a few cilia projecting from the anterior pole of the larva

capillary seta: hairlike bristle that may be ornamented

chevron: V-shaped teeth laterally on the proboscis

ciliated pit: small dorsal ciliated cavity in segment 1, 2, or 3, limiting posteriorly the neurotroch of the spionidae

cirrus: respiratory and tactile appendage of the setiger, without blood vessel

compound seta: jointed bristle most often composed of two parts: the base and the distal part

crest (nuchal crest): mediodorsal zone of the cephalic area in the Spionidae bounded by two ciliated grooves

elytron (pl. elytra): dorsal scale, inserted on the dorsal side of the parapodium, arising from transformation of some dorsal cirri

erpochaeta: creeping stage, moving on or in the sediment using its setae

erposoma: creeping stage, without setae, moving by undulation or contraction of the body

gastrotroch: ventral troch

geniculate seta: seta that is bent but not articulated

hooded hook: seta that is curved distally and covered with a transparent envelope

limbate seta (spatulate seta): seta with a bladelike flattened margin

melanophore: black pigment cell or group of cells

meniscotroch: crescent-shaped area of short cilia in the region anterior to the prototroch; the cilia of the central part are longer, bent, and form a pointed brush

mesotroch: transversal ciliated ring in the middle part of the larval body

metatrochophore: larval planktonic stage with marked segmentation; if parapodia are not yet formed, stage 1; if parapodia are present but not yet functioning, stage 2

monostichious: troch made of one ring of long cilia

nectochaeta: developmental stage bearing functioning parapodia serving locomotion

neuropodium: ventral section of the parapodium

neuroseta: seta of the neuropodium

neurotroch: row of short cilia running along the ventral side of the larva

notopodium: dorsal section of the parapodium

notoseta: seta of the notopodium

palea (pl. paleae): simple and stout seta, often enlarged as an oar (palea of parapodium) or simple and stout seta, not enlarged, but regularly tapering (opercular palea)

palpi: paired projections arising from the prostomium, used for food gathering and tactile purposes

papilla (pl. papillae): conus-like projection

parapodium: lateral segment footlike projections, composed of two parts, neuropodia and notopodia, each bearing a cirrus and setae

peristomium: buccal segment

polystichious: troch made of several rings of long cilia

proboscis: eversible anterior portion of the alimentary tract; with or without papillae

prostomium: preoral lobe that contains the cerebral ganglia and bears the most important sense organs; posteriorly and ventrally, the prostomium is delimited by the peristomium

prototroch: on the equatorial part of the trochophore, a ring of cilia anterior to the mouth; the single ciliated ring of certain trochophores

pygidium: posterior part of the body bearing the anus, always devoid of coelomic cavity

serrated seta: seta or part of seta with one or two edges notched like a saw

seta (pl. setae): slender chitinous structure projecting from the parapodium, used for locomotion and defense

setiger: segment bearing setae

simple seta: unjointed bristle

telotroch: ring of cilia near the anus

troch: ciliated ring of embryonic or larval stage, used in locomotion

trochophore: free pelagic larval stage, top-shaped structure, without visible segmentation, bearing one or two equatorial ciliated rings; in the latter case, the metatroch is situated behind the mouth

uncini: stout spines curved at the distal part; referred to as hooded hooks when the distal part is protected by hoods (see hooded hook); when they are more or less rectangular, with numerous teeth, they are called uncini

Description and Identification of Local Taxa

Key to larvae of polychaete families from the Pacific Northwest (adapted from Bhaud and Cazaux, 1987; Plate and Husemann, 1994)

1a. Larval tube present

    a1. Tube transparent with diameter constant thoughout length, larva bearing 1—several large tentacles, always directed toward body and equaling at least half its length ....
    .... TEREBELLIDAE (p. 73)

    a2. Tube nontransparent and conical, no large tentacles, external paleae directed forward .... PECTINARIDAE (p. 73)

1b. Larva without tube 2

2a. Larva bearing 1 or several stout anterior tentacles ....
.... TEREBELLIDAE (p. 73)

2b. No stout anterior tentacles present 3

3a. Trochophore or metatrochophore with 2 large prototrochal-ventral lobes fused into funnel, resulting in dorsoventral asymmetry .... PECTINARIDAE (p. 73)

3b. Lateral lobes not present 4

4a. Presence of heavy projecting acicular sickle-shaped hooks on notopodium of several segments .... PILARGIDAE (p. 60)

4b. Dorsal sickle-shaped acicular hooks not present.............................. 5

5a. Body and prostomium covered with stout spherical papillae.......
.... SPHAERODORIDAE (p. 63)
5b. Spherical papillae not covering body 6

6a. Larva with 50–100+ segments..
.... TROCHOCHAETIDAE (p. 72)
6b. Larva with <50 segments.. 7

7a. Setae absent 8
7b. Setae present 13

8a. Prostomium broad with marked frontal notch ....
.... TOMOPTERIDAE (p. 64)
8b. Prostomium rounded or conical 9

9a. Prototroch present with no other ciliary bands ....
.... LOPADORHYNCHIDAE (p. 56)
9b. Multiple ciliary bands present, or atrochus ciliation covering
    majority of larva 10

10a. Ciliation present as broad and definitive prototroch,
    metatroch, telotroch, and neurotroch .... EUNICIDAE (p. 49)
10b. Ciliation in distinctly narrow bands or not in defined bands ......
.... 11

11a. Eyespots absent, or if present only in conjunction with
    segmentation .... CIRRATULIDAE (p. 67)
11b. Segmentation absent, eyespots present 12

12a. Cilia arranged in narrow bands at segment definitions, with or
    without short cilia covering rest of larva ....
.... DORVILLEIDAE (p. 49)
12b. Cilia long and nearly covering entire larva...
.... LUMBRINERIDAE (p. 50)

13a. 2 bundles of well-delineated setae 14
13b. Setae not clumped in 2 bundles but distributed on numerous
    parapodia or segments 20

14a. Pair of well-developed tentacles present 15
14b. No such pair of tentacles... 17

15a. 2 long tentacles, covered with adhesive papillae and internal
    blood vessel, without ciliated groove, tentacles motile and often
    coiled in spiral.. MAGELONIDAE (p. 68)
15b. Tentacles not papillated, with 1–2 ciliated grooves, not coiled ..
.... 16

16a. Tentacles with 2 ciliated grooves, tentacles projecting dorsally
    and not laterally, conical prostomium without terminal antenna.
.... AMPHINOMIDAE (p. 46)
16b. 2 laterodorsal tentacles, each with 1 ciliated groove; 2 narrow
    lateral bundles of setae, slightly bent and armed with regular
    denticulated collars .... SABELLARIIDAE (p. 73)

17a. Top-shaped larva with roughly conical episphere 18
17b. Umbrella-, bell- or mushroom-shaped larva 19

18a. Transparent body; 2 lateral bundles of slender setae that are
    smooth or slightly serrated and bright .... OPHELIDAE (p. 51)
18b. Opaque body; 2 lateral bundles of bent larval setae that are
    thick, dark, and coarsely serrated on the convex side;
    transversely striated notosetal spines and paleae; neither
    tentacles or palpi present .... CHRYSOPETALIDAE (p. 53)

19a. Umbrella-shaped larva; main part of larval body a transparent,
    unpigmented umbrella; setal sacs set close together inside
    umbrella .... OWENIIDAE (p. 52)
19b. Mushroom-shaped body with umbrella and foot; umbrella
    opaque, with pigmentation; separate setal sacs set laterally ....
    .... SABELLARIIDAE (p. 73)

20a. Compact, barrel-shaped larva nearly as wide as long 21
20b. Larva clearly longer than wide, formed by series of similar
    segments 23

21a. 1–2 narrow ciliated rings, 1 pygidial cirrus present, adult setae
    present in early planktonic stages ....
    .... CHAETOPTERIDAE (p. 67)
21b. 1–2 broad ciliated bands present; no or 2 anal cirri present .....
    22

22a. Metatrochophore with broad ciliary bands covering most of
    larva; erpochaete with bulbous prostomium ....
    .... LUMBRINERIDAE (p. 50)
22b. Metatrochophore with broad prototroch, remaining bands
    narrow; prostomium rounded or spatulate ....
    .... EUNICIDAE (p. 49)

23a. 1–2 pairs of palps on prostomium 24
23b. No palps on prostomium 25

24a. 1 palp projecting from each side of prostomium; larval setae
    on notopodia and neuropodia on each segment; body opaque
    and often with pigment spots .... SPIONIDAE (p. 68)
24b. 2 palps projecting from each side of prostomium; 3 pairs of
    palps projecting from peristomium; setae on neuropodium ....
    .... LOPADORHYNCHIDAE (p. 56)

25a. 2 distinct nuchal organs; slight pigmentation with transverse
    spots .... OPHELIDAE (p. 51)
25b. Nuchal organs absent 26

26a. Ocular spots large 27
26b. Ocular spots small 30

27a. Jaws and anal cirri present .... NEREIDAE (p. 56)
27b. Jaws and anal cirri absent 28

28a. 2 pairs of anterior tentacles at extreme end of prostomium ....
.... ALCIOPIDAE (p. 53)
29b. Prostomial tentacles absent 29

29a. Prostomial collar extending posteriorly to third segment ....
.... SPIRORBIDAE (p. 66)
29b. Prostomium without collar .... SERPULIDAE (p. 65)

30a. Trochophore and metatrochophore elongated with diameter
not varying substantially along length of body; larva at least 3 to
4 times longer than wide 31
30b. Trochophore and metatrochophore bulky; nectochaeta with
dorsal elytra, lamellar or elongated dorsal cirri 39

31a. Prostomium and pygidium with large zones of short cilia 32
31b. Prostomium and pygidium with only 1 ring of long cilia 35

32a. 2 pairs of eyes; 3 antennae, possibly short; poorly developed
acrotroch, prototroch, and metatroch present, telotroch absent
in early trochophore; peristomium achaetous, remaining
segments uniramous .... SYLLIDAE (p. 64)
32b. No or 1 pair of eyes 33

33a. Ciliated zones reduced to prototroch and telotroch; conical
prostomium with rudiment of the first tentacle; no statocyst ....
.... TEREBELLIDAE (p. 73)
33b. Larva with multiple ciliary bands, including acrotroch (in early
forms), prototroch, metatroch, telotroch, and neurotroch 34

34a. Larva lacking observable mouth; possessing 13 tufts of cilia on
apical end of prostomium; larva with 3 setigerous segments
bearing 1 pair of short and 1 pair of long notopodial setae on
each segment .... SABELLIDAE (p. 65)
34b. Larva with defined mouth; buccal segment lacking setae,
following segment bearing metatroch; first setae on third
segment of nectochaete .... ORBINIIDAE (p. 51)

35a. Parapodia present 36
35b. Parapodia absent 38

36a. Prostomium flat, rectangular, and short, without antennae; 1
unpaired anal cirrus; parapodia developed from first segment
onward .... NEPHTYIDAE (p. 56)
36b. Prostomium conical and annulated, with 4 terminal antennae;
2 anal cirri; parapodia reduced on first and/or second segment.
.... 37

37a. Parapodia uniramous; 0–4 eyes; parapodia of first segment
reduced; erpochaeta bearing chevrons on proboscis; no jaws,
opaque body, brown pigmentation .... GONIADIDAE (p. 54)
37b. Parapodia reduced and uniramous on first and second
segments; parapodia biramous from third segment onward; eyes
absent; proboscis of erpochaeta lacking chevrons; 4 jaws; body
transparent and colorless .... GLYCERIDAE (p. 53)

38a. Each setigerous segment bearing 1 pair of capillary and 1 pair of spatulate setae; prostomium the largest segment .... ARENICOLIDAE (p. 47)

38b. Each setigerous segment bearing 1 pair of capillary setae; posterior segments bearing hooked setae .... CAPITELLIDAE (p. 47)

39a. Trochophore with curved acrotroch and tuft of long, thin cilia pressed against hyposphere at level of prototroch; elytra present on dorsal surface of metatrochophore 40

39b. Trochophore with equatorial ciliated ring; long tuft of cilia absent from hyposphere or present on episphere; metatrochophore without elytra 42

40a. Tentacular cirri present on peristomium .... APHRODITIDAE (p. 53)

40b. Tentacular cirri absent on peristomium 41

41a. 4–5 pairs of elytra; segments lacking elytra bearing dorsal cirri; simple setae; 7–10 larval segments .... POLYNOIDAE (p. 61)

41b. Four pairs of elytra or elytrophores; dorsal cirri absent or present only in third segment; setae of neuropodia compound, of notopodia simple; 5–6 larval segments .... SIGALIONIDAE (p. 63)

42a. Trochophore with ventral menisotroch; metatrochophore with 2 pairs of eyes and segmentation indistinct; nectochaeta with 4–5 antennae, unarmed proboscis, lamellate dorsal cirri, several pairs of smooth and nonarticulate tentacular cirri, compound and spinigerous notosetae and neurosetae .... PHYLLODOCIDAE (p. 58)

42b. Trochophore lacking ventral menisotroch; metatrochophore with >2 pairs of eyes; clearly delineated segments; anterior part broad; nectochaeta with long and cylindrical dorsal cirri, jointed tentacular cirri, compound and bent neurosetae .... HESIONIDAE (p. 55)

Order Amphinomida

Family Amphinomidae (Local Species 2, Local Species with Described Larvae 0). Development of this family includes a rostraria larva (Fig. 2). Amphinomids are called fire worms because of the discomfort caused when coming in contact with the spines of the adult worms (Fauchald, 1977).

Chloeia entypa (Notopygos labiatus)
Chloeia pinnata

Family Euphrosinidae (Local Species 2, Local Species with Described Larvae 0). Euphrosinids are considered closely related to, but distinct from, the Amphinomidae (Fauchald

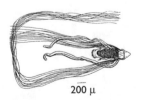

200 μ

Fig. 2. Rostraria larva, family Amphinomidae. (From Mileikovskii, 1961)

1977). Adults are short and thick-bodied with one pair of antennae, tufts of neurosetae, and no palps. Development is not described.

> *Euphrosine bicirrata*
> *Euphrosine hortensis*

Order Capitellida

Family Arenicolidae (Local Species 4, Local Species with Described Larvae 2). Arenicolid worms spawn freely, brood larvae in burrows, or produce benthic egg masses. Larvae are non-feeding. Two local species (*) are briefly pelagic during larval development prior to settling (Strathmann, 1987). *Abarenicola pacifica* develops in gelatinous masses that are brooded within adult tubes. *Branchiomaldane vincenti* undergoes direct development.

> *Abarenicola claparedi oceanica (A. vagabunda oceanica,*
>   *Arenicola pusilla)**
> *Abarenicola pacifica (Arenicola pusilla, in part)*
> *Arenicola marina**
> *Branchiomaldane vincenti (B. simplex, Protocapitella)*

Key to pelagic arenicolid larvae (Fig. 3)

I a. Larva bearing apical tuft, broad prototroch; vague indentations marking segments; brown eyes .... *Arenicola marina*
I b. Larva lacking apical tuft; narrow prototroch; red eyes ....
.... *Abarenicola claparedi*

Family Capitellidae (Local Species 8, Local Species with Described Larvae 2). Intratubular brooding of larvae is quite common among the Capitellidae, though some species are known to broadcast spawn. Likewise, development ranges from direct within the parental tube to lecithotrophic and planktotrophic swimming larvae (Rouse, 1992). *Capitella capitata* appears to have two different developmental strategies: direct development from jelly masses, and a short pelagic phase. *Capitella capitata* may be a complex of six sibling species (see Strathmann, 1987), and therefore it is likely that different sibling species use different developmental strategies. *Heteromastus filiformis* egg masses are attached to the end of the female's tube, and larvae develop to the trochophore stage before release into the plankton (Rasmussen, 1956). Generalization of the development of capitellids based on trochophores is difficult because of generic differences (Lacalli, 1980).

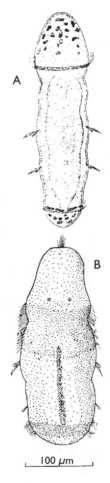

Fig. 3. (A) *Arenicola marina* and (B) *Arenicola claparedi*, Family Arenicolidae. (From Okuda, 1946; Newell, 1948)

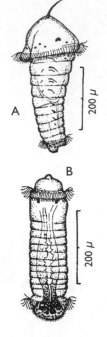

Fig. 4. (A) *Heteromastus filiformis* and (B) *Capitella capitata*, family Capitellidae. (From Lacalli, 1980)

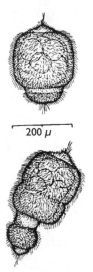

Fig. 5. Example of a pelagic maldanid larva, *Clymenella torquata*; not a local species. (From Lacalli, 1980).

Capitella capitata
Heteromastus filiformis
Heteromastus filobranchus
Mediomastus californiensis
Notomastus giganteus (N. magnus)
Notomastus lineatus
Notomastus magnus
Notomastus tenuis (N. lineatus var. balanoglossi)

Key to pelagic capitellid larvae (Fig. 4)

1a. Long apical tuft present; gut subdivided into several sections; body light green .... *Heteromastus filiformis*

1b. Apical tuft absent or reduced; gut not subdivided ....
.... *Capitella capitata*

Family Maldanidae (Local Species 13, Local Species with Described Larvae 0). Development for the maldanids was presumed to be primarily nonpelagic by Thorson (1946). Rouse (1992) cites two examples of free-spawned lecithotrophic larvae in the genera *Euclymene* and *Axiothella*. Wilson (1983) suggests that *Axiothella rubrocincta* is a sibling pair with one type brooding larvae and the other broadcasting demersal lecithotrophic larvae (Fig. 5).

Asychis disparidentata
Axiothella rubrocincta
Euclymene reticulata
Isocirrus longiceps
Maldane sarsi
Maldanella harai (Maldane robusta)
Nicomache lumbricalis
Nicomache personata
Notoproctus pacificus
Praxillella affinis var. pacifica
Praxillella gracilis
Rhodine bitorquata

Order Cossurida

Family Cossuridae (Local Species 1, Local Species with Described Larvae 0). Adult cossurids are common in sand and deep slope and abyssal muds, where they apparently feed on detritus through a pharynx (Fauchald, 1977). Development is not described.

Cossura modica

Order Ctenodrilida

Family Ctenodrilidae (Local Species 1, Local Species with Described Larvae 0). Fauchald (1977) describes adult worms of this family as small and grub-shaped. They are especially common in areas of aquaculture and may be commensal with *Flabelliderma commensalis* and *Strongylocentrotus purpuratus*. Bhaud and Cazaux (1987) warn that some species of this family are also holoplanktonic, and adults are often confused for larval stages.

Ctenodrilus serratus

Order Eunicida

Family Arabellidae (Local Species 4, Local Species with Described Larvae 0). Arabellids are often parasitic on other animals, especially polychaetes and echiurans. Parasitism may be a larval stage or a lifelong condition. For example, Allen (1952) found larvae through young worms of *Arabella iricolor* present in a single *Diopatra cuprea* host. Descriptions and figures of arabellid development are not known, though Pettibone (1957) indicates that *A. iricolor* spawns throughout the summer.

Arabella iricolor
Drilonereis falcata
Drilonereis filum
Notocirrus californiensis

Family Dorvilleidae (Local Species 4, Local Species with Described Larvae 1). Adult dorvilleids are mainly small and common in shallow water (Fauchald, 1977). The described larva, *Schistomeringos longicornis* (Fig. 6), is free-spawned, lecithotrophic, and has been collected swimming at the water surface (Moore, 1903).

Dorvillea moniloceras
Dorvillea pseudorubrovittata
Protodorvillea gracilis (Dorvillea, P. kefersteini, P. recuperata, Stauronereis)
Schistomeringos longicornis (Dorvillea, Stauronereis, D. rudolphi, D. atlantica, Stauronereis articulata)

Family Eunicidae (Local Species 3, Local Species with Described Larvae 0). Eunicids are large polychaetes, mostly associated with hard substrates and shallow water, mainly carnivorous, burrowers, and tube builders (Fauchald, 1977). *Eunice valens* is known to brood benthic non-feeding larvae (Akesson, 1967). Richards (1967) describes several species that produce benthic egg masses in which the larvae develop.

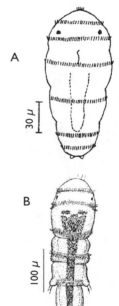

Fig. 6. *Schistomeringos longicornis* larvae, family Dorvilleidae: (A) late trochophore, (B) metatrochophore. (From Richards, 1967)

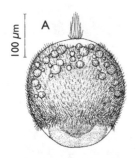

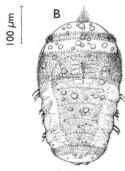

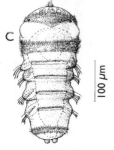

Fig. 7. *Sarsonuphis elegans* larvae, family Onuphidae:
(A) trochophore,
(B) metatrochophore,
(C) nectochaete. (From Blake, 1975a)

Eunice segregata
Eunice valens (E. kobiensis)
Marphysa stylobranchiata

Family Lumbrineridae (Local Species 8, Local Species with Described Larvae 2). Adults brood directly developing larvae in the two local species (*) for which development has been described. Lumbrinerids are mostly free-living burrowers in sand and mud or in algal holdfasts. They occur in shallow and deep water (Fauchald, 1977).

Lumbrineris aff. abyssicola
Lumbrineris bicirrita (L. bifurcata)
Lumbrineris californiensis
Lumbrineris cruzensis
Lumbrineris japonica
*Lumbrineris latreilli**
*Lumbrineris zonata**
Ninoe gemmea

Family Onuphidae (Local Species 7, Local Species with Described Larvae 1). Onuphid development is either lecithotrophic or direct, with only one known example of planktotrophy (*Sarsonuphis elegan,* Fig. 7; Blake, 1975a). *Diopatra cuprea* broods larvae in egg masses attached to the adult tube and produces either nonpelagic (Monro, 1924) or swimming larvae (Allen, 1959).

Diopatra ornata (Onuphis longibranchiata)
Mooreonuphis stigmatis (Nothria)
Nothria occidentalis (Onuphis)
Nothria geophiliformis (Onuphis)
Nothria iridescens (Onuphis)
Sarsonuphis elegans (Onuphis, Nothria)
Sarsonuphis lepta (Onuphis, Nothria, N. abyssalis)

Order Fauveliopsida

Family Fauveliopsidae (Local Species 1, Local Species with Described Larvae 0). This deep-water family is not well known. Adults are smooth-bodied without anterior appendages (Fauchald, 1977).

Fauveliopsis armata

Order Flabelligerida

Family Flabelligeridae (Local Species 3, Local Species with Described Larvae 0). Development has not been described for any local species. Thorson (1946) suggests that *Flabelligera affinis* larvae are non-pelagic.

Flabelligera affinis (F. infundibularis)
Pherusa inflata (Styleroides, Trophonia)
Pherusa plumosa (Stylaroides, P. papillata, P. neopapillata)

Order Ophelida

Family Ophelidae (Local Species 7, Local Species with Described Larvae 2). Ophelids have either direct or pelagic development. *Armandia brevis* is planktotrophic and settles at the 20 setiger stage (Hermans, 1978). *Euzonus mucronata* larvae are lecithotrophic and pelagic.

Armandia brevis (A. bioculata)
Euzonus mucronata (Thoracophelia)
Euzonus williamsi (Thoracophelia)
Ophelia limacina (O. borealis)
Ophelina acuminata (Ammotrypane aulogaster)
Travisia brevis
Travisia pupa (T. carnea)

Key to pelagic ophelid larvae (Fig. 8)

1a. Apical tuft present in metatrochophore; setae long, extending to or past pygidium .... *Euzonus mucronata*
1b. Apical tuft absent; setae not extending past pygidium; 2–3 pairs of setae per setiger .... *Armandia brevis*

100 µm

Family Scalibregmidae (Local Species 3, Local Species with Described Larvae 0). Thorson (1946) collected young polychaetes of this family at 2 m above the sea bottom and therefore assumes that larvae are able to swim or float. Further developmental characteristics are not available.

Asclerocheilus beringianus
Hyboscolex pacificus (Oncoscolex)
Scalibregma inflatum

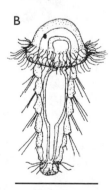

Order Orbiniida

Family Orbiniidae (Local Species 7, Local Species with Described Larvae 4). Three of the four described local species (*) develop in gel masses and spend only a short time swimming as nectochaetes after hatching. *Scoloplos armiger* larvae have been described as nonpelagic (Blake, 1980) and pelagic (Plate and Husemann, 1994).

*Leitoscoloplos elongatus (Scoloplos, Haploscoloplos, L. pugettensis)**

100 µm

Fig. 8. Larvae of (A) *Euzonus mucronata* and (B) *Armandia brevis*, family Ophelidae. (From Dales, 1952; Hermans, 1977)

> *Leitoscoloplos panamensis (Scoloplos, Haploscoloplos, H. alaskensis)*
> *Naineris dendritica (N. laevigata)**
> *Naineris uncinata (N. berkeleyorum)*
> *Phylo felix (Aricia michaelseni, Orbinia)*
> *Scoloplos acmeceps*
> *Scoloplos armiger**

Key to pelagic orbiniid larvae (Fig. 9)

1a. Prototroch band continuous dorsally 2
1b. Prototroch discontinuous dorsally ....*Scoloplos armiger*

2a. Pygidium deeply grooved; prototroch and metatroch of nearly equal width in metatrochophore and nectochaete stage ....
.... *Naineris dendritica*
2b. Pygidium rounded or not grooved; prototroch and metatroch of unequal widths in metatrochophore or nectochate stage 3

3a. Prototroch widest ciliary band in nectochaete stage ....
.... *Leitoscoloplos elongatus*
3b. Metatroch as wide as prototroch in nectochaete state ....
.... *Scoloplos acmeceps*

Family Paraonidae (Local Species 5, Local Species with Described Larvae 0). Development of Oregon species is not known.

> *Aedicira pacifica*
> *Allia ramosa (Aricidea)*
> *Aricidea wassi*
> *Cirrophorus lyra (Paraonis)*
> *Tauberia gracilis (Paraonis, P. ivanovi)*

Order Oweniida

Family Oweniidae (Local Species 2, Local Species with Described Larvae 2). Oweniidae worms live in tubes and spawn or deposit stalked gelatinous egg masses. Trochophores are

Fig. 9. Larvae from the family Orbiniidae.
(A) *Scoloplos armiger;*
(B) *Naineris dendritica,*
(left) metatrochophore,
(right) nectochaete.
(C) *Leitoscoloplos elongatus.* (D) *Scoloplos acmeceps,* (left) metatrochophore, (right) nectochaete. (From Anderson, 1959; Blake, 1980; Giangrande and Petraroli, 1991)

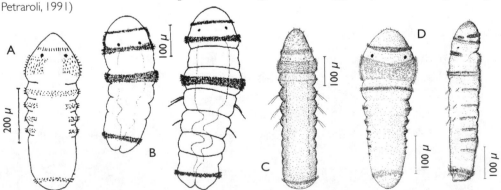

planktotrophic and develop into a characteristic stage called the mitraria, in which they have triangular bodies with undulating ciliated margins and numerous long flotation bristles (Strathmann, 1987).

Myriochele oculata (M. heeri)
Owenia fusiformis (Ammochares)

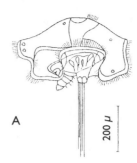

Key to pelagic oweniid larvae (Fig. 10)

1a. Umbrellar lobes evident (except at early stages); prototroch broad with yellow pigment .... *Owenia fusiformis*
1b. Umbrellar margin continuous, lacking lobes; greenish gut, irregular yellow-orange pigment around girdle, red blotches near mouth during metamorphosis .... *Myriochele oculata*

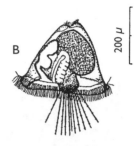

Order Phyllodocidae

Family Alciopidae (Local Species 2, Local Species with Described Larvae 0). Descriptions of larvae of local species are not known, though an undetermined alciopid larva is illustrated in Srikrishnadhas and Ramamoorthi (1975); see Fig. 11. Adults of this family are slender-bodied, exclusively pelagic, and known for their large and complex eyes (Fauchald, 1977). Because alciopids are holoplanktonic, adults are often confused with larval stages (Bhaud and Cazaux, 1987).

Fig. 10. Larvae of (A) *Owenia fusiformis* and (B) *Myriochele oculata*, family Oweniidae. (From Korn, 1960; Lacalli, 1980)

Alciopa reynaudi
Alciopina tenuis (Plotohelmis)

Family Aphroditidae (Local Species 5, Local Species with Described Larvae 0).

Aphrodita japonica
Aphrodita longipalpa
Aphrodita magellanica
Aphrodita parva
Aphrodita refulgida

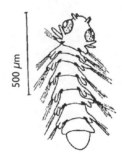

Family Chrysopetalidae (Local Species 2,, Local Species with Described Larvae 1). The larvae of *Paleanotus bellis* (Fig. 12) are known for their sluggish behavior, presence of dorsal paleae, and red coloration of the gut (Blake, 1975b). Cazaux (1968) described similar features for *Chrysopetalum debile*, which may indicate that these characteristics are general for the family.

Fig. 11. General alciopid larva. (From Srikrishnadhas and Ramamoorthi, 1975)

Paleanotus bellis (P. chrysolepis)
Paleanotus occidentale (Chrysopetalum)

Family Glyceridae (Local Species 8, Local Species with Described Larvae 4). Glycerids adults are long, slender-bodied

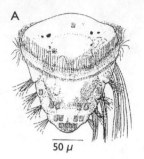

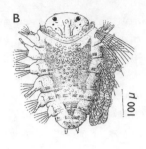

Fig. 12. *Paleanotus bellis,* (A) late metatrochophore, (B) nectochaete, family Chrysopetalidae. (From Blake, 1975b)

with numerous segments, mainly carnivorous, have a long eversible pharynx with four black jaws at the tip, and live in soft sand or mud. Larvae are planktotrophic as trochophores and later, during a benthipelagic stage, they feed on detritus and algae until their jaws are formed (Strathmann, 1987).

> *Glycera americana*
> *Glycera capitata (G. nana)*
> *Glycera convoluta*
> *Glycera gigantea*
> *Glycera oxycephala (G. tenuis)*
> *Glycera robusta*
> *Glycera tesselata*
> *Hemipodus borealis*

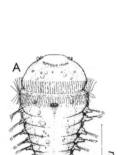

Fig. 13. (A) *Hemipodus borealis,* late metatrochophore, and (B) *Glycera convoluta,* late metatrochophore, family Glyceridae. (From Bhaud and Cazaux, 1987; Plate and Husemann, 1994)

Key to glycerid metatrochophores (Fig. 13)

1a. Metatrochophore conical, narrowing posteriorly; menisotroch present .... *Hemipodus borealis*
1b. Metatrochophore barrel- or cigar-shaped; lacking menisotroch . ... *Glycera convoluta*

Key to glycerid nectochaetes (Fig. 14)

1a. Prototroch present 2
1b. Prototroch absent; weak annulations of prostomium .... *Glycera oxycephala*

2a. Prototroch a double band of cilia 3
2b. Prototroch a single band of cilia .... *Hemipodus borealis*

3a. Prostomium with distinct annulations; body clear with red chromatophores near prototroch; 9 setigers .... *Glycera capitata*
3b. Annulations on prostomium less distinct; 8 setigers; possible brownish pigment near prototroch .... *Glycera convoluta*

Family Goniadidae (Local Species 4, Local Species with Described Larvae 2). Goniadids resemble glycerids in form, but their pharyngeal organs are much larger. Goniadid species may also be distinguished by differences in pharyngeal teeth (Fauchald, 1977).

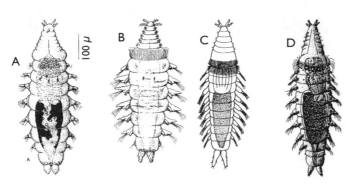

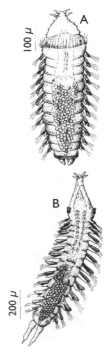

Fig. 14. (A) *Glycera oxycephala,* (B) *Hemipodus borealis,* (C) *Glycera capitata,* and (D) *Glycera convoluta,* family Glyceridae. (From Blake, 1975b; Bhaud and Cazaux, 1987; Plate and Husemann, 1994)

Glycinde armigera
Glycinde polygnatha
Glycinde picta
Goniada brunnea

Key to goniadid nectochaetes (Fig. 15)

1a. Body pale green with red eyes and deep red intestine; prostomium broadly tapering anteriorly .... *Glycinde armigera*

1b. Body with numerous reddish granular pigment markings on each segment and on parapodial lobes of medial and posterior setigers; prostomium narrowly tapered .... *Glycinde polygnatha*

Family Hesionidae (Local Species 2, Local Species with Described Larvae 2). There are few descriptions of hesionid larvae. Direct development and short pelagic lecithotrophic larvae have been described for a few species. *Ophiodromus pugettensis* produces lecithotrophic larvae with a short pelagic phase (Blake, 1975). Adults are common in shallow water and hard substrates; they are fragile and fragment easily (Fauchald, 1977).

Ophiodromus pugettensis (Podarke)
Podarkeopsis brevipalpa (Gyptis arenicola glabra; Gyptis)

Key to hesionid metatrochophores (Fig. 16)

1a. Metatrochophore with several long tentacular cirri, some reaching nearly to posterior margin of body ....
....*Podarkeopsis brevipalpa*

1b. Elongated tentacular cirri, none extending to posterior end of body; 2 lateral antennae on prostomium ....
.... *Ophiodromus pugettensis*

Fig. 15. Larvae of (A) *Glycinde armigera* and (B) *G. polygnath,* family Goniadidae. (From Blake, 1975b)

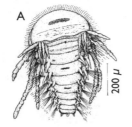

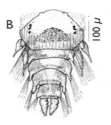

Fig. 16. Larvae of (A) *Podarkeopsis brevipalpa* and (B) *Ophiodromus pugettensis,* family Hesionidae. (From Blake, 1975b)

Fig. 17. (A) *Pelagobia longicirrata* nectochaete and cirri structure and (B, C; not to scale) *Lopadorhyncus uncinatus* metatrochophore and nectochaete, family Lopadorhynchidae. (From Muss, 1953; Akesson, 1966)

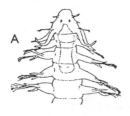

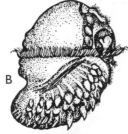
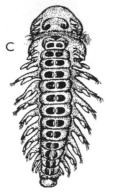

Family Lopadorhynchidae (Local Species 2, Local Species with Described Larvae 1). This is an exclusively pelagic family with short-bodied adults (Fauchald, 1977). Distinction between the local species is based on characteristics of the setae.

> *Lopadorhyncus uncinatus (L. varius)*
> *Pelagobia longicirrata*

Key to lopadorhynchid larvae (Fig. 17)

1a. Dorsal and ventral cirri long and digitiform ....
.... *Pelagobia longicirrata*
1b. Dorsal and ventral cirri thick and lanceolate.......................................
.... *Lopadorhyncus uncinatus*

Family Nephtyidae (Local Species 12, Local Species with Described Larvae 1). Worms of the genus *Nephtys* are predatory, free-spawn, and have pelagic development. Typical trochophores have a large dome-shaped episphere, one pair of eyes, and a barrel-shaped trunk with simple setae. Larvae are predatory as well (Strathmann, 1987). Lacalli (1980) suggests that the pigmentation pattern of irregular ruby red to red-brown bands on the larval tegument near the prototroch and pygidium is typical of larvae of the genus. *Nephtys caeca* (Fig. 18) has brownish pigmentation on the episphere and developing prostomium and pygidium, an olive-colored gut, and no blue pigmentation typical of other larvae in this family.

> *Nephtys assignis*
> *Nephtys caeca*
> *Nephtys caecoides*
> *Nephtys californiensis*
> *Nephtys cornuta cornuta*
> *Nephtys cornuta franciscana*
> *Nephtys ferruginea*
> *Nephtys longosetosa*
> *Nephtys paradoxa*
> *Nephtys punctata*
> *Nephtys rickettsi (N. discors)*
> *Nephtys schmitti*

Family Nereidae (Local Species 17, Local Species with Described Larvae 5). Nereid embryos develop in the plankton or in gelatinous benthic egg masses. Larvae do not feed until lipid drops in the gut are depleted and feeding structures have developed (Strathmann, 1987). Nereids either hatch as nectochaetes or proceed rapidly to the nectochaete stage if hatched at an earlier stage of development (Lacalli, 1980).

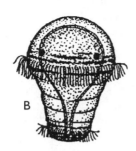
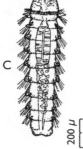

Fig. 18. (A, B) *Nephtys caeca* trochophore and metatrochophore and (C) general *Nephtys* nectochaete, family Nephtyidae. (From Lacalli, 1980)

Ceratonereis paucidentata
Cheilonereis cyclurus
Micronereis nanaimoensis (M. variegata, M. bodegae,
    Phyllodocella)
Nereis brandti (Neanthes)
Nereis eakini
Nereis grubei (N. callaona, N. mediator)
Nereis limnicola (N. lighti, N. diversicolor, N. japonica,
    Neanthes)
Nereis natans
Nereis neoneanthes
Nereis pelagica
Nereis procera
Nereis vexillosa
Nereis virens (Neanthes)
Nereis zonata
Nicon moniloceras (Platynereis)
Perinereis monterea
Playtnereis bicanaliculata (P. dumerili var. *agassizi)*

Key to nereid nectochaetes (Fig. 19)

1a. Tentacular cirri present 2
1b. Tentacular cirri absent .... *Micronereis nanaimoensis*

2a. Prostomium rounded 3
2b. Prostomium blunt or indented slightly at anterior margin ....
.... *Nereis limnicola*

3a. Pygidium conical or bilobed; telotroch present but
    discontinuous 4
3b. Pygidium rounded with distinct and continuous telotroch ....
.... *Nereis pelagica*

4a. Eyes posterior to prototroch at setiger stage 3, yolk reserves
    minimal; pygidium lacks posterior cleft .... *Nereis virens*
4b. Eyes anterior to or at level of prototroch at setiger stage 3,
    abundant blue-green oil droplets in gut; posterior cleft in
    pygidium .... *Platynereis bicanaliculata*

Fig. 19. Larvae of local species from the family Nereidae. (A) *Micronereis nanaimoensis* (B) *Nereis limnicola,* (C) *Nereis pelagica,* (C) *Nereis virens* (D) *Platynereis bicanaliculata.* (From Dales, 1950; Berkeley and Berkeley, 1953; Korn, 1960; Blake, 1975b; Plate and Husemann, 1994)

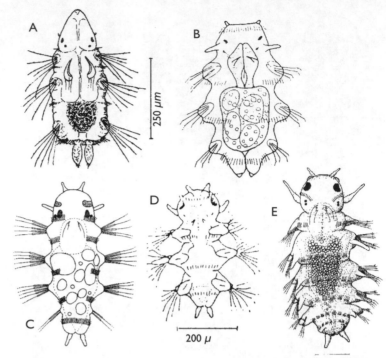

Family Pholoididae (Local Species 1, Local Species with Described Larvae 0). Though development of this family is not known, elytra may be seen in larval forms since the family is part of the scale worm group.

   Pholoides aspera

Family Phyllodocidae (Local Species 19, Local Species with Described Larvae 6). Female phyllodocids spawn near the bottom or deposit strings of small eggs in benthic gelatinous masses. Larvae hatch and spend several weeks in the plankton. Nectochaetes are large and predatory (Strathmann, 1987). No illustrations of local *Phyllodoce* larvae are available, but descriptions of the genera from Bhaud and Cazaux (1987) are included in the key.

   Anaitides groenlandica (Phyllodoce)
   Anaitides hartmanae (Phyllodoce)
   Anaitides medipapillata (Phyllodoce)
   Anaitides mucosa (Phyllodoce)
   Anaitides multiseriata (Phyllodoce)
   Anaitides williamsi (Phyllodoce)
   Eteone californica
   Eteone longa
   Eteone pacifica (E. bistriata, E. maculata, E. spitsbergensis var.
      pacifica)
   Eulalia bilineata
   Eulalia levicornuta

Eulalia nigrimaculata (Bergstroemia, Eumida, Genetyllis,
  Phyllodoce)
Eulalia quadrioculata (E. aviculiseta)
Eulalia sanguinea (Eumida)
Eulalia viridis
Notophyllum imbricatum
Notophyllum tectum (Hesperophyllum)
Phyllodoce castanea (Genetyllis)
Phyllodoce polynoides (Paranaitis)

Key to phyllodocid metatrochopores (Fig. 20)

1a. 2–3 pairs of tentacular cirri 5
1b. 4 pairs of tentacular cirri 2

2a. Well-developed proboscis; cirri are foliaceous .... *Phyllodoce*
2b. Dorsal cirri just budding, proboscis not well developed 3

3a. Prototroch a distinct double band of cilia 4
3b. Prototroch with 1 band of cilia facing anteriorly on ventral
  surface; larva grayish olive green .... *Anaitides groenlandica*

4a. Prototroch with thick, long anterior cilia band and thin posterior
  band; uniform dark green color with deep red eyes ....
  .... *Anaitides williamsi*
4b. Bands of prototroch of equal thickness; larva light green; eyes
  red ....*Anaitides mucosa*

5a. 2 pairs of tentacular cirri; anal cirri not distinct in
  metatrochophore .... *Eteone longa*
5b. 3 pairs of tentacular cirri; anal cirri distinct .... *Eulalia sanguinea*

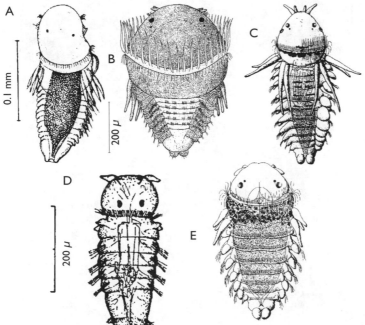

Fig. 20. (A) *Anaitides groendlandica*, (B) *A. williamsi*, (C) *A. mucosa*, (D) *Eteone longa*, and (E) *Eulalia sanguine*, family Phyllodocidae. (From Thorson, 1946; Blake, 1975b; Lacalli, 1980; Bhaud and Cazaux, 1987; Plate and Husemann, 1994)

Key to phyllodocid nectochaetes (Fig. 21)

1a. Rounded or oval prostomium 2
1b. Triangular prostomium, no dorsal cirri on setiger 1; pigmented
     prostomium and pygidium .... *Eteone longa*

2a. Oval dorsal cirri 3
2b. Rounded dorsal cirri 4

3a. 1 pair of red eyes; 1 pair of rounded anal cirri; slight olive-brown
     to dark green gut; olive-brown pygidium .... *Eulalia viridis*
3b. 2 pairs of red eyes; 1 pair of tapering anal cirri; body greenish
     with yellow-white pigmentation posterior to prototroch; dark
     brown gut .... *Eulalia sanguinea*

4a. Dorsal cirri foliaceous and not glandular .... *Phyllodoce*
4b. Dorsal cirri broadly rounded, palmlike, and glandular; first 2
     pairs of dorsal cirri digitataed 5

5a. Prototroch strongly present or reduced 6
5b. Prototroch absent; body grayish olive green; anal cirri spaced
     slightly apart and bulbous .... *Anaitides groenlandica*

6a. Dorsal ciliation beginning on setiger 6 .... *Anaitides williamsi*
6b. Dorsal ciliation beginning prior to setiger 6 .... *Anaitides mucosa*

Family Pilargiidae (Local Species 4, Local Species with
Described Larvae 0). Adults of this family are found in
moderately coarse mixed sediments at shelf depths (Fauchald,
1977). Development of local species has not been described,
but an unidentified *Ancistrosyllis* (Fig. 22) is described in Blake
(1975b).

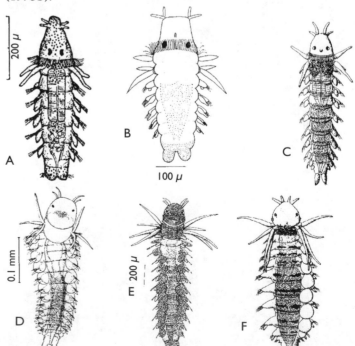

Fig. 21. (A) *Eteone longa*,
(B) *Eulalia viridis*, (C) *E.
sanguinea*, (E) *A. williamsi*,
and (F) *A. mucosa*, family
Phyllodocidae. (From
Thorson, 1946; Blake,
1975b; Lacalli, 1980;
Bhaud and Cazaux, 1987;
Plate and Husemann,
1994)

Ancistrosyllis aff. groenlandica
Pilargus berkeleyi
Sigambra tentaculata
Syhelmis aff. klatti

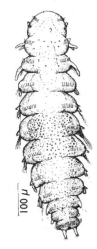

Figuere 22. *Ancystrosyllis* sp. nectochaete. (From Blake, 1975b)

Family Polynoidae (Local Species 21, Local Species with Described Larvae 5). This is the most common family of the scale worm complex. Those polynoids with smooth elytra are often commensal, whereas those with heavily ornamented elytra are free-living (Fauchald, 1977). Some species brood early embryos beneath dorsal elytra; other species spawn freely. Larvae are pelagic for long periods (Strathmann, 1987). *Harmothoe brevisetosa* larvae develop in the plankton, as do *H. imbricata* and *Lepidonotus squamatus*. *Harmothoe extenuata* and *Harmothoe multisetosa* larvae are brooded for a period of time between the parapodia (Pettibone, 1954, 1963).

Arctonoe fragilis
Arctonoe pulchra
Arctonoe vittata
Byglides macrolepida (Antinoella, Antinoe)
Eunoe nodosa
Eunoe oerstedi (E. barbata)
Eunoe senta (Gattyana)
Halosydna brevisetosa
Harmothoe extenuata (Lagisca, H. triannulata, L. rarispina)
Harmothoe fragilis
Harmothoe imbricata (H. hartmanae)
Harmothoe lunulata (Malmgrenia, M. nigralba)
Harmothoe multisetosa (Lagisca)
Harmothoe tenebricosa (H. pellucelytris)
Hermadion truncata
Hesperonoe adventor
Hesperonoe complanata
Lepidasthenia longicirrata
Lepidonotus squamatus (L. caekorus)
Polynoe canadensis (Enipo)
Polynoe gracilis (Enipo, Enipo cirrata)

Key to polynoid late trochophores (Fig. 23)

1a. Episphere flattened and careened orally .. *Halosydna brevisetosa*
1b. Episphere domed 2

2a. Flat lenticular body, colorless tegument; ventral side of intestine green, diameter at prototroch 160 μm ...... *Harmothoe extenuata*
2b. Spherical lenticular body 3

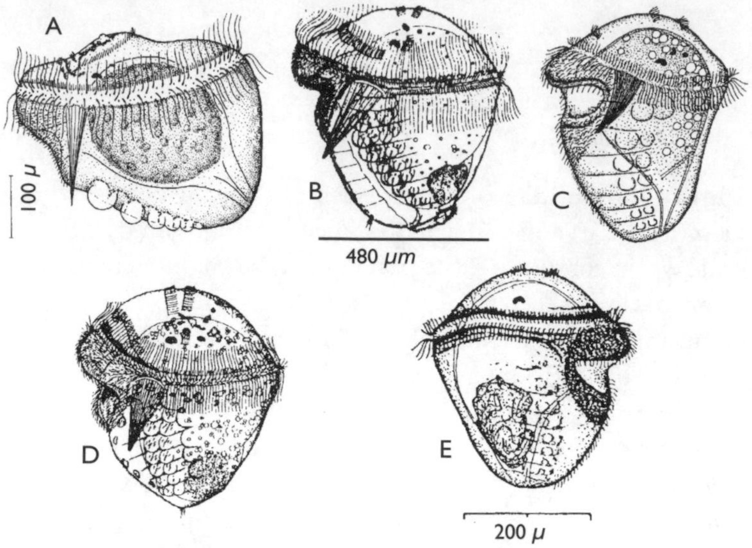

Fig. 23. Late trochophores of species in the family Polynoidae. (A) *Halosydna brevisetosa,* (B) *Harmothoe extenuata,* (C) *Harmothoe imbricata,* (D) *Harmothoe lunulata,* and (E) *Lepidonotus squamatus.* (From Blake, 1975b; Lacalli, 1980; Bhaud and Cazaux, 1987; Plate and Husemann, 1994)

Fig. 24. Metatro-chophores of species in the family Polynoidae. (A) *Halosydna brevisetosa,* (B) *Harmothoe extenuata,* (C) *Harmothoe imbricata,* (D) *Harmothoe lunulata,* and (E) *Lepidonotus squamatus.* (From Blake, 1975b; Lacalli, 1980; Bhaud and Cazaux, 1987; Plate and Husemann, 1994)

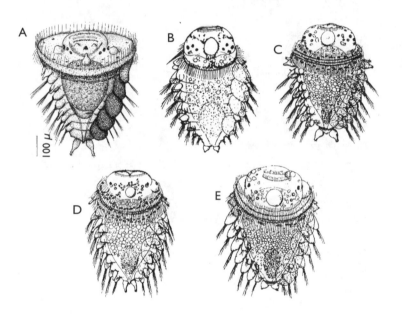

3a. Ventral side of intestine brown, diameter 180 μm .... *Harmothoe imbricata*
3b. Ventral side of intestine purplish violet, diameter 240 μm .... *Harmothoe lunulata*
3c. Ventral side of intestine greenish blue, diameter 200 μm ....*Lepidonotus squamatus*

Key to polynoid metatrochophores (Fig. 24)

1a. 8 segments ....*Lepidonotus squamatus*
1b. 9 segments 2

2a. Prostomium broad and flattened ....*Halosydna brevisetosa*
2b. Prostomium rounded 3

3a. 4 pairs of elytra 4
3b. 5 pairs of elytra, ventral side of intestine light yellow-green ....
.... *Harmothoe lunulata*

4a. Pale red-brown pigmentation on prostomium, peristomium, and
mouth .... *Harmothoe extenuata*
4b. Pale yellow spots, ventral side of intestine purple-brown ....
.... *Harmothoe imbricata*

Key to polynoid nectochaetes (Fig. 25)

1a. 8 segments, 4 pairs of elytra, short dorsal cirri....................................
....*Lepidonotus squamatus*
1b. More than 8 segments, 5 pairs of elytra 2

2a. Dorsal cirri almost twice as long as neuropodia 3
2b. Dorsal cirri only slightly longer than neuropodia; light red-
brown pigment on prostomium and mouth ....
.... *Harmothoe extenuata*

3a. Elytra with papillae; whitish intestine.............. *Harmothoe imbricata*
3b. Elytra without papillae; purple-brown intestine ....
.... *Harmothoe lunulata*

Family Sigalionidae (Local Species 4, Local Species with Described Larvae 1). Larvae of this family are quite similar to those of the Polynoidae. They are somewhat smaller, the elytra are less developed, and they have compound setae rather than simple setae (Lacalli, 1980). *Sthenelais* lack all dorsal elytra; *Pholoe* have dorsal elytra (Plate and Husemann, 1994). Blake (1975b) notes that nectochaetes of *Pholoe caeca* (Fig. 26) feed on algae in culture, though the yolky gut seen in early larval stages suggests that they are lecithotrophic as early larvae. Other members of this family are also described as planktotrophic.

> *Neoleanira areolata (Leanira calcis)*
> *Pholoe caeca (Pholoe tuberculata, P. minuta)*
> *Sthenelais berkeleyi (S. fusca)*
> *Sthenelais verruculosa*

Family Sphaerodoridae (Local Species 4, Local Species with Described Larvae 0). Development mode has been described only for *Sphaerodoropsis minuta*. Mileikovskii (1967) states that development is direct and nonpelagic for this species.

> *Sphaerodoropsis biserialis (Sphaerodorum, Sphaerodoridium)*
> *Sphaerodoropsis sphaerulifer (Ephesia, Sphaerodoridium)*
> *Sphaerodorum papillifer (Ephesia)*
> *Sphaerodoropsis minuta (Sphaerodorum, Ephesiella,*
> *Sphaerodoridium)*

Fig. 25. Nectochaetes of species in the family Polynoidae. (A) *Harmothoe imbricata,* (B) *Harmothoe lunulata,* (C) *Lepidonotus squamatus,* and (D) *Harmothoe extenuata.* (From Blake, 1975b; Lacalli, 1980; Bhaud and Cazaux, 1987; Plate and Husemann, 1994)

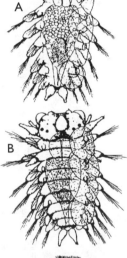

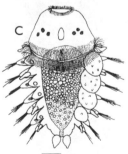

100 µm

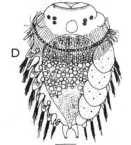

100 µm

Fig. 26.
(A) Metatrochophore
and (B) nectochaete of
Pholoe caeca, family
Sigalionidae. (From Blake,
1975b)

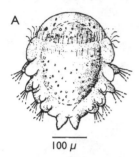

100 μ

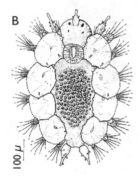

Family Syllidae (Local Species 25, Local Species with Described Larvae 1). Several species are known to have direct development (*), and Pocklington and Hutcheson (1983) suggest that all *Exogone* species are direct developers (Fig. 27).

> *Amblysyllis sp. (A. lineata var. alba)**
> *Autolytus cornutus (A. prismaticus)*
> *Autolytus varius*
> *Ehlersia cornuta (Syllis, S. heterochaeta, S. alternata, S.*
>   *oerstedi, Langerhansia)*
> *Eusyllis assimilis*
> *Eusyllis blomstrandi*
> *Exogone lourei (E. uniformis)**
> *Exogone naidina (E. gemmifera)**
> *Exogone verugera**
> *Haplosyllis spongicola*
> *Odontosyllis parva*
> *Odontosyllis phosphorea (O. phosphorea var. nanaimoensis)*
> *Pionosyllis gigantea*
> *Sphaerosyllis californiensis*
> *Sphaerosyllis hystrix**
> *Sphaerosyllis pirifera*
> *Syllis adamantea (S. spenceri, Typosyllis)*
> *Syllis alternata (Typosyllis)*
> *Syllis armillaris (Typosyllis)*
> *Syllis elongata (Typosyllis)*
> *Syllis hyalina (Typosyllis)*
> *Syllis pulchra (Typosyllis)*
> *Syllis variegata (Typosyllis)*
> *Trypanosyllis gemmipara*
> *Trypanosyllis ingens*

Family Tomopteridae (Local Species 3, Local Species with Described Larvae 0). There are no descriptions of local tomopterid larvae. Adults of this family are holoplanktonic, transparent, and flattened (Fauchald, 1977).

> *Tomopteris cavalli*
> *Tomopteris pacifica (T. elegans, T. renata)*
> *Tompoteris septentrionalis*

Family Typhloscolecidae (Local Species 2, Local Species with Described Larvae 0). This is a poorly known family. Adults are holoplanktonic, transparent, and fusiform (Fauchald, 1977).

> *Sagitella kowalevskii*
> *Travisiopsis lobifera*

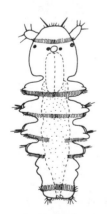

Fig. 27. *Autolytus* sp., family Syllidae (From Plate and Husemann, 1994)

Order Sabellida

Family Sabellidae (Local Species 25, Local Species with Described Larvae 2). Patterns of development in this family include brooded lecithotrophs, direct developers, freely spawned eggs developing into lecithotrophic larvae, and those that develop in gel masses (reviewed by McEuen et al., 1983).

Chone ecaudata (C. minuta, C. gracilis)
Chone gracilis
Chone infundibuliformis (C. teres)
Chone magna
Chone mollis
Demonax media (Sabella, S. aulaconota, Distylia rugosa, Parasabella, P. maculata, Potamilla californica)
Euchone analis
Euchone sp. cf. hancocki
Euchone incolor (E. rosea, E. trisegmentata, E. barnardi)
Eudistylia polymorpha
Eudistylia vancouveri (E. tenella, E. plumosa, E. abbreviata)
Fabricia brunnea (F. sabella)
Fabricia oregonia
Fabricia sabella (F. dubia)
Fabriciola berkeleyi (F. pacifica, F. sabella)
Megalomma splendida (Branchiomma burrardum)
Myxicola aesthetica
Myxicola infundibulum
Oriopsis gracilis
Potamilla intermedia (Pseudopotamilla, P. reniformis)
Potamilla neglecta
Potamilla occelata (Pseudopotamilla)
Sabella crassicornis
Schizobranchia insignis

Key to pelagic sabellidae larvae (Fig. 28)

1a. 3 setigers at metatrochophore stage; otocysts absent; telotroch absent or reduced .... *Demonax media*
1b. 3–4 setigers; otocysts and telotroch present ....
.... *Chone infundibuliformis*

Family Serpulidae (Local Species 7, Local Species with Described Larvae 1). Adults of this family build calcareous tubes (Fig. 29).

Apomatus geniculatus
Apomatus timsi
Crucigera irregularis
Crucigera zygophora

Fig. 28. Larvae
(A) *Demonax media* and
(B) *Chone infundibuli-*
formis, family Sabellidae.
(From Okuda, 1946;
McEuen et al., 1983)

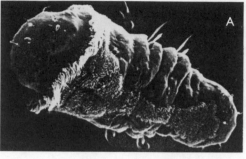

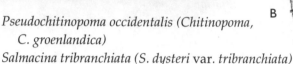

Pseudochitinopoma occidentalis (Chitinopoma,
    C. groenlandica)
Salmacina tribranchiata (S. dysteri var. tribranchiata)
Serpula columbiana (S. vermicularis)

Family Spirorbidae (Local Species 11, Local Species with Described Larvae 2). All spirorbid genera found in Pacific Northwest waters brood their larvae, either within the adult tube or externally (Fauchald, 1977). Larvae have a characteristic complex of gland cells in the ventral and lateral ectoderm of the thorax.

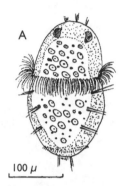

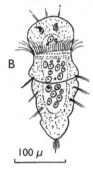

Circeis amoricana (C. spirillum, Spirorbis rugatus)
Circeis spirillum (Dexiospira, Spirorbis)
Janura rugata (Spirorbis, Dexiospira)
Paradexiospira violacea
Paradexiospira vitrea (Laeospira, Eulaeospira, Spirorbis,
    S. variabilis, S. semidentatus, S. racemosus)
Pileolaria langerhansi
Pileolaria potswaldi (Laeospira, Spirorbis, S. moerchi)
Pileolaria quadrangularis
Protolaeospira eximia (Spirorbis, S. ambilateralis)
Sinistrella media (Spirorbis, Laeospira, Romanchella)
Spirorbis bifurcatus

Key to pelagic spirorbid larvae (Fig. 30)

1a. Body and pygidium of metatrochophore tapering narrowly;
    collar extending over setiger 1 .... *Circeis spirillum*
1b. Pygidium flanged; collar extending over setiger 2 .... *Pileolaria potswaldi*

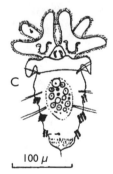

Fig. 29. (A) Late
trochophore,
(B) metatrochophore,
and (C) late
metatrochrophore larvae
of *Salmancia tribranchiata,*
family Serpulidae. (From
Rullier, 1960)

Order Spionida

Family Acrocirridae (Local Species 1, Local Species with Described Larvae 0). Little is known about the larval development in this family. It may be similar to that in the Cirratulidae and Flabelligeridae (Banse, 1969).

Acrocirrus heterochaetus

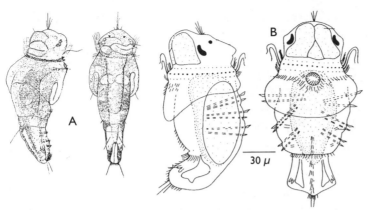

Fig. 30. Larvae of species in the family Spirorbidae. (A) *Circeis spirillum*, lateral and dorsal views, (B) *Pileolaria potswaldi*, lateral and ventral views. (From Okuda, 1946; Potswald, 1978)

Family Apistobranchidae (Local Species 1, Local Species with Described Larvae 0). Development in this family has not been described. Adults are tube dwellers but can exist outside tubes in loosely constructed burrows. They are common in shelly sands and muds (Fauchald, 1977).

Apistobranchus ornatus

Family Chaetopteridae (Local Species 4, Local Species with Described Larvae 2). Chaetopterid larvae are separated by comparison of the modified setae on the fourth anterior segment.

Chaetopterus variopedatus
Mesochaetopterus taylori
Phyllochaetopterus prolifica
Spiochaetopterus costarum (Telepsavus)

Key to chaetopterid genera (Fig. 31)

1a. 1 pair of setae on setiger 4 2
1b. Several setae on setiger 4 3

2a. Modified setae with enlarged heart-shaped distal section ....
.... *Spiochaetopterus*
2b. Setae not enlarged distally .... *Phyllochaetopterus*

3a. Distal part of setae blade-shaped .... *Mesochaetopterus*
3b. Distal part of setae with lateral tip .... *Chaetopterus*

Family Cirratulidae (Local Species 11, Local Species with Described Larvae 1). Development in this family includes many species that produce lecithotrophic larvae (*) that hatch from eggs laid in a gel mass as well, as species that are direct developers (†). Christie (1985) notes that all larvae of this family are large and yolky (Fig. 32).

Caulleriella alata
Caulleriella hamata (Tharyx)
*Chaetozone setosa**

A

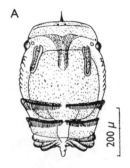

200 μ

B

130 μ

Fig. 31. (A) *Chaetopterus variopedatus* and (B) *Spiochaetopterus costarum*, family Chaetopteridae. (From Bhaud and Cazaux, 1987)

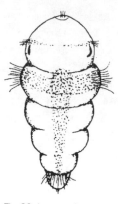

Fig. 32. Larva of
Dodecaceria fewkesi,
family Cirratulidae. (From
Berkeley and Berkeley,
1954)

Cirratulus cirratus cirratus† *(Chaetozone berkeleyorum,*
    Caulleriella gracilis)
Cirratulus spectabilis (C. cirratus cingulatus, C. robustus)
Cirriformia spirabranchia
*Dodecaceria choncharum**
Dodecaceria fewkesi
Dodecaceria fistulicola
Tharyx multifilis
Tharyx parvus (T. multifilus var. *parvus)*

Family Magelonidae (Local Species 1, Local Species with
Described Larvae 0). An unknown magelonid, possibly
Magelona cerae, is often found in plankton samples taken from
Coos Bay, Oregon (K. Johnson, pers. comm.). Adults are
common in sandy substrates and move through the sediment
with a shovel-like prostomium and thread-like body. Larvae
of this family have characteristically long larval tentacles that
are flexible and often coiled. *Magelona alleni* illustrates
characteristics of the family (Fig. 33), but it is not a species
known to be in local waters.

Magelona sp.

Family Spionidae (Local Species 27, Local Species with
Described Larvae 18). The spionids are a large and diverse
group of polychaetes. Adult habitat and development modes
vary across the family. A large number of larvae of local species
have been described.

Boccardia columbiana (Polydora)
Boccardia polybranchia
Boccardia proboscidea (B. californica)
Boccardiella hamata (Boccardia, B. uncata, Polydora)
Laonice cirrata (Nerine)
Paraprionospio pinnata (Prionospio, P. ornata)

Fig. 33. Late
metatrochophore of
Magelona alleni, family
Magelonidae. (From
Wilson, 1982)

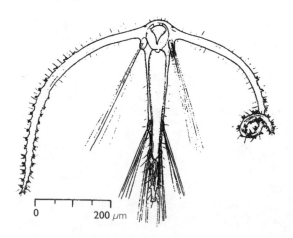

0   200 μm

Polydora alloporis
Polydora armata
Polydora brachycephala
Polydora cardalia
Polydora commensalis
Polydora giardi
Polydora ligni
Polydora pygidialis (P. ciliata)
Polydora socialis (P. caeca, P. caeca var. *mangna, P. plena)*
Polydora spongicola (P. ciliata var. *spongicola)*
Polydora websteri
Prionospio lighti (Minuspio cirrifera)
Prionospio steenstrupi (P. malmgreni)
Pseudopolydora kempi (Neopygospio laminifera, Polydora,
   Polydora kempi japonica)
Pygospio elegans
Scolelepis foliosa (Nerine)
Scolelepis squamata (S. acuta)
Spio filicornis
Spiophanes berkeleyorum (S. cirrata)
Spiophanes bombyx
Streblospio benedicti

Key to spionid nectochaetes (Fig. 34)

1a. Branchiae absent; setiger 1 with 1–2 large curved neuropodial
   spines in addition to normal capillaries .... *Spiophanes bombyx*
1b. Branchiae present; setiger 1 without specialized setae 2

2a. Setiger 5 modified, with specialized setae 11
2b. Setiger 5 not modified, without specialized setae 3

3a. Prostomium distally pointed (may appear conical with rounded
   apex in extremely contracted specimens), with or without
   subdistal lateral horns 4
3b. Prostomium not distally pointed, with distal lateral or frontal
   horns, broadly rounded or incised on anterior margin 5

4a. Prostomium bell-shaped, orange-pigmented; gut blue-green ....
   .... *Scolelepis squamata*
4b. Prostomium triangular and lacking orange pigment; gut blackish-
   brown .... *Scolelepis foliosa*

5a. Branchiae limited to middle and posterior setigers except for
   single pair on setiger 2 in males .... *Pygospio elegans*
5b. Branchiae beginning on setiger 1 or 2 and continuing for
   variable number of setigers 6

6a. Branchiae concentrated in anterior setigers 1—2, absent
   posteriorly 7
6b. Branchiae present over most of body 10

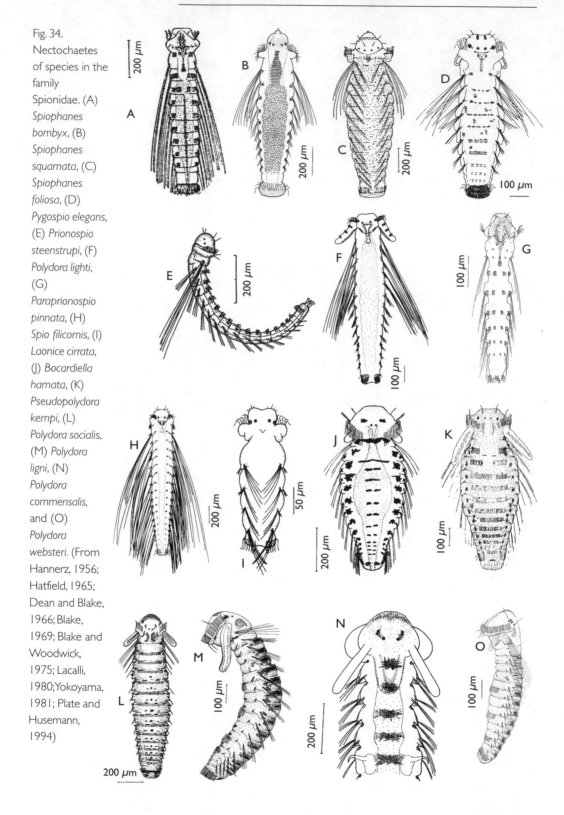

Fig. 34. Nectochaetes of species in the family Spionidae. (A) *Spiophanes bombyx*, (B) *Spiophanes squamata*, (C) *Spiophanes foliosa*, (D) *Pygospio elegans*, (E) *Prionospio steenstrupi*, (F) *Polydora lighti*, (G) *Paraprionospio pinnata*, (H) *Spio filicornis*, (I) *Laonice cirrata*, (J) *Bocardiella hamata*, (K) *Pseudopolydora kempi*, (L) *Polydora socialis*, (M) *Polydora ligni*, (N) *Polydora commensalis*, and (O) *Polydora websteri*. (From Hannerz, 1956; Hatfield, 1965; Dean and Blake, 1966; Blake, 1969; Blake and Woodwick, 1975; Lacalli, 1980; Yokoyama, 1981; Plate and Husemann, 1994)

7a. Branchiae beginning on setiger 1 9
7b. Branchiae beginning on setiger 2 8

8a. Peristomial palps short; 1 pair of anal cirri in early larvae, 2 pairs
    of anal cirri present in later larvae; eyes dark red .... *Prionospio*
    steenstrupi
8b. Persitomial palps differentiate only in late larval stages; 1 pair of
    anal cirri in late larvae .... *Prionospio lighti*

9a. 1 pair of cirriform branchiae; dorsal collar across setiger 2 ....
    .... *Streblospio benedicti*[a]
9b. 3 or more pairs of branchiae; no dorsal collar on setiger 2 ....
    .... *Paraprionospio pinnata*

10a. Branchiae beginning on setiger 1; hooded hooks only in
    neuropodia .... *Spio filicornis*
10b. Branchiae beginning on setiger 2; prostomium broad, bluntly
    rounded or squared on anterior margin; branchiae free from
    dorsal lamellae .... *Laonice cirrata*

11a. Branchiae beginning on setiger 2 12
11b. Branchiae beginning on setigers 6–12 13

12a. Major spines of setiger 5 of one type: simple, falcate, with
    smaller companion setae .... *Bocardiella hamata*
12b. Major spines of setiger 5 of two types: first with expanded
    ends bearing cusps or bristles, second simple, falcate ....
    .... *Boccardia proboscidea*[b]

13a. Setiger 5 slightly to moderately modified, usually with
    prominent parapodia; major spines of two types: first simple,
    acicular or falcate, second pennoned with both types usually
    arranged in U- or J-shaped row. Hooded hooks with secondary
    tooth placed close to main fang .... *Pseudopolydora kempi*
13b. Setiger 5 greatly modified, with reduced parapodia; major
    spines of one or two types in curved row, neither J- or U-
    shaped; hooded hooks with prominent angle between teeth 14

14a. Larvae with distinct ventral chromatophores (black, yellow-
    green, or iridescent) 15
14b. Larvae without ventral pigment, or if present not in distinct
    chromatophores 16

15a. Ventral pigment black .... *Polydora socialis*
15b. Ventral pigment yellow-green or iridescent .... *Polydora lighti*

16a. Single dorsal row of melanophores present, at least posterior
    to segment 3 .... *Polydora commensalis*
16b. Multiple rows of dorsal melanophores present 17

17a. Prostomium with yellow-brown pigment; gastrotrochs on
    segments 7, 10, 13, 15 .... *Polydora websteri*
17b. Edges of prostomium with black pigment; gastrotrochs on
    segments 3, 5, 7, 10, 13, 15, 17, 19 .... *Polydora pygidialis*

[a]Levin (1991) illustrates lecithotrophic pelagic larvae that lack swimming setae. [b]Development of this species is described in Woodwick (1977).

Fig. 35. Representative
trochochaetid larva,
Disoma multisetosum.
(From Hannerz, 1961)

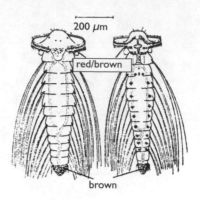

Family Trochochaetidae (Local Species 1, Local Species with Described Larvae 0). This is a small family (nine species) of non-tubiculous worms that live in soft substrates in mainly shallow water. Adults are cylindrical with a slightly flattened body (Fauchald, 1977). An image of *Disoma multisetosum* is included for reference (Fig. 35).

Trochochaeta multisetosa (T. franciscanum)

Order Sternaspida

Family Sternaspidae (Local Species 1, Local Species with Described Larvae 0). This small family has approximately ten recognized species. Adults are common in sand and mud at all depths, but usually at 100–200 m. Adults are rarely found in large numbers, are burrowers, and have dark yellow or reddish chitinized shields (Fauchald, 1977).

Sternaspis scutata (S. fossor, S. affinis)

Order Terebellidae

Family Ampharetidae (Local Species 14, Local Species with Described Larvae 0). Thorson (1946) suggests that several species (*) in this family are direct developers. Nyholm (1950), however, describes *Melinna cristata* as a free-spawned lecithotrophic larva. This is a fairly large, deep-water family (Fauchald, 1977).

Amage anops
*Ampharete acutifrons (A. grubei)**
*Ampharete finmarchica (A. arctica)**
Ampharete goesi goesi
Amphicteis mucronata
Amphicteis scaphobranchiata
Amphisamytha bioculata
*Anobothrus gracilis (Ampharete gagarae)**
Asabellides lineata (Pseudosabellides)

Asabellides sibrica
Hobsonia florida (Amphicteis gunneri floridus)*
Lysippe annectens
Melinna cristata (M. denticulata)
Schistocomus hiltoni

Family Pectinariidae (Local Species 3, Local Species with Described Larvae 3). Trochophores of this family are easily confused with nephtyid larvae (Lacalli, 1980). Pectinariids construct and live in fragile tusk-shaped tubes (Fauchald, 1977).

Amphictene moorei (A. auricoma, Pectinaria)
Cistenides granulata (C. brevicoma, Pectinaria)
Pectinaria californiensis (P. belgica)

Key to pectinariid metatrochophores (Fig. 36)

1a. Prototroch and telotroch pronounced 2
1b. Cilia of prototroch and telotroch short or reduced ....
.... *Pectinaria californiensis*

2a. Eyes red, body surface with regular rows of round dark red to brown spots; lobes of oral hood blunt and indented ....
.... *Cistenides granulata*
2b. Oral hood winglike and large, neurotroch present, oral cilia long
.... *Amphictene moorei*

Family Sabellariidae (Local Species 4, Local Species with Described Larvae 0). Larval morphology is quite similar among sabellariid species (Ecklebarger, 1975, 1977). Opercular characteristics are useful for distinguishing genera according to Bhaud and Cazaux (1987). The key below is modified from Bhaud and Cazaux (1987) for the two local genera. *Sabellaria alveola* (Fig. 37) is not a local species.

Idanthyrsus armatus
Idanthyrsus ornamentatus
Sabellaria cementarium
Sabellaria gracilis

Key to sabellariid genera

1a. Opercular peduncles fused; lacking opercular hooks ... *Sabellaria*
1b. Opercular peduncles not completely fused; opercular hooks present .... *Idanthyrsus*

Family Terebellidae (Local Species 20, Local Species with Described Larvae 2). Terebellids are common shallow-water polychaetes and are found in a diversity of environments. They inhabit permanent tubes from which they extend a crown of elongated extensile tentacles that they use to capture small food

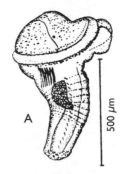

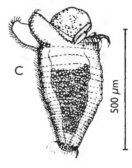

Fig. 36. (A) *Pectinaria californiesis*, (B) *Cistenides granulata*, and (C) *Amphictene moorei*, family Pectinariidae. (From Thorson, 1946; Korn, 1960; Lacalli, 1980)

A — 500 µm

B — 200 µm

C — 500 µm

width 8 μm; length 600 μm

Fig. 37. Larvae of *Sabellaria alveola*, family Sabellaridae; this not a local species. (From Bhaud and Cazaux, 1987)

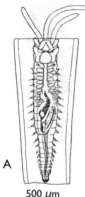

A

500 μm

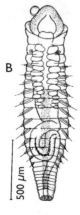

B

Fig. 38. (A) *Lanice conchilega* and (B) *Amphitrite cirrata*, outside tube, family Terebellidae. (From Thorson, 1946)

particles from the surface of the substrate or water (Fauchald, 1977).

Amphitrite cirrata (A. palmata)
Artacama coniferi
Eupolymnia heterobranchia (E. crescentis)
Lanice chonchilega
Loimia medusa
Neoamphitrite robusta (Amphitrite, A. scionides dux)
Neoleprea spiralis
Pista brevibranchiata (P. fasciata)
Pista cristata
Pista elongata
Pista fasciata
Pista fimbriata (P. brevibranchiata)
Pista moorei
Pista pacifica
Polycirrus californicus (P. caliendrum)
Scionella japonica
Thelepus cincinnatus (T. hamatus)
Thelepus crispus
Thelepus hamatus
Thelepus setosus

Key to terebellidae larvae (Fig. 38)

1a. Larva with peristomial tentacles; colorless with pale olive-green intestine and faint brownish anal pigment; tentacles yellow-brown; each parapodium with 2–3 setae adhering at distal ends; eyes yellow-brown .... *Lanice conchilega*
1b. No tentacles present; anal segment yellowish-brown; 1 pair of setae per setiger; eyes red .... *Amphitrite cirrata*

Family Trichobranchidae (Local Species 3, Local Species with Described Larvae 0). Developmental patterns for *Artacamella* and *Trichobranchus* are not known. *Terebellides stroemi* develops directly in egg masses (Thorson, 1946).

Artacamella hancocki
Terebellides stroemi
Trichobranchus glacialis

References

Akesson, B. (1967). The embryology of the *polychaete Eunice kobiensis*. Acta Zool. (Stockholm). 43:135–99.

Allen, M. J. (1952). An example of parasitism among polychaetes. Nature (London) 169–97.

———— (1959). Embryological development of the polychaetous annelid *Diopatra cuprea* (Bosc). Biol. Bull. 116:339–61.

Anderson, D. T. (1959). The embryology of the polychaete *Scoloplos armiger*. Quart. J. Microsc. Sci. 100:89–166.

Banse, K. (1969). Acrocirridae n. fam. (Polychaeta sedentaria). J. Fish. Res. Bd. Can. 26:2595–2620.

Berkeley, E. and C. Berkeley (1953). *Micronerieis nanaimoensis* sp.n.; with some notes on its life history. J. Fish. Res. Bd. Can. 10:84–95.

———— (1954). Notes on the life history of the polychaete *Dodecaceria fewkesi* (nom. nov.). J. Fish. Res. Bd. Can. 11:326–34.

Bhaud, M. and C. Cazaux (1987). Description and identification of polychaete larvae; their implications in current biological problems. Oceanis 13(6):596–753.

Blake, J. A. (1969). Reproduction and larval development of *Polydora* from northern New England (Polychaeta: Spionidae). Ophelia 7:1–63.

———— (1975a). The larval development of Polychaeta from the Northern California coast. II. *Nothria elegans* (Family Onuphidae) Ophelia 13:43–61.

———— (1975b). The larval development of Polychaeta from the Northern California coast. III Eighteen species of Errantia. Ophelia 14:23–84.

———— (1980). The larval development of Polychaeta from the Northern California coast. IV. *Leitoscoloplos pugettensis* and *Scoloplos acmeceps* (Family Orbiniidae). Ophelia 19(1):1-18.

———— and K. H. Woodwick (1975). Reproduction and larval development of *Pseudopolydora paucibranchiata* (Okuda) and *Pseudopolydora kempi* (Southern) (Polychaeta: Spionidae). Biol. Bull. 149:109–27.

Cazaux, C. (1968). Développement larvaire de *Chaetopterus variopedatus* (Renier). Actes Soc. Linné. Bordeaux. 102:1–31.

Christie, G. (1985). The reproductive cycles of two species of *Pholoe* (Polychaeta: Sigalionidae) off the Northumberland Coast. Sarsia 67:283–92.

Dales, R. P. (1950). The reproduction and larval development of *Nereis diversicolor* (O. F. Müller) J. Mar. Biol. Ass. U.K. 30:113–17.

———— (1952). The larval development and ecology of *Thoracophelia mucronata* (Treadwell). Biol. Bull. 102:232–42.

Dean, D. and J. A. Blake (1966). Life-history of *Boccardia hamata* (Webster) on the east and west coasts of North America. Biol. Bull. 130:316–30.

Eckelbarger, K. J. (1975). Developmental studies of the post-settling stages of *Sabellaria vulgaris* (Polychaeta: Sabellariidae). Mar. Biol. 30:137–49.

———— (1977). Larval development of *Sabellaria floridensis* from Florida and *Phragmatopoma californica* from southern California (Polychaeta: Sabellariidae) with a key to the Sabellariid larvae of Florida and a review of development in the family. Bull. Mar Sci. 27:241–55.

Fauchald, K. (1977). The polychaete worms. Definitions and keys to the orders, families and genera. Nat. Hist. Mus. Los Angeles County, Sci. Ser. 28:1–190.

Giangrande, A. and A. Petraroli (1991). Reproduction, larval development and post-larval growth of *naineris laevigata* (Polychaeata, Orbiniidae) in the Mediterranean Sea. Mar. Biol. 111:129–37.

Hannerz, L. (1956). Larval development of the polychaete families Spionidae Sars, Disomidae Mesnil, and Poecilochaetidae n. fam. in the Gullmar Fjord (Sweden). Zool. Bidr. Uppsala 31:1–204.

——— (1961). Polychaeta: Larvae of Families: Spionidae, Disomidae, Poecilochaetidae. Cons. Int. Expl. Mer. 1–12.

Hatfield, P. A. (1965). *Polydora commensalis* (Andrews)—Larval development and observations on adults. Biol. Bull. 128:356–78.

Heffernan, P. and B. F. Keegan (1988). The larval development of *Pholoe minuta* (Polychaeta: Sigalionidae) in Galway Bay, Ireland. J. Mar. Biol. Assn. U.K. 68:339–50.

Hermans, C. O. (1978). Metamorphosis in the Opheliid Polychaete *Armandia brevis*. In: Settlement and Metamorphosis of Marine Invertebrate Larvae, F-S. Chia and M. Rice (eds.), pp. 113–126. Elsevier, New York.

Korn, H. (1960). Introduction to the Polychaete larvae. II. Catalogue of Marine Larvae. Universidade de Sao Paulo Instituto Oceanografico, Sao Paulo.

Lacalli, T. C. (1980). A guide to the marine flora and fauna of the Bay of Fundy: Polychaete larvae from PassamaquoddyBay. Canadian Technical Reports, Fisheries and Aquatic Sciences. 940: 1–26.

Levin, L. A. (1984). Multiple patterns of development in *Streblospio benedicti* Webster (Spionidae) from three coasts of North America. Biol. Bull. 166:494–508.

McEuen, F. S., B. L. Wu, and F.-S. Chia (1983). Reproduction and development of *Sabella media*, a sabellid polychaete with extratubular brooding. Mar. Biol. 76:301–9.

Mileikovskii, S. A (1961). Assignment of two rostraria-type polychaete larvae from the plankton of the northwest Atlantic to species *Amphinome pallasi* Quatrefages 1865 and *Chloenea atlantica* McIntosh 1885 (Polychaeta, Errantia, Amphinomorpha). Dokl. Biol. Sci. (Transl.) 141:1109–12.

——— (1967). Larval development of polychaetes of the family Sphaerodoridae and some considerations of its systematics. Dokl. Biol. Sci. (Transl.) 177:851–54.

Monroe, C. C. A. (1924). On the post-larval stage in *Diopatra cuprea* Bose, a polychaetous annelid of the family Eunicidae. Ann. Mag. Natur. Hist. Ser. 9:193-99.

Moore, J. P. (1903). Polychaeta from the coastal slope of Japan and from Kamchatka and Bering Sea. Proc. Acad. Nat. Sci. Philadelphia 55:401-90.

Muus, B. J. (1953). Polychaeta Families: Aphroditidae, Phyllodocidae, and Alciopidae. Conseil International pour L'Exploration de la Mer. 1–5.

Okuda, S. (1946). Studies on the development of Annelida Polychaeta. I. J. Fac. Sci. Hokkaido Univ. Ser. 6. 9:115–219.

Pettibone, M. H. (1954). Marine polychaete worms from Point Barrow, Alaska, with additional records from the North Atlantic and North Pacific. Proc. U.S. Nat. Mus. 103:203–356.

———. (1957). Endoparasitic polychaetous annelids of the family Arabellidae with descriptions of new species. Biol. Bull. 113:170–87.

——— (1963). Marine polychaete worms of the New England Region. I. Aphroditidae-Trochochaetidae. Bull. U.S. Nat. Mus. 227:1–356

Plate, S. and E. Husemann (1991). An alternative mode of larval development in *Scoloplo armiger* (O. F. Muller, 1776) (Polychaeta, Orbiniidae) Helgolander Meersunters. 45:487–92.

Pocklington, P. and M. S. Hutcheson (1983). New record of viviparity for the dominant benthic invertebrate *Exogone hebes* (Polychaeta: Syllidae) from the Grand Banks of Newfoundland. Mar. Ecol. Prog. Ser. 11:239–44.

Potswald, H. E. (1978). Metamorphosis in *Spirorbis* (Polychaeta). In: Settlement and Metamorphosis of Marine Invertebrate Larvae, F-S. Chia and M. Rice (eds.), pp. 127–43. Elsevier, New York.

Rasmussen, E. (1956). Faunistic and biological notes on marine invertebrates III. The reproduction and larval development of some polychaetes from the Isefjord, with some faunistic notes. Biol. Medd. Kg. Dan. Vidensk. Selsk. 23:1–84.

Richards, T. L. (1967). Reproduction and development of the polychaete *Stauronereis rudolphi*, including a summary of development in the superfamily Eunicea. Mar. Biol. 1:124–33.

Rouse, G. W. (1992). Oogenesis and larval development in *Micromaldane* spp. (Polychaeta: Capitellida: Maldanidae). Invert. Reprod. and Dev. 21(3):215–30.

Rullier, F. (1960). Developpement de *Salmacina dysteri* (Huxley). Cah. Biol. Mar. 1:37–46.

Smith, P. R. and F-S. Chia. (1985). Larval development and metamorphosis of *Sabellaria cementarium* Moore, 1906 (Polychaete: Sabellariidae). Can. J. Zool. 63:1037-49.

Srikrishnadhas, B. and K. Ramamoorthi (1975). Studies on some Polychaete larvae of Porto Novo waters. Bull. Dept. Mar. Sci. Univ. Chochin. 7(4):733–49.

Strathmann, M. F. (1987). Reproduction and Development of Marine Invertebrates of the Northern Pacific Coast. University of Washington Press, Seattle, pp 138–95.

Thorson, G. (1946). Reproduction and larval development of Danish marine bottom invertebrates, with special reference to the planktonic larvae in the Sound (Oresund). Medd. Komm. Dan. Fisk.-Havundersogelser. Ser. Plankton 4(1):1–523.

Wilson, D. P. (1982). The larval development of three species of *Magelona* (Polychaeta) from localities near Plymouth. J. Mar. Biol. Ass. U.K. 62:385–401.

Wilson, W. H. (1983). Life-history evidence for sibling species in *Axiothella rubrocincta* (Polychaeta: Maldanidae). Mar. Biol. 76:297–300.

Woodwick, K. H. (1977). Lecithotrophic larval development in *Boccardia proboscidea* hartman. In: Essays on polychaetous annelids, D. J. Reish and K. Fauchald (eds.), pp. 347–371. Allan Hancock Found.

Yokoyama, H. (1981). Larval development of a spionid polychaete *Paraprionospio pinnata* (Ehlers). Publ. Seto Mar. Biol. Lab. XXVI (1/3):157–70.

Zottoli, R. A. (1974). Reproduction and larval development of the ampharetid polychaete *Amphicteis floridus*. Trans. Amer. Micros. Soc. 93:78–89.

Sipuncula:
The Peanut Worms

Kevin B. Johnson

The sipunculids are a small phylum of worm-shaped marine animals. They are characterized by a lack of segmentation, and the body is divided into two sections: a thick, frequently bulbous, posterior trunk and a thinner introvert. The introvert can be retracted, or introverted, into the trunk (Rice, 1980). With the introvert retracted, many species take on the shape of a peanut, hence, their common name, "peanut worms."

Worldwide there are around 320 species of sipunculids. They are found in all marine habitats. Kosloff (1996) lists five species present in the Pacific Northwest (Table 1) and Morris et al. (1980) list six species in the intertidal zone of the California coast. Local species are found in a variety of intertidal habitats (e.g., wedged between or under rocks, in boring clams' holes and mussel beds, amongst the roots of surfgrass, and burrowed into mud or sand). They are generally deposit feeders.

Table 1. Species in the phylum Sipuncula from the Pacific Northwest

Family Phascolosomatidae
Phascolosoma agassizii

Family Golfingiidae
Golfingia margaritacea
Themiste pyroides
Themiste dyscrita
Thysanocardia nigra
(=Golfingia pugettensis)

Reproduction and Development

Most sipunculans reproduce sexually and, in all but one known case, are dioecious. Asexual reproduction has been reported for two species (Rice, 1970). Parthenogenesis, the development of unfertilized eggs into viable adults, has been observed in the sipunculan *Themiste lageniformis* (Pilger, 1987). In general, gametes are spawned into the water column, where fertilization and development (indirect) take place. Exceptions include several species that undergo direct development on the benthos. Additional details regarding reproductive patterns and strategies in the phylum Sipuncula are summarized in Brusca and Brusca (1990).

Of the species with indirect development, all but *T. lageniformis* produce a simple trochophore larva. In some species the trochophore, short-lived and lecithotrophic, quickly metamorphoses directly into a juvenile peanut worm. In others the trochophore develops to the pelagosphera stage, which itself may be either lecithotrophic or planktotrophic. The four most common patterns of development exhibited by the Sipuncula are categorized in a numerical system devised by Rice (1975b). This system is presented in Table 2. Except for species with planktotrophic pelagosphera larvae, sipunculid larvae are pelagic only briefly and, consequently, are almost never observed in plankton samples (M. Rice, pers. comm.).

The pelagosphera larva, vermiform with a single pair of anterior eye spots, bears an enlarged metatroch and completes metamorphosis by elongating and settling to become a juvenile benthic peanut worm.

Identification of Local Taxa

Sipunculan species with known geographic ranges encompassing the Pacific Northwest are *Phascolosoma agassizii, Golfingia margaritacea, Thysanocardia nigra, Themiste pyroides,* and *T. dyscrita* (Table 1; see Rice, 1980; Austin, 1985; Kozloff, 1993, 1996). Most of these species undergo direct and benthic development, or their larvae are lecithotrophic and only briefly planktonic.

Only one local species, *P. agassizii,* exhibits a long-lived, planktotrophic pelagosphera larval stage, so pelagosphera larvae observed in local plankton samples are most likely of

Table 2. Sipunculans with described development

Species	Development[1]	References[2]
Golfingia minuta	Direct (I)	Akesson, 1958[a]
Themiste pyroides	Direct (I)	Rice, 1967[a]
Phascolion cryptus	Direct (I)	Rice, 1975a[a]; Rice, 1975b[a]
Themiste lageniformis	Indirect (lacks trochophore)	Pilger, 1987[e]
Phascolion strombi	Indirect (II)	Akesson, 1958[a,c]
Phascolopsis gouldi	Indirect (II)	Gerould, 1907[d]; Rice, 1975a[c]
Themiste alutacea	Indirect (III)	Rice, 1975a[a]; Rice, 1975b[a]
Golfingia vulgaris	Indirect (III)	Gerould, 1907 [d]; Rice, 1975a[c]
Golfingia elongata	Indirect (III)	Akesson, 1961[c]; Rice, 1975a[c]
Golfingia pugettensis	Indirect (III)	Rice, 1967[a]; Rice, 1975a[a]
Aspidosiphon sp.	Indirect (IV)	Rice, 1976a[a,b]; Rice, 1981[b]
Phascolosoma agassizii	Indirect (IV)	Rice, 1967[a]; Rice, 1973[a]
Phascolosoma antillarum	Indirect (IV)	Rice, 1975b[e]
Phascolosoma perlucens	Indirect (IV)	Rice, 1975a[d,a]; Rice, 1975b[a,d]
Phascolosoma varians	Indirect (IV)	Rice, 1975b[a]
Paraspidosiphon fischeri	Indirect (IV)	Rice, 1975b[a]
Siphonosoma sp.	Indirect (IV)	Rice, 1976[a,b]
Siphonosoma cumanense	Indirect (IV)	Rice 1981[d,a]; Rice, 1988[a,b]
Sipunculus nudus	Indirect (IV)	Hatschek, 1883[c]; Rice, 1975a[c]; Rice, 1988[a,b]
Sipunculus polymyotus	Indirect (IV)	Rice, 1975a[a]
Sipunculus sp.	Indirect (IV)	Rice, 1976a[a,b]; Rice, 1981[b]
Golfingia misakiana	Indirect (IV)	Rice, 1981[b] (tentative species I.D.)

[1]Numerals I–IV represent development categories in the Sipuncula as defined by Rice (1975b): I, direct development; II, indirect development through a trochophore stage only; III, indirect development with a lecithotrophic pelagosphera larva; IV, indirect development with a planktotrophic pelagosphera larva.
[2]Description type provided: [a]photographs, [b]scanning electron micrographs, [c]line drawings, [d]other illustrations, [e]verbal description.

Fig. 1. Dorsal views of the pelagosphera larva of *Phascolosoma agassizii*. Prominent metatrochal ciliary band and everted head are easily visible when the animal is moving. May be >1 mm in length. ((From *A Guide to Marine Coastal Plankton and Marine Invertebrate Larvae*, Second Edition, by DeBoyd L. Smith and Kevin B. Johnson. Copyright 1996 by DeBoyd L. Smith and Kevin B. Johnson. Reprinted by permission of Kendall/Hunt Publishing Company.))

Fig. 2. (A) Planktotrophic pelagosphera larva of *Phascolosoma agassizii*, tail retracted. (B) Partially retracted head of *P. agassizii* pelagosphera. The region anterior to the metatroch folds within the body when retracted.

this species (Figs. 1, 2). The metatrochal band (Fig. 2B) is responsible for the relatively fast swimming speeds of pelagosphera larvae (mm/sec). If disturbed, pelagosphera larvae may retract their heads, cease swimming, and sink.

Because pelagosphera larvae are able to retract both head and tail, they may appear as nondescript, blimplike creatures, difficult to recognize as the larvae of sipunculans. The pelagosphera of *P. agassizii* in Fig. 2A has its tail completely retracted and the head partially retracted. If observed alive, however, distinct pelagosphera features such as the retractable head and large metatrochal band are quickly apparent. Photographs of the pelagosphera larva of *P. agassizii*, along with behavioral observations, are available in Rice (1973). Less detailed photographs of *P. agassizii* development are available in Rice (1967), where the development of *P. agassizii* is compared to two other Pacific Northwest sipunculans, *Golfingia pugettensis* and *Themiste pyroides*.

Additional Literature

Visual descriptions of larval development in a variety of genera and species are available to researchers interested in learning more about development of sipunculans in other geographic regions. Table 2 summarizes the available publications that provide descriptions of the early development of larval sipunculans. A review of morphology and behavior of pelagosphera larvae is given by Jägersten (1963), who, incidentally, observes that the large-mouthed head, typical of many pelagosphera larvae, is reminiscent of a hippopotamus. Most pelagosphera larvae share a similar basic morphology. Using a combination of published descriptions, knowledge of local species distributions, and known modes of development, however, it may be possible to determine the specific identity of field-caught pelagosphera larvae.

A
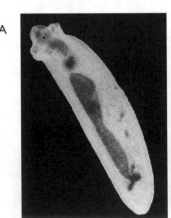

B

Acknowledgments

I thank Mary E. Rice of the Smithsonian Marine Station at Link Port, Florida, for comments on this manuscript and sharing her invaluable knowledge of local species and their development.

References

Akesson, B. (1958). A study of the nervous system of the Sipunculoideae with some remarks on the development of two species *Phascolion strombi* Montagu and *Golfingia minuta* Keferstein. Undersökningar över Öresund 38, 1–249.

——— (1961). The development of *Golfingia elongata* Keferstein (Sipunculidea) with some remarks on the development of neurosecretory cells in sipunculids. Arkiv for Zoologi 13(23): 511–31.

Austin, W. C. (1985). An Annotated Checklist of Marine Invertebrates in the Cold Temperate Northeast Pacific, Volume 2. Khoyatan Marine Laboratory, Cowachin Bay, B.C.

Brusca, R. C. and G. J. Brusca (1990). Invertebrates. Sinauer Associates, Inc., Sunderland.

Gerould, J. H. (1907). The development of *Phascolosoma*. Zool. Jb. 23(1):77–185.

Hatschek, B. (1883). Ueber Entwicklung von *Sipunculus nudus*. Arb. Zool. Inst. Univ. Wien Zool. Stat. Triest 5, 61–140.

Jägersten, G. (1963). On the morphology and behaviour of pelagosphaera larvae (Sipunculoidea). Zool. Bidr. Uppsala 36:27–35.

Kozloff, E. N. (1993). Seashore Life of the Northern Pacific Coast. University of Washington Press, Seattle.

——— (1996). Marine Invertebrates of the Pacific Northwest. University of Washington Press, Seattle.

Pilger, J. F. (1987). Reproductive biology and development of *Themiste lageniformis*, a parthenogenic Sipunculan. Bull. Mar. Sci. 41(1):59–67.

Rice, M. E. (1967). A comparative study of the development of *Phascolosoma agassizii, Golfingia pugettensis,* and *Themiste peroides* with a discussion of developmental patterns in the sipuncula. Ophelia 4:143–71.

——— (1970). Asexual reproduction in a sipunculan worm. Science 167, 1618–1620.

——— (1973). Morphology, behavior, and histogenesis of the pelagosphaera larva of *Phascolosoma agassizii* (Sipuncula). Smith. Contri. Zool. 132:1–51.

——— (1975a). Sipuncula. In: Reproduction of Marine Invertebrates II, A. C. Giese and J. S. Pearse (eds.) Academic Press, San Francisco.

——— (1975b). Observations on the development of six species of Caribbean Sipuncula with a review of development in the phylum. In: Proceedings of the International Symposium on the Biology of the Sipuncula and Echiura. M. E. Rice and M. Todorovic (eds.). Naucno Delo Press, Belgrade.

——— (1976). Larval development and metamorphosis in Sipuncula. Amer. Zool. 16: 563–71.

———— (1980). Sipuncula and Echiura. In: Intertidal Invertebrates of California, Morris et al. (eds.). Stanford University Press, Stanford.

———— (1981). Larvae adrift: patterns and problems in life histories of sipunculans. Amer. Zool. 21:605–19.

———— (1988). Observations on development and metamorphosis of *Siphonosoma cumanense* with comparative remarks on *Sipunculus nudus* (Sipuncula, Sipunculidae). Bull. Mar. Sci. 42(1):1–15.

8

Echiura and Pogonophora:
The Coelomate Worms

Alan L. Shanks

Phylum Echiura

Echiurans are dioecious protostomes. There is marked sexual dimorphism in some species. For example, male *Bonellia* are the size of a settling larva and reside in the nephridial sack of the female (Brusca and Brusca, 1990). There is no evidence of asexual reproduction. At least one species *(Urechis caupo)* is capable of parthenogenesis (Stephano and Gould, 1995). Fertilization is external and mass spawning events have been observed (Brusca and Brusca, 1990). Echiurans have at times been placed within the annelid phylum. Recent molecular evidence suggests that they may be derived from annelids and some authors have suggested that they should be included as a family (the Echiuridae) within the annelids (McHugh, 1997).

The initial larval stage is a trochophore (Fig. 1A), which begins feeding at about 40 hours post-fertilization. The prototroch and preprototroch develop into the adult probosus while the postprototroch develops into an elongated truck (Fig. 1B). The trunk superficially appears to be segmented, but the segmentation is due to bands of epithelial cells. Larval development takes two to three months (Gould 1967; Brusca and Brusca, 1990). Locally, the larvae of *Urechis caupo* are the only echiuran larvae that have been described (Fig. 1). Table 1 lists species observed locally.

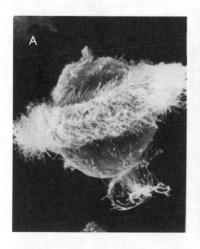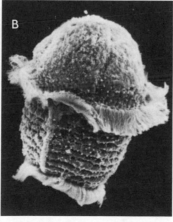

Fig. 1. Larvae of *Urechis caupo*. (A) Trochophore. (B) 15-day-old larva. Note the apparent segmentation of the trunk due to bands of epithelial cells. (From Brusca and Brusca, 1990, Fig. 9)

Table 1. Species in the phylum Echiura from the Pacific Northwest (from Kozloff, 1996)

Order Bonelloinea Family Bonelliidae
Nellobia eusoma

Order Echiuroinea Family Echiuridae
Arhynchite pugettensis
Echiurus echiurus **subsp.**
  Alaskanus

Order Xenopneusta Family Urechidae
Urechis caupo

Fig 2. Larval Pogonophora in the order Athecanephria; no larvae from the order Thecanephria have been described. (A, B) *Oligobrachia webbi*. (C) *Oligobrachia dogieli*. (D) *Siboglinum fiordicum*; there are two local *Siboglinum* species. Scales = 0.1 mm. (From Bakke, 1990, Fig. 5)

Phylum Pogonophora

The pogonophorans are a fairly recently discovered group, and our knowledge of them is still rather fragmentary. They are worm-shaped with generally long and thin bodies; they lack both mouth and gut. As with the echiurans, there is some discussion that the pogonophorans are derived from the annelids. They are sedentary tube dwellers, and most species are found in deep water. Some of the smaller species can be found in water as shallow as 30 m (Southward, 1975; Bakke, 1990; McHugh, 1997).

Essentially all of our knowledge of spawning and larval development has been gained from studies of species in the order Athecanephria. In this order, males release spermatophores with long thin filaments (Southward, 1975; Bakke, 1990). These presumably drift with the current, perhaps with the filament acting in an analogous fashion to the byssus thread in thread drifting bivalves (Titman and Davies, 1976). It is not clear how the female worms obtain the spermatophores. Each spermatophore contains enough sperm to fertilize the entire spawn of a female (Bakke, 1990). The site of fertilization is unclear; it occurs either within the female or in her tube. Larval development occurs in the tube, in front of the female (Southward, 1975; Bakke, 1990). At the end of the brooding period, larvae are worm-shaped and have two bands of cilia (Fig. 2). Short setae are present just anterior to the posterior ciliary band.

On leaving the maternal tube, the larvae swim upward briefly (probably for less than half an hour) before settling down to the sediment. Swimming is via the ciliary bands. The larvae swim in a helical path rotating along the longitudinal axis. Given their brief pelagic phase and weak swimming ability (Southward, 1975), the larvae are probably most common near the bottom. After settling to the bottom they

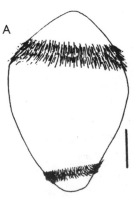

crawl around briefly before burrowing into the sediment, where metamorphosis occurs (Bakke, 1990).

No species in the order Thecanephria has been observed brooding young. There are no descriptions of their larvae. Their eggs are relatively small. Given the lack of evidence of brooding and the small egg size, they may have a free-living larval stage (Southward, 1975; Bakke, 1990). Table 2 lists species observed locally.

References

Bakke, T. (1990). Pogonophora. In: Reproductive Biology of Invertebrates vol. IV, Part B, G. K. Adiyodi and R. G. Adiyodi (eds.), pp. 37–48. John Wiley and Sons, New York.

Brusca, R. C. and G. J. Brusca (1990). Invertebrates. Sinauer Associates, Inc., Sunderland.

Gould, M. C. (1967). Echiuroid Worms: *Urechis*. In: Methods in Developmental Biology, F. H. Wilt and N. K. Wessells (eds.), pp. 163–72. Thomas Y. Crowell Co., New York.

Kozloff, E. N. (1996). Marine Invertebrates of the Pacific Northwest. University of Washington Press, Seattle.

McHugh, D. (1997). Molecular evidence that echiurans and pogonophorans are derived from annelids. Proc. Nat. Acad. Sci. USA 94:8006–9.

Southward, E. C. (1975). Pogonophora. In: Reproduction of Marine Invertebrates Vol. II, A. C. Giese and J. S. Pearse (eds.), pp. 129–56. Academic Press, New York.

Stephano, J. L. and M. C. Gould (1995). Parthenogenesis in *Urechis caupo* (Echiura). 1. Persistance of functional maternal asters following activation without meiosis. Dev. Biol. J. 167:104–17.

Titman, C. W. and P. A. Davies (1976). The dispersal of young post-larval bivalve molluscs by byssus threads. Nature 262:386–87.

Table 2. Shallow-water species in the phylum Pogonophora from the Pacific Northwest (from Kozloff, 1996)

Order Athecanephria
Family Siboglinidae
Siboglinum fedotovi
Siboglinum pusillum

Order Thecanephria
Family Polybrachiidae
Galathealinum brachiosum
Heptabrachia ctenophora
Polybrachia canadensis
Lamellisabella coronata
Lamellisabella zachsi

9

Mollusca: Gastropoda

Jeffrey H. R. Goddard

Gastropods include the snails and slugs and the less familiar free-swimming pteropods ("sea butterflies") and heteropods. They are the most speciose class in the second-largest phylum of animals. Approximately 400 species are known from the Pacific Northwest or have distributions encompassing the Pacific Northwest (Goddard, 1984, 1990, 1997; Austin, 1985). Type of development is known for a significant number of these species, but illustrations and specific descriptions of their larvae, especially at later stages of development, are lacking for the vast majority. Identification of larvae is therefore usually possible only to higher taxonomic levels.

Gastropods are now divided into two, rather than the traditional three, subclasses, the Prosobranchia and Heterobranchia (Table 1). The prosobranchs include most of the familiar limpets, abalone, snails, and slipper shells. The heterobranchs include the opisthobranchs (sea slugs and allies, including the pteropods), pulmonates (land and freshwater snails and slugs), the small, ectoparasitic pyramidellid snails, the pulmonate-like gymnomorphs, and some lesser known families not known from local waters and formerly allied with the Prosobranchia (see Bieler, 1992; Gosliner, 1996; Ponder and Lindberg, 1997, for recent reviews of gastropod higher classification).

The prosobranchs include more species than the heterobranchs, but repeated evolutionary loss of the shell in the latter subclass has resulted in greater morphological and ecological diversity. Of the local gastropods, 63% (255 species) are prosobranchs; the remainder are heterobranchs, mostly opisthobranchs. Families and genera of local gastropods are listed in Table 2.

Morphology

Gastropods have a distinct head with sensory tentacles and eyespots, and they crawl, burrow, or swim by means of a broad muscular foot. The foot supports and carries an overlying visceral mass that is covered, in turn, by a layer of tissue called the mantle and a calcified, protective shell secreted by the mantle. The foot of most species also carries an operculum, a shield-like structure used for closing off the shell aperture after

Table 1. Higher classification of gastropods known from the Pacific Northwest.

Phylum Mollusca
  Class Gastropoda
    Subclass Prosobranchia
      Order Patellogastropoda (= Archaeogastropoda, in part; includes most limpets)
      Vetigastropoda (= Archaeogastropoda, in part; keyhole limpets, abalone, turban snails, etc.)
      Caenogastropoda (= Meso- and Neogastropoda; periwinkles, whelks, drills, moon snails)
    Subclass Heterobranchia
      Superorder Pyramidellidacea (pyramidellid snails)
      Opisthobranchia (sea slugs, bubble snails, pteropods)
      Gymnomorpha (gymnomorph slugs)
      Pulmonata (lunged snails)

Table 2. Families and genera of the class Gastropoda known from the Pacific Northwest (from Austin, 1985; Behrens, 1990; Kozloff, 1996)[1]

Patellogastropoda
Acmaeidae: *Acmaea*
Lottidae: *Discurria, Lottia, Tectura*
Lepetidae: *Cryptobranchia, Iothia, Lepeta*

Vetigastropoda
Fissurellidae: *Arginula, Craniopsis, Diodora, Fissurellidea, Puncturella*
Haliotidae: *Haliotis*
Scissurellidae: *Anatoma*
Trochidae: *Bathybembix, Calliostoma, Halistylus, Liruloria, *Margarites, Solariella, Tegula, Tricolia*

Caenogastropoda
Lacunidae: *Lacuna*
Littorinidae: **Littorina*
Rissoidae: *Alvania, Onoba*
Barleeidae: *Barleeia*
Assimineidae: **Assiminea*
Turritellidae: *Tachyrynchus, Turritellopsis*
Vermetidae: *Dendropoma, Vermetus*
Caecidae: *Fartulum, Micranellum*
Cerithiidae: **Bittium*
Potamididae: *Batillaria* (introduced)
Cerithiopsidae: *Cerithiopsis*
Hipponicidae: **Hipponix*
Calyptraeidae: **Crepidula, Crepipatella*
Trichotropididae: **Trichotropis*
Naticidae: *Calinaticina, Natica, Neverita, Polinices*
Marseniidae: *Lamellaria, Marsenina, Marseniopsis*
Velutinidae: *Velutina*

Cymatiidae: *Fusitriton*
Epitoniidae: *Nitidiscala* (=*Epitonium*), *Opalia*
Janthinidae: *Janthina* (pelagic)
Eulimidae: *Balcis, Eulima*
Entoconchidae: *Enteroxenos, Thyonicola*
Muricidae: **Ceratostoma, *Ocenebra, *Trophonopsis,*
Nucellidae: **Acanthina, *Nucella*
Buccinidae: **Buccinium, *Searlesia*
Neptuneidae: *Ancistrolepis, *Beringius, *Colus, Exilioidea, *Neptunea, *Plicifusus*
Columbellidae: *Alia, *Amphissa, Mitrella*
Nassariidae: **Nassarius*
Fusinidae: **Fusinus*
Olividae: **Olivella*
Marginellidae: *Granulina*
Cancellariidae: *Cancellaria*
Turridae: *Ophiodermella, Pseudomelatoma, Taranis, Clathromangelia, Kurtzia, Kurtziella, Oenopota, Cymakra, Antiplanes*

Heteropoda
Carinariidae: *Carinaria*
Atlantidae: *Atlantia*
Pterotracheidae: *Pterotrachea*

Pyramidellacea
Pyramidellidae: *Iselica, Odostomia, Turbonilla*

Opisthobranchia, Cephalaspidea
Acteonidae: *Rictaxis, Microglyphis*
Haminoeidae: *Haminaea*
Retusidae: *Volvulella*
Diaphanidae: *Diaphana*
Philinidae: *Philine*
Aglajidae: *Aglaja*
Gastropteridae: *Gastropteron*
Cyclichnidae: *Cylichna, Acteocina*
Runcinidae: **Runcina*

[1]Genera with representatives in Oregon known (or suspected, based on the development of congeners from other regions) to have direct development and therefore lack a larval stage are preceded by an asterisk (*). See the above sources for lists of species known from the Pacific Northwest.

Table continues

Table 2. Families and genera of the class Gastropoda known from the Pacific Northwest (continued)

Opisthobranchia, Anaspidea
Aplysiidae: *Aplysia*, **Phyllaplysia*

Opisthobranchia, Notaspidea
Pleurobranchidae: *Berthella*, *Pleurobranchia*

Opisthobranchia, Sacoglossa
Stiligeridae: *Placida*, *Stiliger*
Elysidae: *Elysia*
Hermaeidae: *Alderia*, *Hermaea*

Opisthobranchia, Nudibranchia, Doridacea
Corambidae: *Corambe*, *Doridella*
Goniodorididae: *Ancula*, *Hopkinsia*
Onchidorididae: *Acanthodoris*, *Onchidoris*, *Adalaria*,
    Diaphorodoris
Notodorididae: *Aegires*
Polyceratidae: *Crimora*, *Laila*, *Polycera*, *Triopha*
Chromodorididae: *Cadlina*
Actinocyclidae: *Hallaxa*
Aldisidae: *Aldisa*
Rostangidae: *Rostanga*
Archidorididae: *Archidoris*
Discodorididae: *Anisodoris*, *Diaulula*, *Geitodoris*

Opisthobranchia, Nudibranchia, Dendronotacea
Tritoniidae: *Tochuina*, *Tritonia*
Dendronotidae: *Dendronotus*
Dotoidea: *Doto*
Tethyidae: *Melibe*

Opisthobranchia, Nudibranchia, Arminacea
Arminidae: *Armina*
Dironidae: *Dirona*
Zephyrinidae: *Janolus*

Opisthobranchia, Nudibranchia, Aeolidacea
Flabellinidae: *Chamylla*, *Flabellina*
Cumanotidae: *Cumanotus*
Eubranchidae: *Eubranchus*
Tergipedidae: *Catriona*, *Cuthona*, *Tenellia*
Fionidae: *Fiona*
Facelinidae: *Hermissenda*
Aeolidiidae: *Aeolidia*, *Cerberilla*

Opisthobranchia, Thecosomata, Euthecosomata
Limacinidae: *Limacina*
Cavoliniidae: *Cavolinia*, *Clio*, *Creseis*, *Diacria*, *Styliola*,
    Cuvierina

Opisthobranchia, Thecosomata, Pseudothecosomata
Cymbuliidae: *Corolla*

Opisthobranchia, Gymnosomata
Clionidae: *Cliona*
Pneumodermatidae: *Pneumoderma*, *Pneumodermopsis*
Cliopsidae: *Cliopsis*

Gymnomorpha
Onchidiidae: **Onchidella*

Pulmonata
Melampidae: **Mysotella* (= *Ovatella*)
Siphonariidae: **Siphonaria*, *Williamia*
Trimusculidae: *Trimusculus*

withdrawl of the body. Adult shells vary from the spiraling coils of whelks and periwinkles, to the cap-shaped shells of limpets and abalone, to the sessile, calcareous tubes secreted by the wormlike vermetid snails. The shell and operculum are reduced or lost altogether in postlarval slugs and free-swimming forms.

All gastropods are characterized by an unusual developmental phenomenon called torsion, a 180° twisting of the viscera, mantle, and shell relative to the head and foot. Torsion results in a forward placement of the gill, anus, and reproductive openings, all of which are housed in a cavity formed by the overhanging mantle and shell. Gastropods lacking shells generally lack a mantle cavity and also show varying degrees of detorsion. The latter is manifested externally by the position of the anus, gill, and reproductive openings on the right side or posterior end of the organism and can impart a superficial bilateral symmetry to the body.

Adult gastropods exhibit a wide range of feeding modes, ranging from generalist, grazing herbivory, to specialized carnivory, to suspension feeding with mucus nets. Except for some highly specialized ectoparasites, suctorial predators, and suspension-feeders, all use a tooth-studded, ribbon-like radula in feeding. The radula is unique to the Mollusca, and in different groups of gastropods its teeth are variously modified for rasping, grasping, pulling, piercing, or harpooning prey. Acting in concert with secretions from the foot, the radula is also used by members of some taxa for drilling through the shells and skeletons of their prey.

Individuals of most prosobranchs are either male or female, whereas most heterobranchs are hermaphroditic. Fertilization can be external or internal, and eggs are freely spawned into the water column or deposited in a wide variety of benthic or even pelagic egg masses and capsules. Individuals of some species brood their egg capsules, especially in the mantle cavity or under the foot. Reproduction and development of many Pacific Northwest species are summarized by Strathmann (1987), and comparative data on the development of most opisthobranchs known from the Pacific Northwest are presented in Appendix A (pages 118-22).

Larval Forms

Most gastropods hatch from their egg coverings as one of two types of larval forms, veligers and trochophores. A minority (although this can include entire clades) bypass a planktonic larval stage in their life cycles and hatch as crawl-away juveniles (e.g., species of *Nucella*). Though not strictly planktonic, hatchlings of many of these "directly developing" species are capable of significant dispersal via drifting in the water column and may be caught in plankton tows near adult habitat (Martel and Chia, 1991). The same can also apply to post-larval stages of species with planktonic development. Identification of post-larval and juvenile stages is better accomplished using adult characters and is beyond the scope of this chapter. Keen and Coan (1974) and Kozloff (1996) provide keys to adult gastropods known from the Pacific Northwest.

Veliger Larvae

Most gastropods hatch from benthic or pelagic egg capsules or egg masses as veliger larvae (Figs. 1, 2). Gastropod veligers are distinguished by their univalve, usually coiled shells and an anterior, round to multilobed velum, the primary organ of

Fig. 1. External morphology of a veliger larva, ventral view. (Modified from Fretter and Graham, 1962, Fig. 237A)

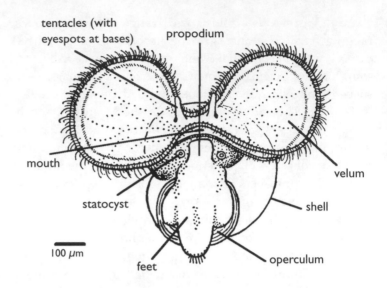

tentacles (with eyespots at bases)

propodium

mouth

velum

statocyst

shell

100 μm

feet

operculum

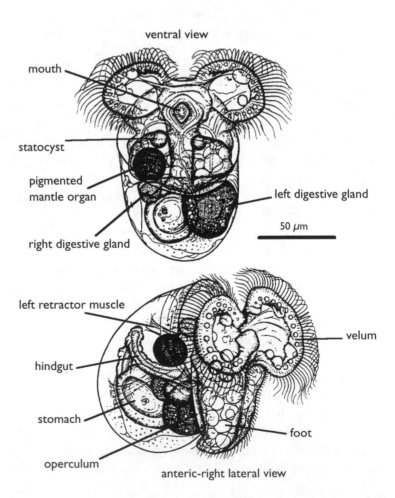

ventral view

mouth

statocyst

pigmented mantle organ

left digestive gland

right digestive gland

50 μm

left retractor muscle

velum

hindgut

stomach

foot

Fig. 2. Internal anatomy of a veliger larva. (Modified from Rasmussen, 1951, Fig. 14)

operculum

anteric-right lateral view

propulsion and food particle capture. The velar lobes of planktotrophic species especially are delicate, extensible, and edged with two powerful circlets of cilia that bound a ciliated food groove (Hyman, 1967). Most veliger larvae also have an operculum, carried on the back of the foot, which is used to close the shell aperture after withdrawal of the body into the shell. Except for the velum, the body plan of most gastropod veligers resembles that of a typical adult prosobranch with a coiled shell. Although there is not a close correspondence between larval and adult appearance which might allow specific identification of one based solely on knowledge of the other (Fretter and Pilkington, 1970), larval shells do persist as "protoconchs" at the apex of the shells of many adult gastropods. Thus, larval shells can be identified by comparison with the protoconch of an identified juvenile or adult specimen (Thorson, 1946; Robertson, 1971; Thiriot-Quiévreux, 1980).

The veliger stage can last from days to months, depending on larval feeding mode (i.e., lecithotrophic vs. planktotrophic), taxon (especially at the species level), and environmental factors such as food supply and presence of settlement cues. Larval life ends with settlement and metamorphosis into a post-larva or juvenile, a process often triggered by specific environmental cues. In the absence of these cues, meta-morphosis can be delayed for long periods. The larvae of holoplanktonic species obviously do not settle to the bottom like those of benthic species, and metamorphosis for many of these is a more gradual process.

A major component of metamorphosis in both holo-planktonic and meroplanktonic species is the irreversible loss or absorption of the velum. After metamorphosis the foot takes over as the organ of propulsion, and its development, especially of the propodium, is one of the best indicators of metamorphic competence.

Trochophore Larvae

Some of the more primitive prosobranchs hatch from their egg coverings as trochophores (Fig. 3), a developmental stage shared with other coelomate protostomes (e.g., polychaete annelids) and one that most gastropods pass through as encapsulated embryos. Gastropod trochophores do not feed on particulate matter (they may take up dissolved organic matter) and swim by means of the prototroch, a band of ciliated cells encircling the body (see Fig. 3). Prosobranchs hatching as trochophore larvae include all of the Patellogastropoda (except those hatching as crawl-away juveniles) and some of the Vetigastropoda (e.g., haliotids and some of the trochids). The

Fig. 3. Gastropod
trochophore larva,
ventral view. (Modified
from Kessel, 1964, Fig. 6)

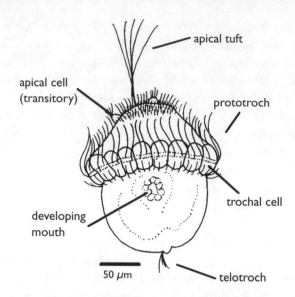

apical tuft

apical cell
(transitory)

prototroch

trochal cell

developing
mouth

50 μm

telotroch

trochophore stage is short in gastropods, lasting only a few hours, and grades into the veliger stage as shell secretion and development of the foot and velum progress.

Polytrochous Larvae

Veliger larvae of gymnosomatous pteropods, the shell-less "sea-butterflies," develop into fusiform, polytrochous larvae before metamorphosing into juveniles. Polytrochous larvae (see 11 in the key) lack both shell and velum and rely on three ciliary rings for propulsion. This stage grades into the juvenile stage as the swimming wings (specialized lobes of the foot) enlarge and replace the ciliary rings as the primary means of propulsion.

Identification of Local Taxa

The key in this chapter is largely pictorial and based on gross morphological features such as shell shape and sculpture, shape of the velum, and, to a lesser degree, color pattern. Identification is best accomplished using live material, but many diagnostic characters are apparent in specimens relaxed in 7.5% magnesium chloride and fixed in 4% formalin buffered with borax (see Strathmann, 1987, Chap. 1, pp. 228–29). Most of the illustrations used in the key were obtained from the primary literature; sources of these are listed in Appendix B (page 123).

For many taxa identification of larvae is not possible beyond the level of family or even order, either because the larvae of most local species have not been described, or because the larvae of closely related species are not sufficiently differentiated to permit identification based on qualitative morphological features (the latter applies especially to young larvae). In many cases, diagnoses are based on descriptions of close relatives from other parts of the world. Many of these descriptions give more detail than could be included in this chapter, and workers are encouraged to examine the primary literature. The most useful references, especially for the larvae of the prosobranchs, include Thorson (1946), Fretter and Pilkington (1970), Richter and Thorson (1975), Pilkington (1976), and the works of M. Lebour and C. Thiriot-Quiévreux.

If a larva keys to a group of opisthobranchs, the comparative data provided in Appendix A (especially data on type and size of the shell) may help to narrow the choices.

Nomenclature used in the keys generally follows Austin (1985), Kozloff (1996), and Behrens (1990), and these sources should be consulted for lists of gastropod species found in the region.

Key to Marine Gastropod Larvae of Oregon

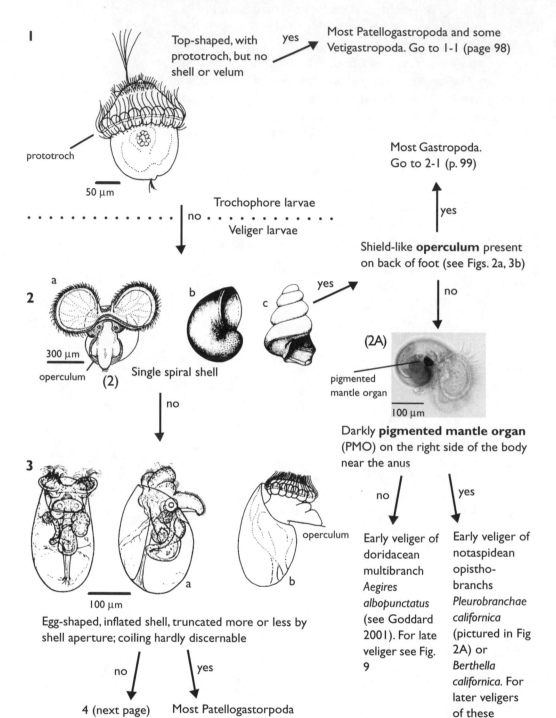

1

Top-shaped, with prototroch, but no shell or velum — **yes** → Most Patellogastropoda and some Vetigastropoda. Go to 1-1 (page 98)

prototroch

50 μm

Trochophore larvae
· · · · · · · · · · · · · · **no** · · · · · · · · · · · · · ·
Veliger larvae

Most Gastropoda. Go to 2-1 (p. 99)

yes

Shield-like **operculum** present on back of foot (see Figs. 2a, 3b)

2

a

300 μm

operculum **(2)**

b

c — **yes** →

Single spiral shell

no

no

(2A)

pigmented mantle organ

100 μm

Darkly **pigmented mantle organ** (PMO) on the right side of the body near the anus

no   **yes**

Early veliger of doridacean multibranch *Aegires albopunctatus* (see Goddard 2001). For late veliger see Fig. 9

Early veliger of notaspidean opistho-branchs *Pleurobranchae californica* (pictured in Fig 2A) or *Berthella californica.* For later veligers of these species, see Figs. 8a, 8b (p. 96)

3

a

b

operculum

100 μm

Egg-shaped, inflated shell, truncated more or less by shell aperture; coiling hardly discernable

no   **yes**

4 (next page)

Most Patellogastorpoda and some Nudibranchia. Go to 3-1 (p. 112)

4

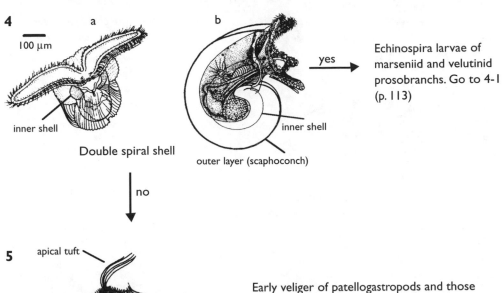

100 μm

inner shell

inner shell

outer layer (scaphoconch)

Double spiral shell

yes → Echinospira larvae of marseniid and velutinid prosobranchs. Go to 4-1 (p. 113)

no

5

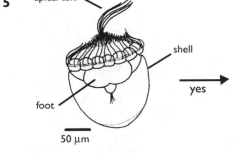

apical tuft

shell

foot

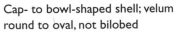

50 μm

yes → Early veliger of patellogastropods and those vetigastropods hatching as trochophores. The former possess an apical tuft of non-motile cilia; the latter do not. This stage is brief in both groups and rapidly develops into a fully developed veliger (for later veligers of these two groups, see 3-1 and 2-4.1 to 4-2)

Cap- to bowl-shaped shell; velum round to oval, not bilobed

no

6

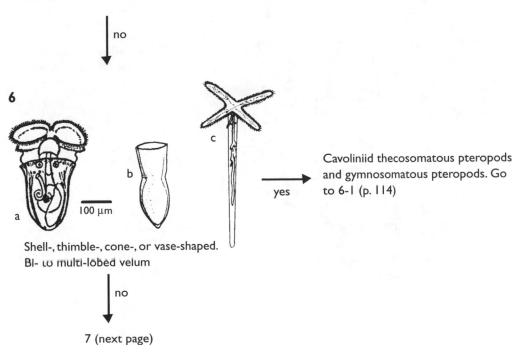

a

b

c

100 μm

yes → Cavoliniid thecosomatous pteropods and gymnosomatous pteropods. Go to 6-1 (p. 114)

Shell-, thimble-, cone-, or vase-shaped. Bi- to multi-lobed velum

no

7 (next page)

7

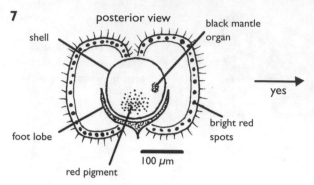

posterior view

shell

black mantle organ

bright red spots

foot lobe

red pigment

100 μm

yes →

Late veliger of the cephalaspidean opisthobranch *Gastropteron pacificum*

Note: The foot lobes of this species grow to completely envelope the shell and can be used for swimming. This species also has a distinctive shell (see Fig. 2-12)

Coiled shell present but covered or obscured by lateral lobes of the **foot**. Bright red spots around edge of velum

↓ no

8

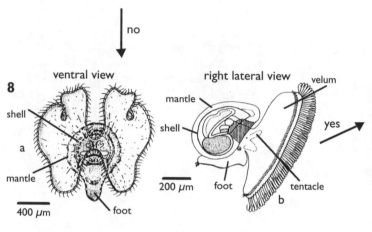

ventral view

shell

a

mantle

400 μm

foot

right lateral view

velum

mantle

shell

200 μm

foot

tentacle

b

yes →

Late veliger, notaspidean opisthobranchs (2 spp. in 2 genera in Oregon)

Note: See Tsubokawa and Okutani (1991) for a description of mantle growth in these larvae; except for their lack of an operculum, the early larvae are similar to those of other opisthobranchs with paucispiral shells and pigmented mantle organs (PMOs) (see Figs. 2-8b, 2-15, 2-9-8)

Coiled shell present but mostly or completely covered by the **mantle**; no operculum

↓ no

9

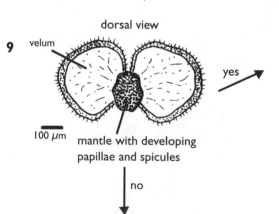

dorsal view

velum

100 μm

mantle with developing papillae and spicules

yes →

Late veliger of doridacean nudibranch *Aegires*.

Note: This type of veliger is known only for the North Atlantic nudibranch *Aegires punctilucens* (Thiriot-Quiévreux, 1977), but is likely also found in the northeastern Pacific *A. albopunctatus* (Goddard, 2001)

↓ no

10 (next page)

10   ventral view

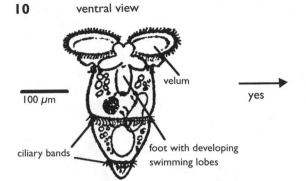

velum

100 μm

ciliary bands

foot with developing
swimming lobes

yes →

Late veliger of gymnosomatous
pteropods (4 species in 4 genera off
Oregon)

Note: *Clione limacina,* pictured at left, is the
most abundant gymnosome nearshore and is
occasionally found in Oregon bays

No shell or operculum. Velum present.
Ciliary bands around fusiform body; foot
with developing swimming lobes

Veliger larvae

· · · · · · · · · |no ·

Polytrochus larvae

11

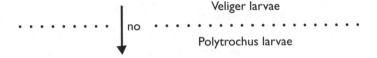

ventral view

right foot lobe

right foot lobe

left foot
lobe

100 μm

ciliary bands

yes →

Polytrochous larvae.
Last larval stage of
gymnosomatous
pteropods (4 species in
4 genera off Oregon)

Note: *Clione limacina,*
pictured at different stages
of development at left, is the
most abundant gymnosome
nearshore and is occasionally found in Oregon bays

No shell, operculum, or velum. Small foot with
developing swimming lobes or "wings"; three ciliary
bands around body

From 1 (top-shaped, with prototroch, but no shell or velum)

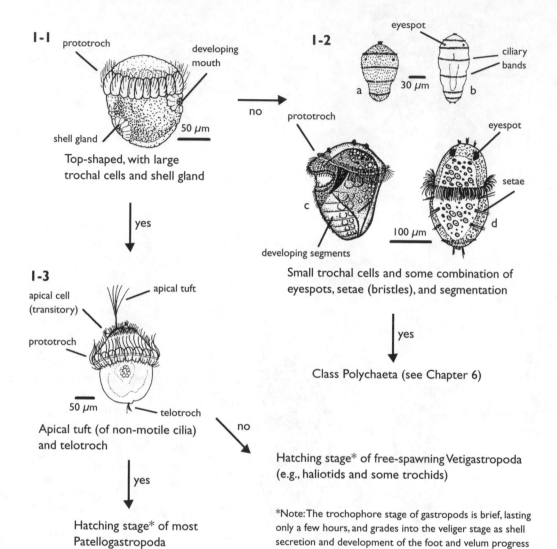

1-1 prototroch — developing mouth

shell gland — 50 μm

Top-shaped, with large trochal cells and shell gland

no →

1-2 eyespot — ciliary bands

a   30 μm   b

prototroch

c — developing segments

eyespot — setae

100 μm   d

Small trochal cells and some combination of eyespots, setae (bristles), and segmentation

↓ yes

Class Polychaeta (see Chapter 6)

↓ yes

1-3
apical cell (transitory) — apical tuft

prototroch

50 μm — telotroch

Apical tuft (of non-motile cilia) and telotroch

no ↘

Hatching stage* of free-spawning Vetigastropoda (e.g., haliotids and some trochids)

↓ yes

Hatching stage* of most Patellogastropoda

*Note: The trochophore stage of gastropods is brief, lasting only a few hours, and grades into the veliger stage as shell secretion and development of the foot and velum progress

From 2 (single spiral shell)

2-1

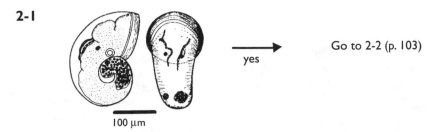

100 μm

yes → Go to 2-2 (p. 103)

Shell planispiral (coiling in one plane) or nearly so

no ↓

2-3

apex

a

200 μm   aperture   100 μm   b

yes → Most prosobranch gastropods.
Go to 2-4 (p. 106)

Shell dextrally coiled and smooth to
elaborately sculptured. Head tentacles usually
present, with eyespots at their bases. Except
some epitoniids, the shell is **hydrophilic** and
does not get trapped at the air-water interface

Note: Coiling direction of inflated, paucispiral shells
(Fig. 2-3b) may be difficult to determine

no ↓

2-5

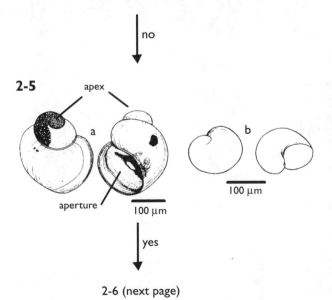

apex

a

aperture   100 μm   b

100 μm

yes ↓

2-6 (next page)

**Shell sinistrally (counter
clockwise) coiled** and generally
unsculptured. Head tentacles usually
lacking, but if present they are
usually separate from and anterior
to the eyespots. Velum usually
bilobed. Most heterobranchs
(starting on section 2-8) and a few
prosobranchs (next 2 sections, 2-6,
2-7). The former have hydrophobic
shells, the latter hydrophilic ones

Note: Coiling direction of inflated,
paucispiral shells (Fig. 2-5b) may be difficult
to determine

2-6

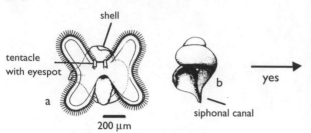

Tentacles with eyespots; shell multiwhorled, with well-developed siphonal canal

Turrid prosobranchs *Antiplanes voyi* and *A. perversa.*

Note: The larvae of these 2 species have not been described and Figs. 2-6a and b are used to depict diagnostic features only. In addition, direction of coiling of the larval shells of these 2 species is assumed to be the same as the adults' (sinistral); it is possible that their coiling direction changes at settlement, a condition known as heterostrophy (see Hadfield and Strathmann, 1990; and Box 1, p. 102)

yes

no

2-7

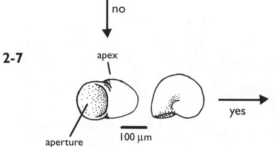

Velum round to only slightly bilobed. Inflated, paucispiral shell of about one whorl. Larvae semiopaque, owing to brown to greenish yolk reserves, and develop head and foot tentacles at metamorphosis. Shell coiling changes direction at metamorphosis, becoming dextral (see Box 1, p. 102)

Margarites pupillus and possibly a few other trochid prosobranchs

Note: *Margarites pupillus* has brownish yolk and a shell finely pitted and 238 μm wide (Hadfield and Strathmann, 1990). The larval shells of most trochids are superficially similar to that of *M. pupillus* but are dextrally coiled (see section 2-4). Most trochids have non-feeding, pelagic, lecithotrophic development and settle after about a week in the plankton (Hickman, 1992)

yes

no (most heterobranchs)

2-8

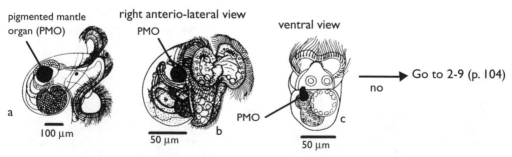

Body with **pigmented mantle organs** (PMOs) located on the right side, near the anus

Go to 2-9 (p. 104)

no

yes

2-10 (next page)

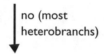

2-10 Black pigment on visceral mass → **yes** → Late veliger, aglajid opisthobranch *Melanochlamys diomedea* and possibly some sacoglossan opisthobranchs.

Note: Hatching *Melanochlamys* has a nearly black PMO and an inflated, paucispiral shell. PMOs are not known in the larvae of the few local species of sacoglossans that have been examined

no ↓

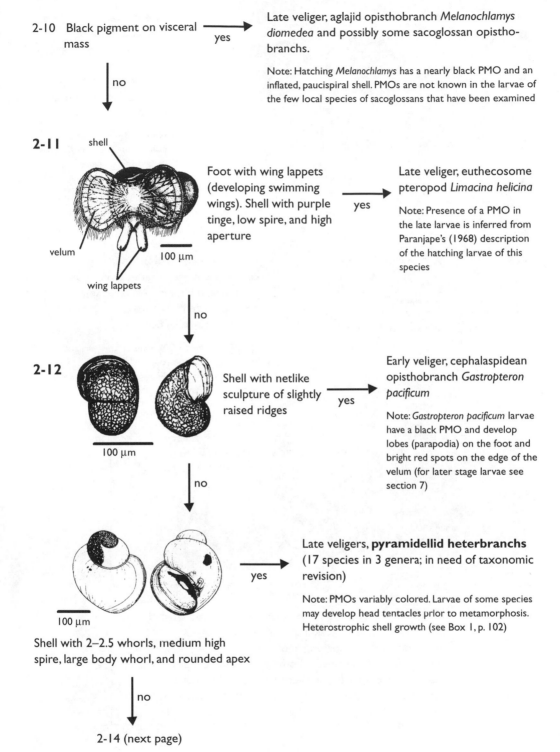

2-11

shell

velum

100 µm

wing lappets

Foot with wing lappets (developing swimming wings). Shell with purple tinge, low spire, and high aperture → **yes** → Late veliger, euthecosome pteropod *Limacina helicina*

Note: Presence of a PMO in the late larvae is inferred from Paranjape's (1968) description of the hatching larvae of this species

no ↓

2-12

100 µm

Shell with netlike sculpture of slightly raised ridges → **yes** → Early veliger, cephalaspidean opisthobranch *Gastropteron pacificum*

Note: *Gastropteron pacificum* larvae have a black PMO and develop lobes (parapodia) on the foot and bright red spots on the edge of the velum (for later stage larvae see section 7)

no ↓

100 µm

Shell with 2–2.5 whorls, medium high spire, large body whorl, and rounded apex

→ **yes** → Late veligers, **pyramidellid heterbranchs** (17 species in 3 genera; in need of taxonomic revision)

Note: PMOs variably colored. Larvae of some species may develop head tentacles prior to metamorphosis. Heterostrophic shell growth (see Box 1, p. 102)

no ↓

2-14 (next page)

2-14

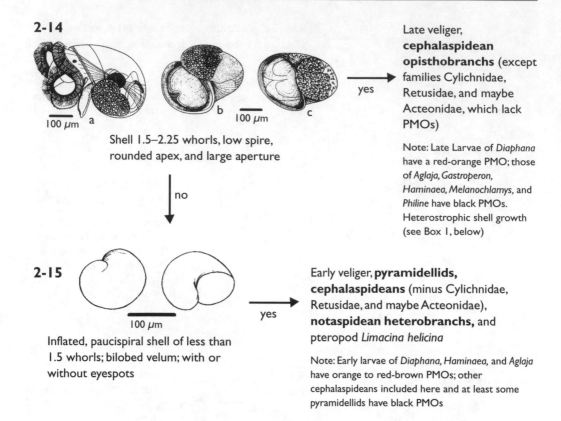

Shell 1.5–2.25 whorls, low spire, rounded apex, and large aperture

yes →

Late veliger, **cephalaspidean opisthobranchs** (except families Cylichnidae, Retusidae, and maybe Acteonidae, which lack PMOs)

Note: Late Larvae of *Diaphana* have a red-orange PMO; those of *Aglaja, Gastroperon, Haminaea, Melanochlamys,* and *Philine* have black PMOs. Heterostrophic shell growth (see Box 1, below)

↓ no

2-15

Inflated, paucispiral shell of less than 1.5 whorls; bilobed velum; with or without eyespots

yes →

Early veliger, **pyramidellids, cephalaspideans** (minus Cylichnidae, Retusidae, and maybe Acteonidae), **notaspidean heterobranchs,** and pteropod *Limacina helicina*

Note: Early larvae of *Diaphana, Haminaea,* and *Aglaja* have orange to red-brown PMOs; other cephalaspideans included here and at least some pyramidellids have black PMOs

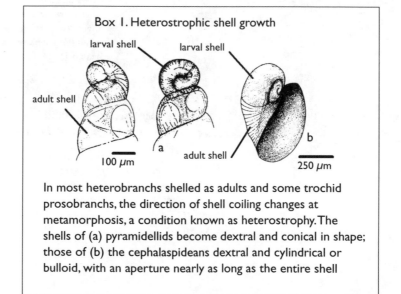

Box 1. Heterostrophic shell growth

In most heterobranchs shelled as adults and some trochid prosobranchs, the direction of shell coiling changes at metamorphosis, a condition known as heterostrophy. The shells of (a) pyramidellids become dextral and conical in shape; those of (b) the cephalaspideans dextral and cylindrical or bulloid, with an aperture nearly as long as the entire shell

From 2-1 (shell planispiral, or nearly so). Also see double spiral shells, section 4-2

Late veliger of pterotracheid heteropod prosobranch *Pterotrachea coronata*

Note: Heteropods are visual predators and in the larval stage develop complex eyes with a lens and retina (see Thiriot-Quiévreux, 1973)

yes ↑

2-2   dorsal view   **2-2-1**

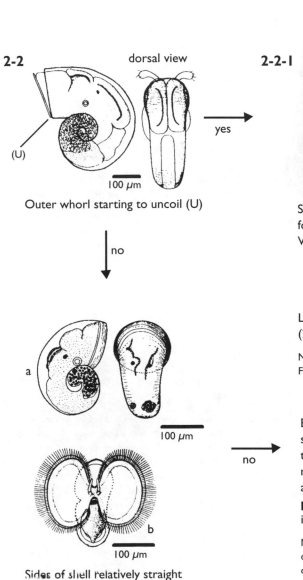

(U)

100 μm

Outer whorl starting to uncoil (U)

yes →

100 μm

Shell thin, with accordion-like transverse folds, especially on the outer whorl. Velum with 4 lobes

no ↓

no ↓

Late veliger of **caecid prosobranchs** (2 species in 2 genera in Oregon)

Note: Velum of caecids is bilobed (see Fig. 2-2-2b) and the shell is smooth

a

100 μm

b

100 μm

Sides of shell relatively straight (not inflated or rounded)

no →

Early veliger, **Calyptraeidae** (slipper shells; 5 species in 2 genera; some of these, as known for *Crepidula adunca*, may lack a planktonic phase and hatch as crawl-away juveniles) or **pterotracheid heteropods** (2 species in one genus)

Note: The calyptroids have a bilobed velum and develop simple eyespots; the pterotracheids develop a 4-lobed velum and complex eyespots with a lens and retina

yes ↓

Early veliger, **caecid prosobranchs** (2 species in 2 genera)

From 2-8 (without pigmented mantle organs)

2-9   left lateral view

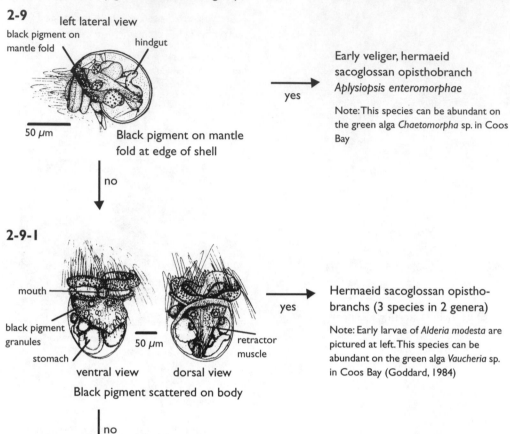

black pigment on mantle fold

hindgut

50 μm

Black pigment on mantle fold at edge of shell

yes →

Early veliger, hermaeid sacoglossan opisthobranch *Aplysiopsis enteromorphae*

Note: This species can be abundant on the green alga *Chaetomorpha* sp. in Coos Bay

↓ no

2-9-1

mouth

black pigment granules

stomach

50 μm

retractor muscle

ventral view   dorsal view

Black pigment scattered on body

yes →

Hermaeid sacoglossan opisthobranchs (3 species in 2 genera)

Note: Early larvae of *Alderia modesta* are pictured at left. This species can be abundant on the green alga *Vaucheria* sp. in Coos Bay (Goddard, 1984)

↓ no

2-9-2   right lateral view

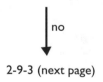

red line

red spots

200 μm

foot   propodium

4 to 6 red spots on perivisceral membrane, and a red line on the edge of the mantle

yes →

Late veliger, anaspidean opisthobranch *Aplysia californica*

Note: Red pigment appears just before settlement The shell at this stage is 400 μm long, with 2.25 whorls, and propodium and eyespots are present. Adults of this species occur infrequently in Oregon bays, and have not been found in Coos Bay. Hatching larvae have opaque white grains in the larval kidney, an apparently unique trait

↓ no

2-9-3 (next page)

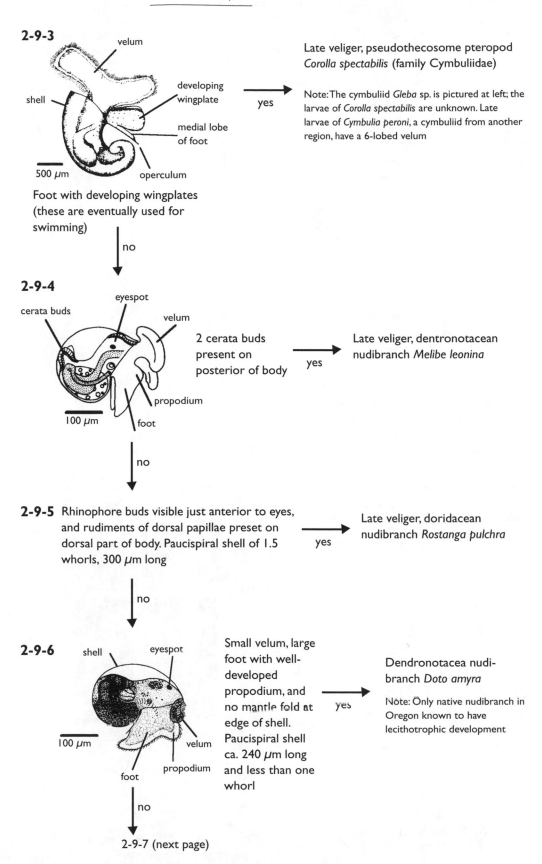

2-9-3

velum

developing wingplate

shell

medial lobe of foot

500 μm

operculum

Foot with developing wingplates (these are eventually used for swimming)

yes

Late veliger, pseudothecosome pteropod *Corolla spectabilis* (family Cymbuliidae)

Note: The cymbuliid *Gleba* sp. is pictured at left; the larvae of *Corolla spectabilis* are unknown. Late larvae of *Cymbulia peroni*, a cymbuliid from another region, have a 6-lobed velum

no

2-9-4

eyespot

cerata buds

velum

propodium

100 μm

foot

2 cerata buds present on posterior of body

yes

Late veliger, dentronotacean nudibranch *Melibe leonina*

no

2-9-5 Rhinophore buds visible just anterior to eyes, and rudiments of dorsal papillae preset on dorsal part of body. Paucispiral shell of 1.5 whorls, 300 μm long

yes

Late veliger, doridacean nudibranch *Rostanga pulchra*

no

2-9-6 shell

eyespot

100 μm

velum

foot

propodium

Small velum, large foot with well-developed propodium, and no mantle fold at edge of shell. Paucispiral shell ca. 240 μm long and less than one whorl

yes

Dendronotacea nudibranch *Doto amyra*

Note: Only native nudibranch in Oregon known to have lecithotrophic development

no

2-9-7 (next page)

2-9-7 Eyespots and propodium present, mantle withdrawn from edge of shell, all at a shell size of 168 μm

→ *yes* → Late veliger, doridacean nudibranch *Doridella steinbergae*

Note: Only nudibranch known to be metamorphically competent at such a small size

↓ *no*

2-9-8

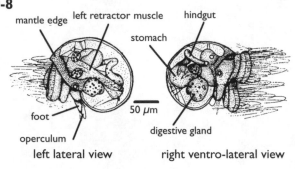

mantle edge left retractor muscle hindgut

stomach

foot

operculum

50 μm

digestive gland

left lateral view   right ventro-lateral view

Clear, generally smooth, paucispiral shell. Bilobed velum. Viscera relatively transparent owing to lack of yolk reserves

↓ *yes*

Early veliger, most nudibranchs (ca. 45 species in Oregon; all but *Doto amyra* have planktotrophic development) and the cephalaspideans of the families Cylichnidae, Retusidae, and possibly Acteonidae (8 species in 4 genera in Oregon)

• •

From 2-3 (shell dextrally coiled)

Calliostoma (5 species in Oregon)

↑ *yes*

2-4

100 μm

Shell **inflated paucispiral** of 1.25–1.5 whorls. **Round to only slightly bilobed velum**. Body semiopaque owing to yellow, green, or brown yolk reserves

→ *yes*

2-4-1

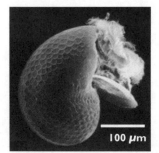

100 μm

↓ *no*

2-4-3 (next page)

↓ *no*

2-4-2 (next page)

2-4-2

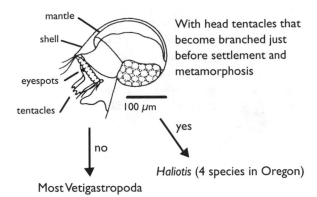

mantle

shell

eyespots

tentacles

100 μm

With head tentacles that become branched just before settlement and metamorphosis

yes

no

Haliotis (4 species in Oregon)

Most Vetigastropoda

2-4-3

outer whorl

300 μm

Inner whorl planispiral and unsculptured; **outer whorl expanded laterally**, resulting in large aperture, low shell height, and beginning of limpet-like form

yes

Late veliger, Calyptraeidae (slipper shells; 5 species in 2 genera in Oregon [*Crepidula adunca* has direct development])

no

2-4-4

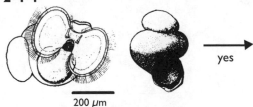

200 μm

Up to 2.5 **strongly inflated shell whorls**; apex roundly blunted

no

2-4-6 (next page)

yes

2-4-5

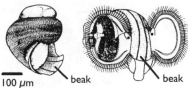

100 μm beak

beak

Body whorl with distinct **tongue-shaped beak**

yes

Cerithiidae (5 species of *Bittium*)

no

Turritellidae (3 species in 2 genera)

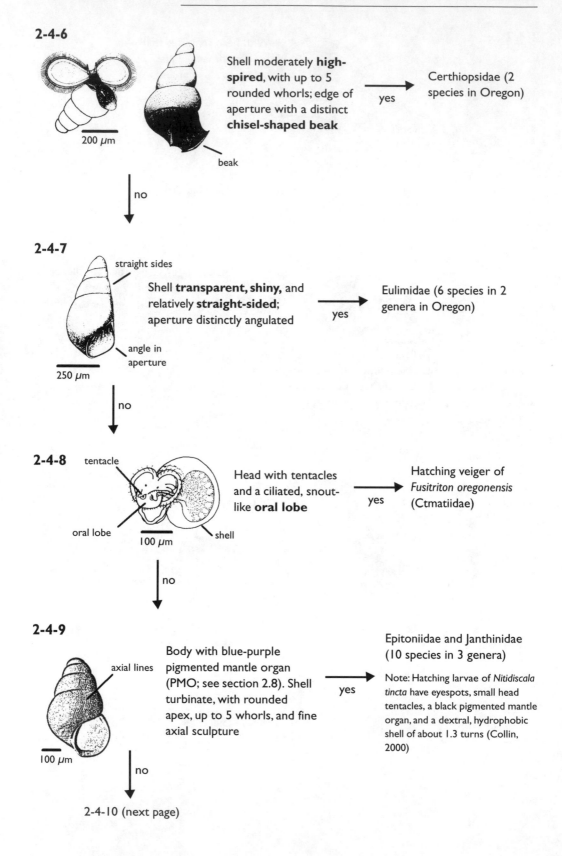

2-4-6

200 μm

beak

Shell moderately **high-spired**, with up to 5 rounded whorls; edge of aperture with a distinct **chisel-shaped beak**

yes → Certhiopsidae (2 species in Oregon)

no

2-4-7

straight sides

angle in aperture

250 μm

Shell **transparent, shiny,** and relatively **straight-sided**; aperture distinctly angulated

yes → Eulimidae (6 species in 2 genera in Oregon)

no

2-4-8

tentacle

oral lobe   shell

100 μm

Head with tentacles and a ciliated, snout-like **oral lobe**

yes → Hatching veiger of *Fusitriton oregonensis* (Ctmatiidae)

no

2-4-9

axial lines

100 μm

Body with blue-purple pigmented mantle organ (PMO; see section 2.8). Shell turbinate, with rounded apex, up to 5 whorls, and fine axial sculpture

yes →

Epitoniidae and Janthinidae (10 species in 3 genera)

Note: Hatching larvae of *Nitidiscala tincta* have eyespots, small head tentacles, a black pigmented mantle organ, and a dextral, hydrophobic shell of about 1.3 turns (Collin, 2000)

no

2-4-10 (next page)

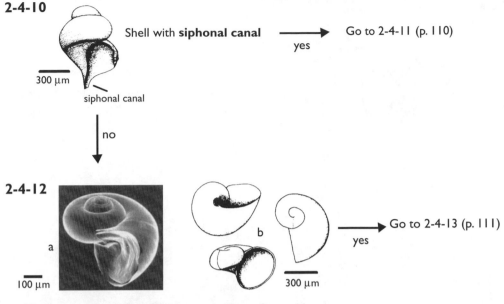

2-4-10

Shell with **siphonal canal** ⟶ yes ⟶ Go to 2-4-11 (p. 110)

300 μm

siphonal canal

no

2-4-12

a

100 μm

b

300 μm

⟶ yes ⟶ Go to 2-4-13 (p. 111)

Shell **globose** (rounded) with **nearly flat spire** and **large aperture**; no apertural beak or siphonal canal

no

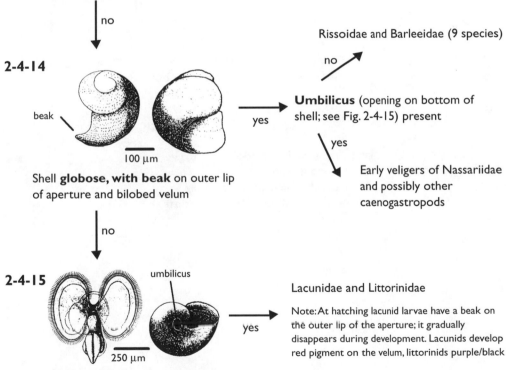

2-4-14

Rissoidae and Barleeidae (9 species)

no

beak

yes ⟶ **Umbilicus** (opening on bottom of shell; see Fig. 2-4-15) present

100 μm

yes

Early veligers of Nassariidae and possibly other caenogastropods

Shell **globose, with beak** on outer lip of aperture and bilobed velum

no

2-4-15

umbilicus

Lacunidae and Littorinidae

yes ⟶ Note: At hatching lacunid larvae have a beak on the outer lip of the aperture; it gradually disappears during development. Lacunids develop red pigment on the velum, littorinids purple/black

250 μm

Globose shell with **umbilicus** but **no beak**; velum bilobed

From 2-4-10 (shell with siphonal canal)

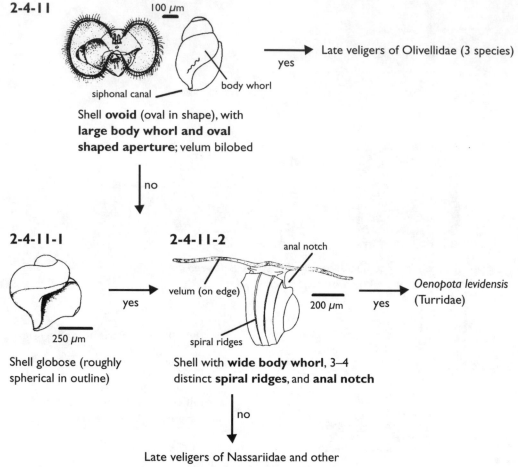

2-4-11

100 μm

body whorl

siphonal canal

yes → Late veligers of Olivellidae (3 species)

Shell **ovoid** (oval in shape), with **large body whorl and oval shaped aperture**; velum bilobed

no ↓

2-4-11-1

250 μm

Shell globose (roughly spherical in outline)

yes →

2-4-11-2

anal notch

velum (on edge)

spiral ridges

200 μm

Shell with **wide body whorl**, 3–4 distinct **spiral ridges**, and **anal notch**

yes → *Oenopota levidensis* (Turridae)

no ↓

Late veligers of Nassariidae and other caenogastropods, including species of Columbellidae, Turridae, and possibly Cancellariidae and some of the Muricidae

From 2-4-12 (shell globose, with flattened spire; no beak or siphonal canal)

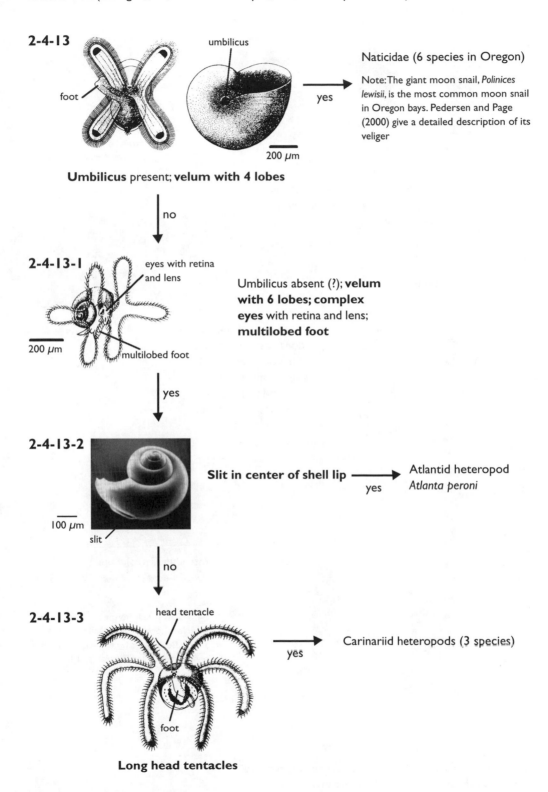

2-4-13

umbilicus

foot

200 μm

Naticidae (6 species in Oregon)

Note: The giant moon snail, *Polinices lewisii*, is the most common moon snail in Oregon bays. Pedersen and Page (2000) give a detailed description of its veliger

yes

Umbilicus present; **velum with 4 lobes**

no

2-4-13-1

eyes with retina and lens

200 μm

multilobed foot

Umbilicus absent (?); **velum with 6 lobes; complex eyes** with retina and lens; **multilobed foot**

yes

2-4-13-2

100 μm

slit

Slit in center of shell lip ⟶ Atlantid heteropod *Atlanta peroni*

yes

no

2-4-13-3

head tentacle

foot

Carinariid heteropods (3 species)

yes

Long head tentacles

From 3 (egg-shaped, inflated shells, coiling hardly discernable)

3-1

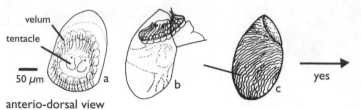

velum
tentacle
50 μm
a
b
c

yes →

anterio-dorsal view

Velum round to oval, with large trochal cells.
Tentacles or tentacle buds with eyespots at bases.
Shell with granulated or wavy surface sculpture
(marked in figure c above)

Patellogastropoda

Note: Shell sculpture has not
been described for local species.
Diagnosis based on descriptions
of NW Pacific congeners in Amio
(1963)

↓ no

3-2

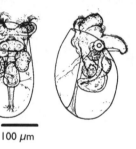

100 μm

Shell smooth and transparent
and does not grow during
larval development. Shell
hydrophobic (repels water)
and is prone to getting trapped
in the air-water interface.
Velum bilobed. No head
tentacles

↓ yes

3-3

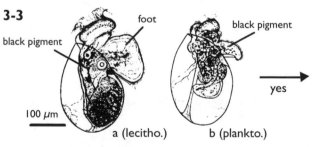

black pigment

foot

black pigment

100 μm

a (lecitho.)   b (plankto.)

Epidermal black pigment

yes →

Introduced aeolid nudibranch
Tenellia adspersa. Shell 195–228 μm
long

Note: This estuarine species can produce
both lecithotropic and planktotrophic
larvae. Both types have epidermal black
pigment; both are illustrated at left

↓ no

Nudibranchs of the families Dendronotidae, Tergipididae,
Eubranchidae, and Fionidae (18 species in 5 genera)

From 4 (double spiral shell). Outer shell (scaphoconch) thin, transparent, not calcified. Inner shell calcified

4-1

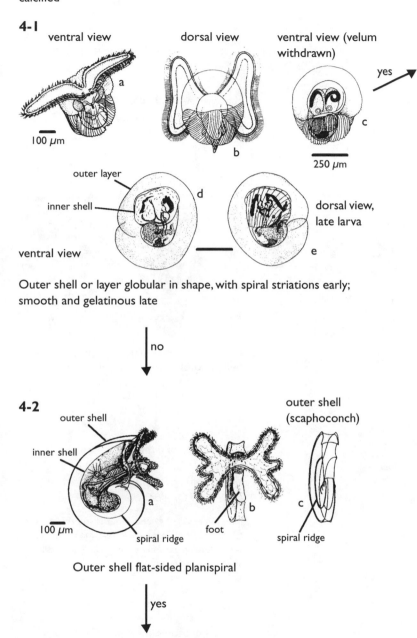

ventral view   dorsal view   ventral view (velum withdrawn)

100 μm

a

b

250 μm

c

outer layer

inner shell

ventral view

d

dorsal view, late larva

e

Outer shell or layer globular in shape, with spiral striations early; smooth and gelatinous late

Velutina velutina and *V. plicatilis*

yes

Note: The inner shell of late veliger acquires spiral ridges and eventually becomes thick, white, and opaque. Hatching larvae of *Velutina plicatilis* described and depicted by Strathmann (1987) (Fig. 4-1a at left) are virtually identical to young larvae of *V. velutina* described by Thorson (1946) (Fig. 4-1c at left). One, possibly 2, additional species of *Velutina* occur in Oregon; their larvae are unknown

no

4-2

outer shell

inner shell

100 μm

a

spiral ridge

outer shell (scaphoconch)

foot

b

c

spiral ridge

Outer shell flat-sided planispiral

yes

Marseniid prosobranchs (4 species in 3 genera)

Note: Diagnosis based on descriptions of congeners from other parts of the world (e.g., Fretter and Pilkington, 1970; Pilkington, 1976)

From 6 (shells thimble-, cone-, or vase-shaped)

6-1

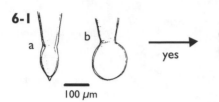

100 μm

yes → Cavoliniid euthecosome pteropods. 6 species in 3 genera (*Clio, Diacria*, and *Cuverina*)

Shell conical with a swollen end

no ↓

Late veliger, euthecosome pteropods of genus *Cavolinia* (5 species)

yes ↑

6-2

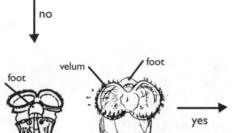

velum   foot
foot
a   b
100 μm
shell

Shell thimble- to cone-shaped with a rounded end

yes →

6-3

left foot lobe   velum   right foot lobe

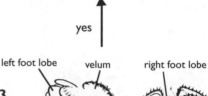

foot
a   100 μm   b

Foot with 2 developing swimming lobes (wings); these expand as the velum shrinks

no ↓

Early veliger of either euthecosome pteropods *Cavolinia* (5 species) or gymnosome pteropods (4 species in 4 genera off Oregon)

Note: The shell of gymnosomes is straight; in most species of *Cavolinia* it curves. Also, the swimming wings develop early in cavoliniids and not until after shell loss in gymnosomes

no ↓

6-4

velum   foot lobes

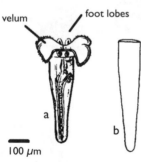

a
b
100 μm

Shell a straight narrow cone with a rounded end

yes ↘

Cavoliniid euthecosome pteropod *Creseis virgula*

no ↓

6-5 (next page)

6-5

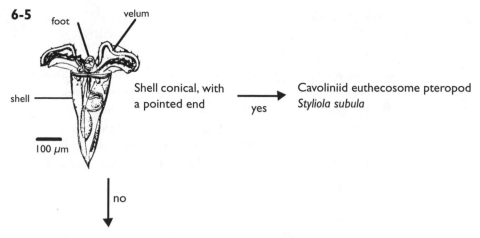

Shell conical, with a pointed end → yes → Cavoliniid euthecosome pteropod *Styliola subula*

no

6-6

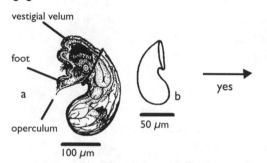

Shell with sinuous branching ridges

yes →

Entoconchid prosobranch *Enteroxenos parastichopoli*

Note: Entoconchids are endoparasites of holothuroids (sea cucumbers). They have not been reported from Oregon waters but are likely present, given their occurrence in Washington and the occurrence of their hosts in Oregon. The larvae apparently do not swim (see Lützen, 1979) and are therefore unlikely to be taken in plankton samples

no

Entoconchid prosobranchs (see note above right) of the genus Thyonicola (2 species known from the Pacific Northwest)

Note: The shells of members of this genus are unsculptured (Lützen, 1979)

Appendix A. Comparative data on the development of opisthobranch molluscs reported from Oregon

This appendix provides a summary of the major developmental features of the larvae of opisthobranch molluscs known from Oregon. It is based on a survey of the published literature and my own observations. Hurst (1967) and Strathmann (1987) provide data on additional species of opisthobranchs known from the Pacific Northwest. Mode of development is known for many species and can be determined by morphological examination of hatching veligers or inferred from data on egg size and embryonic period (see table note 2). A glance at the data table will, however, show that both the duration of the larval period and specific morphological characteristics of metamorphically competent veligers are known for only a few species. This is because most species from our region have fairly long planktotrophic development and require considerable skill, care and facilities to rear through metamorphosis (Hadfield and Switzer-Dunlap, 1984; Strathmann, 1987). Moreover, the larval stages of few, if any, local opisthobranchs have been identified and described by rearing through metamorphosis specimens collected from the plankton. The following data, therefore, comprise an incomplete framework for identifying the larvae of these diverse and morphologically flamboyant organisms.

The type or mode of development of most of the opisthobranchs known from Oregon is given in the table. Almost all hatch as free-swimming planktotrophic veliger larvae with a paucispiral shell, operculum, no eyespots, and a small foot lacking a propodium (see Fig. 2). These larvae feed and grow in the plankton for weeks or months and become competent to settle and metamorphose only after they have acquired eyespots, a propodium and sufficient tissue mass and lipid reserves to fuel the transformation into a functional juvenile. Settlement and metamorphosis in many species is triggered by chemical cues emanating from the prey of the adult slugs (Thompson, 1976; Hadfield and Switzer-Dunlap, 1984).

Two species (*Doto amyra* and the introduced *Tenellia adspersa*) hatch from larger eggs as lecithotrophic veligers competent to metamorphose within a few days of hatching, and one or two more (*Phyllaplysia taylori* and probably *Runcina macfarlandi*) lack a larval stage entirely and hatch as crawl-away juveniles. *Tenellia adspersa* (and possibly also the sacoglossan *Alderia modesta*; see table note 6) has variable developmental mode (known as poecilogony) and also hatches as planktotrophic larvae.

The Oregonian biogeographic province, which stretches from Point Conception, California, to Vancouver Island, British Columbia, has the highest proportion (97%) of nudibranchs with planktotrophic development known from any region in the world. This appears to reflect the suitability, at all but the highest latitudes, of this mode of development for nudibranchs from regions with slow currents, high primary production, and geographically extensive adult habitat with weak gradients in physical factors such as temperature and salinity (Goddard, 1992).

Values given for each species are in most cases means (or ranges in means) derived from at least one source. An additional 22 species of opisthobranchs (17 benthic species and five pelagic) have been recorded from Oregon but are not included here because of lack of information on their development. Species list compiled from Goddard (1984, 1990, 1997, and unpublished observations).

Taxa	Egg diameter (µm)	Shell size at hatching (µm)	Shell type[1]	Eyes at hatching	Dev. type[2]	Min. larval period (days)[3]	Shell size at meta- morph. (µm)	Refs.[4]
Cephalaspidea								
Diaphana californica	73	123	I	no	P	–	–	33
Gastropteron pacificum	95	158	I	no	P	–	–	6, 37
Philine auriformis	–	125	I	no	P	–	–	42
Melanochlamys diomedea	98	180	I	no	P	>40	–	6, 37
Rictaxis punctocaelatus	–	137	I	no	P	–	–	44
Runcina macfarlandi	–	–	–	–	D[5]	none[5]	–	–
Anaspidea								
Aplysia californica	81	125-135	I	no	P	34	400	18, 22, 44
Phyllaplysia taylori	144-157	250-300	I	yes	D	none	250-280	19, 44, 47
Notaspidea								
Berthella californica	93	153	I	yes	P	–	–	33
Pleurobranchaea californica	–	150-215	I	no	P	–	–	44, 45
Sacoglossa								
Alderia modesta	62-80	90-130	I	no	P[6]	35	300-340	3, 5, 8, 37, 48
Aplysiopsis enteromorphae	66-70	109-113	I	no	P	–	–	33, 37
Elysia hedgpethi	68	100-105	I	no	P	–	–	9, 37, 44
Hermaea vancouverensis	63	114	I	no	P	–	–	44
Placida dendritica	47-72	82-127	I	no	P	–	–	9, 20, 37, 44
Stiliger fuscovittatus	95	150	I	no	P	–	–	37

table continues

Taxa	Egg diameter (μm)	Shell size at hatching (μm)	Shell type[1]	Eyes at hatching	Dev. type[2]	Min. larval period (days)[3]	Shell size at meta-morph. (μm)	Refs.[4]
Thecosomata								
Limacina helicina	75[7]	~75	I	no	P	>30	300	10, 24
Gymnosomata								
Clione limacina	85, 110	120-160	3	no	P	14[8]	280-360	1, 17, 39
Nudibranchia: Doridacea								
Acanthodoris brunnea	80	130-150	I	no	P	—	—	37
Acanthodoris hudsoni	67-70	127	I	no	P	—	—	36, 37
Acanthodoris nanaimoensis	—	133	I	no	P	—	—	6
Acanthodoris rhodoceras	—	112	I	no	P	—	—	44
Adalaria sp.	83	140	I	no	P	—	—	33
Aegires albopunctatus	98-120	154	I	yes	P	—	—	37, 46
Aldisa cooperi	110	—	I	—	P	—	—	35
Aldisa sanguinea	90-100	163	I	no	P	—	—	35, 41
Ancula pacifica	59	104	I	no	P	—	—	33
Anisodoris lentiginosa	90	154	I	no	P	36	241	30
Anisodoris nobilis	83	153	I	no	P	—	—	33
Archidoris montereyensis	81-90	154-169	I	no	P	—	—	4, 6, 41
Archidoris odhneri	96	186-189	I	no	P	—	—	6, 41
Cadlina flavomaculata	85	140	I	no	P	—	—	44
Cadlina luteomarginata	90-94	—	—	—	P	—	—	27, 41
Cadlina modesta	92	157	I	no	P	—	—	33
Crimora coneja	73	116-119	I	no	P	—	—	33, 41
Diaphorodoris lirulatocauda	63	115	I	no	P	—	—	33
Diaulula sandiegensis	83	130-153	I	no	P	—	—	6, 37, 41
Doridella steinbergae	75-85	142	I	no	P	25	168	26
Geitodoris heathi	73-79	102,144	I	no	P	—	—	33, 37, 41
Hallaxa chani	81-83	131-152	I	no	P	—	—	33, 41
Hopkinsia rosacea	81-82	141	I	no	P	—	—	44
Laila cockerelli	95	142	I	no	P	—	—	33
Onchidoris bilamellata	100	147-165	I	no	P	38	320	6, 37, 38
Onchidoris muricata	76-80	117-137[9]	I	no	P	>49	—	33, 34, 41
Palio zosterae	65-70	101, 150	I	no	P	—	—	37, 41
Polycera atra	68-71	122	I	no	P	—	—	44
Rostanga pulchra	73-80	148-161	I	no	P	35	300	6, 12, 23
Triopha catalinae	75-87	131-134	I	no	P	—	—	6, 33, 37, 41
Triopha maculata	—	—	I	no	P			15
Nudibranchia: Dendronotacea								
Dendronotus albopunctatus	108	—	2	—	P	—	—	16
Dendronotus diversicolor	96	—	—	—	P	—	—	11, 37
Dendronotus frondosus	85-90	230-245	2	no	P	—	230-245	37, 41
Dendronotus iris	110	268-280	2	no	P	—	268-280	6, 37
Dendronotus subramosus	—	—	—	—	P	—	—	41
Doto amyra	152	239	I	yes	L	I	239	33, 43
Doto kya	78	133	I	no	P	—	—	43
Doto form B	70	122	I	no	P	—	—	43
Melibe leonina	86-90	140-152	I	no	P	30	250	6, 32, 37, 41
Tritonia diomedea	87	145	I	no	P	34	329	6,21,37
Tritonia festiva	79	131	I variable		P	—	-	41

Taxa	Egg diameter (µm)	Shell size at hatching (µm)	Shell type[1]	Eyes at hatching	Dev. type[2]	Min. larval period (days)[3]	Shell size at meta- morph. (µm)	Refs.[4]
Nudibranchia: Arminacea								
Armina californica	95-102	160	1	no	P	—	—	6, 37
Dirona albolineata	70	113-129	1	no	P	—	—	6, 37, 41
Dirona aurantia	—	139	1	no	P	—	—	6
Dirona picta	80[10]	—	—	—	P	—	—	7
Janolus fuscus	81	138	1	yes	P	—	—	41
Nudibranchia: Aeolidacea								
Aeolidia papillosa	74	116-138	1	no	P	—	—	6, 29, 37
Catriona columbiana	100-109	274-302[11]	2	yes	P	—	274-302	33, 44
Catriona rickettsi	98-103	291	2	yes	P	—	291	44
Cuthona abronia	95	224	2	yes	P	—	224	40
Cuthona albocrusta	97	270-281	2	yes	P	—	270-281	6, 40
Cuthona cocoachroma	95	257-277	2 variable	P			—	257-277	33, 40
Cuthona divae	107	249	2	no	P	—	249	33
Cuthona fulgens	94	252	2	yes	P	—	252	40
Cuthona lagunae	98	262	2	yes	P	—	262	40
Cumanotus fernaldi	73	119-130	1	no	P	—	—	6, 41
Eubranchus olivaceus	85	244	2 variable	P			—	244	6, 41
Eubranchus rustyus	93	240	2	yes	P	—	240	33
Fiona pinnata	100-150	280	2	no	P	—	280	14, 31, 37, 44
Flabellina fusca	—	133	1	no	P	—	—	6
Flabellina trilineata	60-65	100-110	1	no	P	—	—	13, 37, 41
Hermissenda crassicornis	65	102-119	1	no	P	34	310	6, 25, 29, 41
Tenellia adspersa	72, 103	195-228	2	yes	P or L[12]	variable[13]	195-228	2, 28, 41, 44

[1]Shell type: 1 = sinistral, pauci-spiral shells; generally 0.75 to 1 whorl. 2 = egg-shaped, inflated shells; these do not grow after hatching. 3 = thimble-shaped or hemiellipsoid shell that flares with growth after hatching.

[2]Development type: P = planktotrophic, L = lecithotrophic, D = direct (capsular metamorphic or ametamorphic). When not stated by the original author, I have assigned development type according to criteria described by Thompson (1967), Bonar (1978), Todd (1983), Hadfield and Switzer-Dunlap (1984), and Hadfield and Miller (1987). Development type was inferred for five species (*Aldisa cooperi, Dendronotus albopunctatus, D. diversicolor, D. subramosus,* and *Dirona picta*) using information on egg size, embryonic period, and comparisons with congeners (see Goddard, 1992, pp. 38–41).

[3]Duration of larval period varies with temperature and food supply. Values given are from laboratory studies (see references) using culture temperatures ranging 10–15° C (16° C for *Clione limacina*).

[4]1, Lebour (1931); 2, Rasmussen (1944); 3, Rasmussen (1951); 4, McGowan and Pratt (1954); 5, Hand and Steinberg (1955); 6, Hurst (1967); 7, Marcus and Marcus (1967); 8, Seelemann (1967); 9, Greene (1968); 10, Paranjape (1968); 11, Robilliard (1970); 12, Anderson (1971); 13, Bridges and Blake (1972); 14, Holleman (1972); 15, Mulliner (1972); 16, Robilliard (1972); 17, Lalli and Conover (1973); 18, Kriegstein et al. (1974); 19, Bridges (1975); 20, Clark (1975); 21, Kempf and Willows (1977); 22, Kriegstein (1977); 23, Chia

notes continue

and Koss (1978); 24, Lalli and Wells (1978); 25, Harrigan and Alkon (1978); 26, Bickell and Chia (1979); 27, Dehnel and Kong (1979); 28, Eyster (1979); 29, Williams (1980); 30, Millen (1982); 31, Schmekel and Portmann (1982); 32, Bickell and Kempf (1983); 33, Goddard (1984); 34, Millen (1985); 35, Millen and Gosliner (1985); 36, Goddard (1987); 37, Strathmann (1987); 38, Chia and Koss (1988); 39, Lalli and Gilmer (1989); 40, Goddard (1991); 41, Goddard (1992); 42, Gosliner (1995); 43, Goddard (1996); 44, Goddard (unpublished observations); 45, Chivers (1967); 46, Goddard (in press); 47, Bertsch and Hirshberg (1973); 48, Krug (1998).

[5]Development has not been examined for *Runcina macfarlandi*; however, direct development is considered diagnostic of the genus (Thompson and Brodie, 1988).

[6]Krug (1998) reported that a population of *Alderia modesta* in San Diego, California, produces planktotrophic larvae from eggs 68 μm diameter and lecithotrophic larvae from eggs 105 μm diameter. Hatching larvae of the latter had shells 186 μm long.

[7]Paranjape (1968, p. 323) stated that "the egg diameter was 95–100 μm in the longest dimension, while the diameter of the ovum was 75 μm." I am assuming that by "egg diameter" Paranjape meant "egg capsule."

[8]Duration of veliger larval stage only; gymnosomes have a second, "polytrochous," larval stage that undergoes a gradual metamorphosis into the adult stage.

[9]Hurst (1967) reported an anonymously high value of 186 μm.

[10]Marcus and Marcus (1967) did not specify if this value was obtained from measurements of living or preserved material.

[11]Hurst (1967) reported an anonymously low value of 230 μm.

[12]Embryos from different egg masses hatch as either planktotrophic or lecithotrophic larvae. In addition, Eyster (1979) reported capsular metamorphic development in some *Tenellia adspersa* (as *T. pallida*) from South Carolina.

[13]Hours for the lecithotrophic larvae; unknown for the planktotrophs.

Appendix B. Sources of illustrations used in the key

1a (Kessel, 1964: 6)
2a (Fretter and Graham, 1962: 237a)
2b (Rasmussen, 1951: 15)
2c (Pilkington, 1976: 2A)
2.a (personal)
3a (Rasmussen, 1944: 19)
3b (Kessel, 1964: 11)
4a (Strathmann, 1987: 11.13)
4b (Fretter and Graham, 1964: 245B)
5a (Kessel, 1964: 8)
6a (Lebour, 1931: plate 1, fig. 6)
6b (van der Spoel, 1967: 61B)
6c (Yamaji, 1977: plate 139, fig. 4c)
7 (personal)
8a (Thiriot-Quiévreux, 1967: 2A)
8b (Tsubokawa and Okutani, 1991: 7C)
9 (personal)
10 (Lebour, 1931: plate 1, fig. 9)
11 (Lebour, 1931: plate 1, figs. 10 and 11)
1.1 (after Crofts, 1937: 41a)
1.2a (Chapter 9, p. X)
1.2b (Chapter 9, p. X)
1.2c (Chapter 9, p. X)
1.2d (Chapter 9, p. X)
1.3 (Kessel, 1964: 6)
2.1 (Thorson, 1946: 108A and B)
2.3a (Pilkington, 1976: 2A)
2.3b (Hadfield and Strathmann, 1990: 2B and C)
2.5a (Thorson, 1946: 117A and B)
2.5b (Rasmussen, 1944: 6)
2.6a (modified from: Pilkington, 1976: 11F)
2.6b (modified from: Pilkington, 1976: 11H)
2.7 (modified from: Hadfield and Strathmann, 1990: 2B and C)
2.8a (Thorson, 1946: 152B)
2.8b (Rasmussen, 1951: 14, lower)
2.8c (Thorson, 1946: 152A)
2.11 (Lalli and Gilmer, 1989: 35C)
2.12 (Hurst, 1967: 24.19)
2.13 (Thorson, 1946: 117A and B)
2.14a (Thorson, 1946: 152C)
2.14b (Thorson, 1946: 147B)
2.14c (Thorson, 1946: 145C)
2.15 (Rasmussen, 1944: 6)
Box 1,a (Rasmussen, 1944: 7)
Box 1, b (Thorson, 1946: 144E)
2.2 (Thorson, 1946: 108E and F)
2.2.1 (Lalli and Gilmer, 1989: 14d)
2.2.2a (Thorson, 1946: 108A and B)
2.2.2b (Fretter and Pilkington, 1970: 7b)
2.9 (Thompson, 1976: 41a)
2.9.1 (Thompson, 1976: 100)
2.9.2 (Kriegstein, 1977: 1, stage 6a)

2.9.3 (Lalli and Gilmer, 1989: 45a)
2.9.4 (Bickell and Kempf, 1983: 8C)
2.9.6 (Goddard, 1996: 3)
2.9.8 (Thompson, 1976: 41a and b)
2.4 (Hadfield and Strathmann, 1990: 2B and C)
2.4.1 (Hickman, 1992: 4)
2.4.2 (Leighton, 1974: 1.7)
2.4.3 (Fretter and Pilkington, 1970: 11a and b)
2.4.4 (Fretter and Pilkington, 1970: 32a and b)
2.4.5 (Thorson, 1946: 109E and F)
2.4.6 (Fretter and Pilkington, 1970: 9a and b)
2.4.7 (Fretter and Pilkington, 1970: 5a)
2.4.8 (Strathmann, 1987: 11.11 [top])
2.4.9 (Richter and Thorson, 1975: 43a)
2.4.10 (Pilkington, 1976: 11H)
2.4.12a (Pilkington, 1976: 7A and C)
2.4.12b (Lalli and Gilmer, 1989: 14C)
2.4.14 (Fretter and Pilkington, 1970: 3a and b)
2.4.15 (Fretter and Pilkington, 1970: 12a and c)
2.4.11 (Thiriot-Quiévreux, 1983: 1F and G)
2.4.11.1 (Pilkington, 1976: 11A)
2.4.11.2 (Shimek, 1986: 7)
2.4.13 (Fretter and Pilkington, 1970: 20b and c)
2.4.13.1 (Lalli and Gilmer, 1989: 13B)
2.4.13.2 (Lalli and Gilmer, 1989: 14B)
2.4.13.3 (Lalli and Gilmer, 1989: 13C)
3.1a (Kessel, 1964:12)
3.1b (Kessel, 1964:13)
3.1c (after Amio, 1963: 16h)
3.2a (Rasmussen, 1944: 19)
3.3a (Rasmussen, 1944: 18A)
3.3b (modified from: Rasmussen, 1944: 19)
4.1a (Strathmann, 1987: 11:13)
4.1b (Fretter and Pilkington, 1970: 33)
4.1c (Thorson, 1946: 133A)
4.1d (Thorson, 1946: 133D)
4.1e (Thorson, 1946: 133E)
4.2a (Fretter and Graham, 1962: 245B)
4.2b (Fretter and Graham, 1962: 246A)
4.2c (Fretter and Graham, 1962: 246B)
6.1a (van der Spoel, 1967: 61B)
6.1b (van der Spoel, 1967: 76C)
6.2a (Lebour, 1931: plate 1, fig. 6)
6.2b (Fol, 1875: plate III, fig 30)
6.3a (Fol, 1875: plate III, fig. 39)
6.3b (Fol, 1875: plate III, fig. 37)
6.4a (Fol, 1875: plate VI, fig. 5)
6.4b (after van der Spoel, 1967: 37)
6.5 (Fol, 1875: plate VI, fig. 7)
6.6a (Lützen, 1979: 4A)
6.6b (Lützen, 1979: 41, lower right)
6.7 (Lützen, 1979: 4A and B)

Literature Cited

Amio, M. (1963). A comparative embryology of marine gastropods, with ecological considerations. J. Shimonoseki U.Fish. 12:15–144.

Anderson, E. S. (1971). The association of the nudibranch *Rostanga pulchra* MacFarland, 1905 with the sponges *Ophlitaspongia pennata*, *Esperiopsis originalis*, and *Plocamia karykina*. Ph.D. Dissertation, University of California, Santa Cruz. 161 pp.

Austin, W. C. (1985). An annotated checklist of marine invertebrates in the cold temperate northeast Pacific. Khoyatan Marine Laboratory, Cowichan Bay, B.C. 682 pp.

Bertsch, H. and J. Hirshberg (1973). Notes on the veliger of the opisthobranch *Phyllaplysia taylori*. The Tabulata (Santa Barbara Malacological Society) 6:3–5.

Bickell, L. R. and F.-S. Chia (1979). Organogenesis and histogenesis in the planktotrophic veliger of *Doridella steinbergae* (Opisthobranchia: Nudibranchia). Mar. Biol. 52:291–313.

Bickell, L. R. and S. C. Kempf (1983). Larval and metamorphic morphogenesis in the nudibranch *Melibe leonina* (Mollusca: Opisthobranchia). Biol. Bull. 165:119–38.

Bieler, R. (1992). Gastropod phylogeny and systematics. Ann. Rev. Ecol. Syst. 23:311–38.

Bonar, D. B. (1978). Morphogenesis at metamorphosis in opisthobranch molluscs. In: Settlement and Metamorphosis of Marine Invertebrate Larvae, F.-S. Chia and M. E. Rice (eds.), pp. 177–196. Elsevier, New York. 290 pp.

Bridges, C. B. (1975). Larval development of *Phyllaplysia taylori* Dall, with a discussion of development in the Anaspidea (Opisthobranchiata: Anaspidea). Ophelia 14:161–84.

Bridges, C. and J. A. Blake (1972). Embryology and larval development of *Coryphella trilineata* O'Donoghue, 1921 (Gastropoda: Nudibranchia). Veliger 14:293–97.

Caldwell, M. E. (1981). Spawning, early development and hybridization of *Haliotis kamtschatkana* Jonas. M.S. Thesis, University of Washington, Seattle. 55 pp.

Chia, F.-S. and R. Koss (1978). Development and metamorphosis of the planktotrophic larvae of *Rostanga pulchra* (Mollusca: Nudibranchia). Mar. Biol. 46:109–19.

——— (1988). Induction of settlement and metamorphosis of the veliger larvae of the nudibranch, *Onchidoris bilamellata*. Int. J. Invertebr. Reprod. Dev. 14:53–70.

Chivers, D. D. (1967). Observations on *Pleurobranchaea californica* MacFarland, 1966 (Opisthobranchia, Notaspidea). Proc. Calif. Acad. Sci. 32:515–21.

Clark, K. B. (1975). Nudibranch life cycles in the northwest Atlantic and their relationship to the ecology of fouling communities. Helgol. Wiss. Meeresunters. 27:28–69.

Collin, R. 1997. Hydrophobic larval shells: another character for higher level systematics of gastropods. J. Molluscan Stud. 63:425–30.

——— (2000). Development and anatomy of *Nitidiscala tincta* (Carpenter, 1865) (Gastropoda: Epitoniidae). Veliger 43:302–12.

Collin, R. and J. B. Wise (1997). Morphology and development of *Odostomia columbiana* Dall and Bartsch (Pyramidellidae): implications for the evolution of gastropod development. Biol. Bull. 192:243–52.

Crofts, D. R. (1937). The development of *Haliotis tuberculata*, with special reference to organogenesis during torsion. Phil. Trans. R. Soc. London, Ser. B 228:219–68.

Cumming, R. L. (1993). Reproduction and variable development of an ectoparasitic snail, *Turbonilla* sp., (Pyramidellidae, Opisthobranchia), on cultured giant clams. Bull. Mar. Sci. 52:760–71.

Dehnel, P. A. and D. C. Kong (1979). The effect of temperature on developmental rates in the nudibranch *Cadlina luteomarginata*. Can. J. Zool. 57:1835–44.

Eyster, L. S. (1979). Reproduction and developmental variability in the opisthobranch *Tenellia pallida*. Mar. Biol. 51:133–40.

Fol, P. H. (1875). Études sur le développement des Mollusques. Premier mémoire: Sur le développement des Ptéropodes. Arch. Zool. Exp. Gén. 4:1–214.

Fretter, V. and A. Graham (1962). British prosobranch molluscs. Ray Society: London. 755 pp.

Fretter, V. and M. C. Pilkington (1970). Prosobranchia. Veliger larvae of Taenioglossa and Stenoglossa. Cons. Int. Explor. Mer, Zooplankton sheets 129–32.

Gibson, G. D. and F.-S. Chia (1989). Embryology and larval development of *Haminoea vesicula* Gould (Opisthobranchia: Cephalaspidea). Veliger 32:409–12.

Goddard, J. H. R. (1984). The opisthobranchs of Cape Arago, Oregon, with notes on their biology and a summary of benthic opisthobranchs known from Oregon. Veliger 27:143–63.

——— (1987). Observations on the opisthobranch mollusks of Punta Gorda, California, with notes on the distribution and biology of *Crimora coneja*. Veliger 29:267–73.

——— (1990). Additional opisthobranch mollusks from Oregon, with a review of deep-water records and observations on the fauna of the south coast. Veliger 33:230–37.

——— (1991). Unusually large polar bodies in an aeolid nudibranch: a novel mechanism for producing extra-embryonic yolk reserves. J. Mollus. Stud. Suppl. 57:143–52.

——— (1992). Patterns of development in nudibranch molluscs from the northeast Pacific Ocean, with regional comparisons. Ph.D. Dissertation, University of Oregon, Eugene, Oregon. 237 pp.

——— (1996). Lecithotrophic development in *Doto amyra* (Nudibranchia: Dendronotacea), with a review of developmental mode in the genus. Veliger 39:43—54.

——— (2001). The early veliger larvae of *Aegires albopunctatus* (Nudibranchia: Aegiridae), with morphological comparisons to members of the Notaspidea. Veliger 44:412-14.

Gosliner, T. M. (1995). Introduction and spread of *Philine auriformis* (Gastropoda: Opisthobranchia) from New Zealand to San Francisco Bay and Bodega Harbor. Mar. Biol. 122:249–55.

——— (1996). The Opisthobranchia. In Scott, P. H., J. A. Blake, and A. L. Lissner (eds.), Taxonomic atlas of the Santa Maria Basin and western Santa Barbara channel. Vol. 9: The Mollusca, part 2—Gastropoda, pp. 161–213. U. S. Department of the Interior, Minerals Management Service, Camarillo, California.

Greene, R. W. (1968). The egg masses and veligers of southern California sacoglossan opisthobranchs. Veliger 11:100–4

Hadfield, M. G. and S. E. Miller (1987). On developmental patterns of opisthobranchs. Amer Malacol. Bull. 5:197–214.

Hadfield, M. G. and M. F. Strathmann (1990). Heterostrophic shells and pelagic development in trochoideans: implications for classification, phylogeny and palaeoecology. J. Molluscan Stud. 56:239–56.

Hadfield, M. G. and M. Switzer-Dunlap (1984). Opisthobranchs. In: The Mollusca, vol. 7, Reproduction, A. S. Tompa, N. H. Verdonk, and J. A. M. van den Biggelaar (eds.), pp. 209–350. Academic Press, New York.

Hand, C. and J. Steinberg (1955). On the occurrence of the nudibranch *Alderia modesta* (Lovén, 1844) on the central California coast. Nautilus 69:22–28.

Harrigan, J. F. and D. L. Alkon (1978). Larval rearing, metamorphosis, growth and reproduction of the eolid nudibranch *Hermissenda crassicornis* (Eschscholtz, 1831) (Gastropoda: Opisthobranchia). Biol. Bull. 154:430–39.

Hickman, C. S. (1992). Reproduction and development of trochacean gastropods. Veliger 35:245–72.

Holleman, J. J. (1972). Observations on growth, feeding, reproduction, and development in the opisthobranch *Fiona pinnata* (Eschscholtz). Veliger 15:142–46.

Holyoak, A. R. (1988). Spawning and larval development of the trochid gastropod *Calliostoma ligatum* (Gould, 1849). Veliger 30:369–71.

——— (1988a). Spawning, egg mass formation, and larval development of the trochid gastropod *Margarites helicinus* (Fabricius). Veliger 31:111–13.

Hurst, A. (1967). The egg masses and veligers of thirty northeast Pacific opisthobranchs. Veliger 9:255–88.

Hyman, L. H. (1967). The invertebrates, Vol. VI, Mollusca I. McGraw-Hill, New York. 792 pp.

Høisæter, T. (1989). Biological notes on some Pyramidellidae (Gastropoda: Opisthobranchia) from Norway. Sarsia 74:283–97.

Keen, A. M. and E. Coan. (1974). Marine molluscan genera of western North America: an illustrated key. Second edition. Stanford University Press, Stanford, Calif.

Kempf, S. C. and A. O. Willows (1977). Laboratory culture of the nudibranch *Tritonia diomedea* Bergh (Tritonidae: Opisthobranchia) and some aspects of its behavioral development. J. Exp. Mar. Biol. Ecol. 30:261–76

Kessel, M. M. (1964). Reproduction and larval development of *Acmaea testudinalis* (Müller). Biol. Bull. 127:294–303.

Koppen, C. L., J. R. Glascock, and A. R. Holyoak (1996). Spawning and larval development of the ribbed limpet, *Lottia digitalis* (Rathke, 1833). Veliger 39:241–43.

Kozloff, E. N. (1996). Marine invertebrates of the Pacific Northwest. University of Washington Press, Seattle.

Kriegstein, A. R. (1977). Stages in the post-hatching development of *Aplysia californica*. J. Exp. Zool. 199:275–88.

Kriegstein, A. R., V. Castellucci, and E. R. Kandel (1974). Metamorphosis of *Aplysia californica* in laboratory culture. Proc. Nat. Acad. Sci., USA 71:3654–58.

Kristensen, J. H. (1970). Fauna associated with the sipunculid *Phascolion strombi* (Montagu), especially the parasitic gastropod *Menestho diaphana* (Jeffreys). Ophelia 7:257–76.

Krug, P. J. (1998). Poecilogony in an estuarine opisthobranch: planktotrophy, lecithotrophy, and mixed clutches in a population of the ascoglossan *Alderia modesta*. Mar. Biol. 132:483–94.

LaFollette, P. I. (1979). Observations on the larval development and behavior of *Chrysallida cincta* Carpenter, 1864 (Gastropoda: Pyramidellidae). West. Soc. Malacol. Annu. Rep. 11:31–34.

Lalli, C. M. and R. J. Conover (1973). Reproduction and development of *Paedoclione doliiformis*, and a comparison with *Clione limacina* (Opisthobranchia: Gymnosomata). Mar. Biol. 19:13–22.

Lalli, C. M. and R. W. Gilmer (1989). Pelagic Snails. Stanford University Press: Stanford, California. 259 pp.

Lalli, C. M. and F. E. Wells (1978). Reproduction in the genus *Limacina* (Opisthobranchia: Thecosomata). J. Zool. 186:95–108.

Lebour, M. V. (1931). *Clione limacina* in Plymouth waters. J. Mar. Biol. Assoc. (UK) 17:785–95.

———. (1932). The eggs and early larvae of two commensal gastropods, *Stilifer stylifer* and *Odostomia eulimoides*. J. Mar. Biol. Assoc. (UK) 18:117–19.

———. (1945). The eggs and larvae of some prosobranchs from Bermuda. Proc. Zool. Soc. London. 114:462–89.

Leighton, D. L. (1972). Laboratory observations on the earlygrowth of the abalone, *Haliotis sorenseni*, and the effect of temperature on larval development and settling success. Fish. Bull. 70:373–81.

——— (1974). The influence of temperature on larval and juvenile growth in three species of southern California abalones. Fish. Bull. 72:1137–45.

Lima, G. M. and R. A. Lutz (1990). The relationship of larval shell morphology to mode of development in marine prosobranch gastropods. J. Mar. Biol. Assoc. (UK) 70:611–37.

Lützen, J. (1979). Studies on the life history of *Enteroxenos* Bonnevie, a gastropod endoparasitic in aspidochirote holothurians. Ophelia 18:1–51.

Marcus, Er. and Ev. Marcus (1967). American opisthobranch mollusks. Part 1, Tropical American opisthobranchs; Part 2, Opisthobranchs from the Gulf of California. Studies in Tropical Oceanography, Miami 6:1–256.

Martel, A. and F.-S. Chia (1991). Drifting and dispersal of small bivalves and gastropods with direct development. J. Exp. Mar. Biol. Ecol. 150:131–47.

McCloskey, L. R. (1973). Development and ecological aspects of the echinospira shell of *Lamellaria rhombica* Dall Prosobranchia; Mesogastropoda). Ophelia 10:155–68.

McGowan, J. A. and I. Pratt (1954). The reproductive system and early embryology of the nudibranch *Archidoris montereyensis* (Cooper). Bull. Mus. Comp. Zool., Harvard Univ. 111:261–76.

Millen, S. V. (1982). A new species of dorid nudibranch (Opisthobranchia: Mollusca) belonging to the genus *Anisodoris*. Can. J. Zool. 60:2694–2705.

——— (1985). The nudibranch genera *Onchidoris* and *Diaphorodoris* (Mollusca, Opisthobranchia) in the northeastern Pacific. Veliger 28:80–93.

Millen, S. V. and T. M. Gosliner (1985). Four new species of dorid nudibranchs belonging to the genus *Aldisa* (Mollusca: Opisthobranchia), with a revision of the genus. Zool. J. Linnean Soc. 84:195–233.

Moran, A. L. (1997). Spawning and larval development of the black turban snail *Tegula funebralis* (Prosobranchia: Trochidae). Mar. Biol. 128:107–14.

Morse, D. E., N. Hooker, H. Duncan, and L. Jensen (1979). γ - aminobutyric acid, a neurotransmitter, induces planktonic abalone larvae to settle and begin metamorphosis. Science 204:407–10.

Mulliner, D. K. (1972). Breeding habits and life cycles of three species of nudibranchs from the eastern Pacific. Festivus, San Diego Shell Club 3:1–5.

Paranjape, M. A. (1968). The egg mass and veligers of *Limacina helicina* Phipps. Veliger 10:322–26.

Pedersen, R. V. K. and L. R. Page (2000). Development and metamorphosis of the planktotrophic larvae of the moon snail, *Polinices lewisii* (Gould, 1847) (Caenogastropoda: Naticoidea). Veliger 43:58–63.

Pernet, B. (1997). Development of the keyhole and growth rate in *Diodora aspera* (Gastropoda: Fissurellidae). Veliger 40:77–83.

Pilkington, M. C. (1976). Descriptions of veliger larvae of monotocardian gastropods occurring in Otago plankton hauls. J. Mollusc. Stud. 42:337–60.

Ponder, W. F. and D. R. Lindberg (1997). Towards a phylogeny of gastropod molluscs: an analysis using morphological characters. Zool. J. Linnean Soc. 119:83–265.

Radwin, G. E. and J. L. Chamberlin (1973). Patterns of larval development in stenoglossan gastropods. Trans. San Diego Soc. Nat. Hist. 17:107–18.

Rasmussen, E. (1944). Faunistic and biological notes on marine invertebrates I. The eggs and larvae of *Brachystomia rissoides* (Hanl.), *Eulimella nitidissima* (Mont.), *Retusa truncatula* (Brug.) and *Embletonia pallida* (Alder & Hancock), (Gastropoda marina). Vidensk. Medd. Dan. Naturhist. Foren. København 107:207–33.

———— (1951). Faunistic and biological notes on marine invertebrates II. The eggs and larvae of some Danish marine gastropods. Vidensk. Medd. Dan. Naturhist. Foren. København 113:201–49.

Richter, G. and G. Thorson (1975). Pelagische prosobranchier-larven des Golfes von Neapel. Ophelia 13:109–85.

Robertson, R. (1971). Scanning electron microscopy of larval marine gastropod shells. Veliger 14:1–12.

———— (1985). Four characters and the higher category systematics of gastropods. Amer. Malacol. Bull., Special Edition No. 1:1–22.

Robilliard, G. A. (1970). The systematics and some aspects of the ecology of the genus *Dendronotus*. Veliger 12:433–79.

———— (1972). A new species of *Dendronotus* from the northeastern Pacific, with notes on *Dendronotus nanus* and *Dendronotus robustus* (Mollusca: Opisthobranchia). Can. J. Zool. 50:421–32.

Rodriguez Babio, C. and C. Thiriot-Quiévreux (1974). Gastéropodes de la réegion de Roscoff. Étude particulière de la protoconque. Cah. Bio. Mar. 15:531–49.

Schaefer, K. (1996). Review of data on cephalaspid reproduction, with special reference to the genus *Haminaea* (Gastropoda, Opisthobranchia). Ophelia 45:17–37.

Scheltema, R. S. (1962). Pelagic larvae of New England intertidal gastropods. I. *Nassarius obsoletus* Say and *Nassarius vibex* Say. Trans. Amer. Microscop. Soc. 81:1–11.

Schmekel, L. and A. Portmann (1982). Opisthobranchia des Mittlemeeres, Nudibranchia und Saccoglossa. Springer-Verlag: Berlin. 410 pp.

Seelemann, U. (1967). Rearing experiments on the amphibian slug *Alderia modesta*. Helgol. Wiss. Meeresunters. 15:128–34.

Shimek, R. L. (1986). The biology of the Northeastern Pacific Turridae. V. Demersal development, synchronous settlement and other aspects of the larval biology of *Oenopota levidensis*. Int. J. Invertebr. Reprod. Dev. 10:313–33.

Smith, F. G. W. (1935). The development of *Patella vulgata*. Phil. Trans. Royal Soc. London, Serial B 225:95–125.

Strathmann, M. F. (1987). Reproduction and development of marine invertebrates of the northern Pacific coast. University of Washington Press: Seattle. 670 pp.

Thiriot-Quiévreux, C. (1967). Descriptions de quelques véligères planctoniques de gastéropodes. Vie Milieu, Ser. A, Biol. Mar. 18:303–15.

——— (1969). Caractéristiques morphologiques des véligères planctoniques de gastéropodes de la région de Banyuls-Sur-Mer. Vie Milieu, Ser. B, Oceanog. 20:333–66.

——— (1970). Transformations histologiques lors de la métamorphose chez *Cymbulia peroni* de Blainville (Mollusca, Opisthobranchia). Z. Morph. Tiere 67:106–17.

——— (1973). Heteropoda. Oceanog. Mar. Biol. Ann. Rev. 11:237–61.

——— (1975). Observations sur les larves et les adultes de Carinariidae (Mollusca: Heteropoda) de l'Océan Atlantique Nord. Mar. Biol. 32:379–88.

——— (1977). Véligère planctotrophe du doridien *Aegires punctilucens* (D'Orbigny) (Mollusca: Nudibranchia: Notodorididae): description et métamorphose. J. Exp. Mar. Biol. Ecol. 26:177–90.

——— (1980). Identification of some planktonic prosobranch larvae present off Beaufort, North Carolina. Veliger 23:1–9.

——— (1983). Summer meroplanktonic prosobranch larvae occurring off Beaufort, North Carolina. Estuaries 6:387–98.

Thiriot-Quiévreux, C. and C. Rodriguez Babio (1975). Étude des protoconques de quelques prosobranches de la région de Roscoff. Cah. Bio. Mar. 16:135–48.

Thiriot-Quiévreux, C. and R. S. Scheltema (1982). Planktonic larvae of New England gastropods. V. *Bittium alternatum*, *Triphora nigrocincta*, *Cerithiopsis emersoni*, *Lunatia heros*, and *Crepidula plana*. Malacologia 23:37–46.

Thompson, T. E. (1967). Direct development in a nudibranch, *Cadlina laevis*, with a discussion of developmental processes in Opisthobranchia. J. Mar. Biol. Assoc. (UK) 47:1–22.

——— (1976). Biology of opisthobranch molluscs, Vol. 1. Ray Society: London. 207 pp.

Thompson, T. E. and G. Brodie (1988). Eastern Mediterranean Opisthobranchia: Runcinidae (Runcinacea), with a review of runcinid classification and a description of a new species from Fiji. J. Mollusc. Stud. 54:339–46.

Thorson, G. (1946). Reproduction and larval development of Danish marine bottom invertebrates, with special reference to the planktonic larvae in the sound (Øresund). Medd. Komm. Dan. Fisk. Havunder. Ser. Fisk 4:1–523.

Todd, C. D. (1983). Reproductive and trophic ecology of nudibranch molluscs. Pp. 225-259 in W. D. Russel-Hunter (ed.), The Mollusca, vol. 6, Ecology. Academic Press, New York.

Tsubokawa, R. and T. Okutani (1991). Early life history of *Pleurobranchia japonica* Thiele, 1925 (Opisthobranchia: Notaspidea). Veliger 34:1–13.

Ueno, S. and M. Amio (1994). Swarming of thecosomatous pteropod *Cavolinia uncinata* in the coastal waters of the Tsushima Strait, the western Japan Sea. Bull. Plankton Soc. Jpn. 41:21–29.

van der Spoel, S. (1967). Euthecosomata, a group with remarkable developmental stages (Gastropoda, Pteropoda). J. Noorduijn en Zoon: Gorinchem. 375 pp.

White, M. E., C. L. Kitting, and E. N. Powell (1985). Aspects of reproduction, larval development, and morphometrics in the pyramidellid *Boonea impressa* (=*Odostomia impressa*) (Gastropoda: Opisthobranchia). Veliger 28:37–51.

Williams, L. G. (1980). Development and feeding of larvae of the nudibranch gastropods *Hermissenda crassicornis* and *Aeolidia papillosa*. Malacologia 20:99–116.

Yamaji, I. (1977). Illustrations of the marine plankton of Japan. Hoikusha Publishing Co., Ltd.: Osaka, Japan. 369 pp.

10

Mollusca: Bivalvia

Laura A. Brink

The bivalves (also known as lamellibranchs or pelecypods) include such groups as the clams, mussels, scallops, and oysters. The class Bivalvia is one of the largest groups of invertebrates on the Pacific Northwest coast, with well over 150 species encompassing nine orders and 42 families (Table 1). Despite the fact that this class of mollusc is well represented in the Pacific Northwest, the larvae of only a few species have been identified and described in the scientific literature. The larvae of only 15 of the more common bivalves are described in this chapter. Six of these are introductions from the East Coast. There has been quite a bit of work aimed at rearing West Coast bivalve larvae in the lab, but this has lead to few larval descriptions.

Reproduction and Development

Most marine bivalves, like many marine invertebrates, are broadcast spawners (e.g., *Crassostrea gigas*, *Macoma balthica*, and *Mya arenaria*,); the males expel sperm into the seawater while females expel their eggs (Fig. 1). Fertilization of an egg by a sperm occurs within the water column. In some species, fertilization occurs within the female, with the zygotes then

text continues on page 134

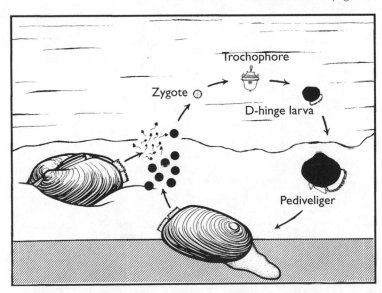

Fig. 1. Generalized life cycle of marine bivalves (not to scale).

Table 1. Species in the class Bivalvia from the Pacific Northwest (local species list from Kozloff, 1996). Species in bold indicate larvae described in this chapter.

Order, Family	Species	Life History[1]	References for Larval Descriptions
Nuculoida			
Nuculidae	*Nucula tenuis*		
	Acila castrensis	FSP	Strathmann, 1987; Zardus and Morse, 1998
Nuculanidae	*Nuculana hamata*		
	Nuculana minuta		
	Nuculana cellutita		
Yoldiidae	*Yoldia amygdalea*		
	Yoldia scissurata		
	Yoldia thraciaeformis		Hutchings and Haedrich, 1984
	Yoldia myalis		
Solemyoida			
Solemyidae	***Solemya reidi***	FSP	Gustafson and Reid, 1986
Arcoida			
Glycymerididae	*Glycymeris subobsoleta*		
	Glycymeris corteziana		
Philobryidae	*Philobrya setosa*		
Mytiloida			
Mytilidae	***Mytilus californianus***	FSV	Lutz et al., 1982; Shaw et al., 1988; Strathmann, 1987; Martel et al., 2000
	Mytilus trossulus (***edulis***)	FSV	Stafford, 1912; Sullivan, 1948; Rees, 1950; Loosanoff et al., 1966, Chanley and Andrews, 1971; Bayne, 1971; Epifanio et al., 1975; De Schweintiz and Lutz, 1976; Le Pennec, 1980
	Crenella decussata		Shaw et al., 1988; Martel et al., 2000
	Lithophaga plumula		
	Adula californiensis	FSV	Morris et al., 1980
	Adula falcata		
	Adula diegensis	FSV	Lough and Gonor, 1971; Strathmann, 1987
	Modiolus modiolus	FSV	Rees, 1950; Chanley and Andrews, 1971; De Schweinitz and Lutz, 1976
	Modiolus rectus		
	Musculista senhousia		
	Musculus niger		
	Musculus discors		
	Musculus taylori		
	Dacrydium pacificum		
	Megacrenella columbiana		
Limoida			
Limidae	*Limatula subauriculata*		
Ostreoida			
Pectinidae	*Patinopecten caurinus*	FSV	Bronson et al., 1984; Strathmann, 1987
	Chlamys hastata	FSV	Strathmann, 1987Hodgson and Bourne, 1988
	Chlamys rubida		
	Chlamys behringiana		
	Hinnites gigantea	FSV	Bronson et al., 1984; Strathmann, 1987
	Delectopecten randolphi		
	Delectopecten vancouverensis		
	Parvamussium alaskensis		

Order, Family	Species	Life History[1]	References for Larval Descriptions
Anomiidae	*Pododesmus cepio*	FSV	Leonard, 1969; Strathmann, 1987
Ostreidae	**Ostrea lurida**	BV	Hori, 1933; Hopkins, 1936; Loosanoff et al., 1966; Strathmann 1987
	Ostrea conchaphilia		
	Crassostrea virginica	FSV	Sullivan, 1948; Loosanoff et al., 1966; Chanley and Andrews, 1971; Epifanio et al., 1975; Lutz et al., 1982
	Crassostrea gigas	FSV	Loosanoff et al., 1966; Epifanio et al., 1975; Le Pennec, 1980; Pauley et al., 1988
Veneroida			
Lucinidae	*Lucina tenuisculpta*		
	Lucinoma annulata		
Thyasiridae	*Thyasira cygnus*		
	Thyasira gouldii	BV or BJ ?	Blacknell and Ansell, 1974, 1975
	Thyasira barbarensis		
	Axinopsida serricata		
	Axinopsida viridis		
	Conchocele bisecta		
Ungulinidae	*Diplodonta orbellus*		
	Diplodonta impolita		
Chamidae	*Chama arcana*		
	Pseudochama exogyra		
Kelliidae	**Kellia suborbicularis**	BV	Lebour, 1938 a,b; Rees, 1950; Strathmann, 1987
	Rhamphidonta retifera		
	Odontogena borealis		
Lasaeidae	*Lasaea subviridis*	BV	Strathmann, 1987
Galeommatidae	*Scintillona bellerophon*		
Montaculidae	*Pseudopythina rugifera*	BV	Strathmann, 1987
	Pseudopythina compressa		
	Mysella tumida	BV	O'Foighil, 1985; Strathmann, 1987
Turtoniidae	*Turtonia minuta*		Matveeva, 1976
Carditidae	*Glans carpenteri*	BJ	Morris et al., 1980
	Miontodiscus prolongatus		
	Cyclocardia ventricosa		
	Cyclocardia crebricostata		
	Crassicardia crassidens		
Astartidae	*Astarte esquimalti*		
	Astarte compacta		
	Astarte undata		
	Tridonta alaskensis		
Cardiidae	*Nemocardium centrifilosum*		
	Clinocardium nuttalli	FSV	Gallucci and Gallucci 1982; Strathmann, 1987
	Clinocardium blandum		
	Clinocardium cliatum		
	Clinocardium californiense		
	Clinocardium fucanum		
	Serripes groenlandicus		
Mactridae	*Mactra californica*		
	Spisula falcata		

table continues

Order, Family	Species	Life History[1]	References for Larval Descriptions
	Tresus nuttalli	FSV	Morris et al., 1980
	Tresus capax	FSV	Bourne and Smith, 1972a,b; Strathmann, 1987
Cultellidae	*Siliqua patula*	FSV	Breese and Robinson, 1981; Lassuy and Simons, 1989; Morris et al., 1980
	Siliqua lucida		
	Siliqua sloati		
Solenidae	*Solen sicarius*		
Tellinidae	*Tellina bodegensis*		
	Tellina carpenteri		
	Tellina modesta		
	Tellina nuculeoides		
	Macoma balthica	FSV	Sullivan, 1948; Strathmann, 1987
	Macoma calottensis		
	Macoma eliminata		
	Macoma nasuta	FSV	Marriage, 1954; Rae, 1978, 1979
	Macoma secta	FSV	Marriage, 1954; Rae, 1978, 1979
	Macoma yoldiformis		
Psammobiidae	*Gari californica*	FSV	Strathmann, 1987
Scrobiculariidae	*Semele rubropicta*		
	Cumingia californica		
Solecurtidae	*Tagelus californianus*		
Corbiculidae	**Corbiucla fluminea**	BJ	Kennedy and Van Huekelem, 1985; Kennedy et al., 1991
Veneridae	*Protothaca staminea*	FSV	Marriage, 1954; Nickerson, 1977; Chew and Ma, 1987; Strathmann, 1987
	Protothaca tenerrina		
	Humilaria kennerlyi		
	Tapes philippinarum	FSV	Bourne, 1982; Strathmann, 1987
	Psephidia ovalis		
	Psephidia lordi	BV	Strathmann, 1987
	Transenella confusa		Gray, 1982
	Transenella tantilla	BG	Strathmann, 1987
	Lyocyma fluctuosa		
	Saxidomus giganteus	FSV	Fraser, 1929; Breese and Phibbs, 1970; Nickerson, 1977; Strathmann, 1987
	Compsomyax subdiaphana		
Petricolidae	*Petricola carditoides*		
	Petricola pholadiromis	FSV	Chanley and Andrews, 1971
Cooperellidae	*Cooperella subdiaphana*		
Thraciidae	*Thracia beringi*		
	Thracia curta		
	Thracia trapezoides		
	Thracia challisiana		
Myoida			
Myidae	**Mya arenaria**	FSV	Stafford, 1912; Sullivan, 1948; Marriage, 1954; Loosanoff et al., 1966; Chanley and Andrews, 1971; Savage and Goldberg, 1976; Lutz et al., 1982
	Mya truncata		
	Cryptomya californica		
	Platyodon cancellatus		
	Sphenia ovoidea		

Order, Family	Species	Life History[1]	References for Larval Descriptions
Hiatellidae	*Hiatella arctica*	FSV	Lebour, 1938b; Sullivan, 1948, Rees, 1950, Savage and Goldberg, 1976
	Panope abrupta	FSV	Marriage, 1954; Goodwin, 1973; Strathmann, 1987
	Panomya chrysis		
Pholadidae	*Barnea subtruncata*		
	Zirfaea pilsbryii		
	Penitella conradi		Wilson and Kennedy, 1984
	Penitella gabbii		
	Penitella penita	FSV	Morris et al., 1980
	Penitella turnerae		
	Netastoma rostrata		
Xylophagaidae	*Xylophaga washingtona*		
Teredinidae	*Teredo navalis*	FSV	Sullivan, 1948; Loosanoff et al., 1966; Chanley and Andrews, 1971; Culliney, 1975; Quayle, 1992
	Bankia setacea	FSV	Quayle, 1953, 1959, 1992; Townsley et al., 1966; Haderlie, 1983
Pholadomyoida			
Lyonsiidae	*Lyonsia californica*	FSV	Strathmann, 1987
	Entodesma pictum		
	Mytilimeria nuttalli	FSV	Yonge, 1952; Strathmann, 1987
	Agriodesma saxicola		
Pandoridae	*Pandora bilirata*		
	Pandora filosa		Thomas, 1994
	Pandora punctata		
	Pandora wardiana		
	Pandora glacialis		
Septibranchida			
Cuspidariidae	*Cardiomya oldroydi*		
	Cardiomya pectinata	BEC	Gustafson et al., 1986; Strathmann, 1987
	Cardiomya planetica		
	Cardiomya californica		
	Plectodon scaber		

[1]Life History: FSV, free-spawning veliger larvae; FSP, free-spawning periclymmna larvae; BV, brooded to veliger; BJ, brooded to juvenile; BEC, benthic egg capsule.

expelled into the surrounding water column. After fertilization of the egg, the zygote first develops into a planktonic trochophore and then into a shelled veliger larva with a ciliated velum used for swimming and respiration. This planktonic larval stage can last from a few days to months, depending on the species. Just prior to metamorphosis the foot develops, at which time the larva is called a pediveliger. After settling on the bottom, loss of the ciliated swimming velum takes place and the post-larval bivalve is able to dig and crawl around using its now fully functional foot. At this stage, these post-larval individuals are referred to as "spat," or juveniles.

In contrast, some bivalve species brood their larvae either within the mantle cavity or attached to the external shell surface (Strathmann, 1987). The developmental stage at which these larvae are released is species specific; some species release their young at the early veliger stage (i.e., *Kellia suborbicularis*, at ca 72 µm; Strathmann, 1987), but others retain the larvae until they are ready to live as juveniles on the bottom (i.e., *Corbicula fluminea*, at ca 210 µm; V. Kennedy, pers. comm.).

Finally, some species in the orders Nuculoida and Solemyoida produce non-feeding (lecithotrophic) larvae with unique morphology, the periclymma larva (see Fig. 3). The larvae are barrel-shaped with prominent apical tufts and are propelled by cilia. The outer cellular body, or test, is ovoid and completely surrounds the shell. The test is cast off at metamorphosis when the juvenile clam begins its benthic existence. Two local species with this mode of development have thus far been described (*Acila castrensis* and *Solemya reidi*).

Occasionally post-settlement-sized "larvae" are found in plankton samples, especially those samples taken near the bottom. These are individuals that have already meta-morphosed and taken up residence on the bottom. Non-planktonic individuals are occasionally found in plankton samples for one of three reasons: (1) They were stirred up from the bottom by the currents. (2) They are byssus thread drifters. Byssus thread drifting occurs when post-settlement bivalves extend a byssus thread and the drag on the thread causes the individual to become resuspended in the water column. This is a common way for post-larval bivalves (and occasionally gastropods) to further their dispersal (Lane et al., 1985; Martel and Chia, 1991; Cummings et al., 1993). (3) Less likely, these individuals have not found suitable settlement sites and are able to delay metamorphosis (Bayne, 1965).

Identification and Description of Local Taxa

Sampling of bivalve larvae is best accomplished with a plankton net that has a mesh size small enough to retain the smallest larvae one wishes to study. Larvae smaller than about 100 μm are nearly impossible to identify without the aid of a scanning electron microscope. The larval shells are surprisingly strong and thus can be sampled either by a towed plankton net or through an electric or gas-powered pump with minimal damage to the shells. For short-term storage (on the order of months), 5% formalin buffered with calcium carbonate works well for preservation; buffering prevents low pH that causes shells to dissolve. If longer storage is necessary (months to years), 70% ethanol is recommended to prevent shell loss.

Because plankton samples are typically kept in buffered formalin, the bodies of the bivalve larvae are often shriveled while the shell has remained intact. For some larvae, shell color is helpful for identification. Depending on how long the sample has been stored, however, this color may not have survived, and shell shape is the primary tool for identification, not shell or body color. Color characteristics can be useful as a backup to help confirm an identification

Use cross-polarized light to view larvae. Place a polarizing filter below the stage on the microscope and a second filter between the sample and the lens of the microscope. Rotate one of the filters until the background becomes dark and the bivalve shells "light up." The crossed polarization causes birefringence due to the microcrystalline aragonitic structure of the larval shells (Gallager et al., 1989) and dramatically aids finding and identifying larvae within a sample.

At all developmental stages (veliger or pediveliger) regardless of size, the shape of the umbo is most useful for identification (Fig. 2). The umbo can take on a variety of shapes and sizes, most of which are species-specific. Prior to the development of the umbo, during the early veliger or D-hinge stage, larvae can be extremely difficult to differentiate and, without the use of scanning electron microscopy, probably cannot be taken down to species. Aside from the umbo, the length and slope of the shoulders can also be helpful in identification. The shoulders typically become prominent during the veliconcha stage.

Periclymma Larvae

There are probably 12 local bivalve species that produce periclymma larvae. The larvae of two of these species have been described, and illustrations of several additional periclymma are available (Fig. 3).

Umbo Stage

Straight-hinge stage

straight-hinge line

posterior   anterior

length

dorsal

umbo

posterior
shoulder

anterior
shoulder

posterior
end

anterior
end

ventral

depth

height

Umbonal Shapes

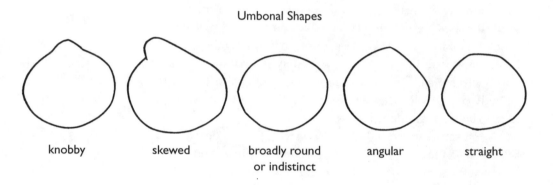

knobby   skewed   broadly round   angular   straight
                                or indistinct

D-shaped larva or straight-hinge larva

Veliconcha

prodissoconch I

prodissoconch II

Early bottom-living stage

prodissoconch

dissoconch

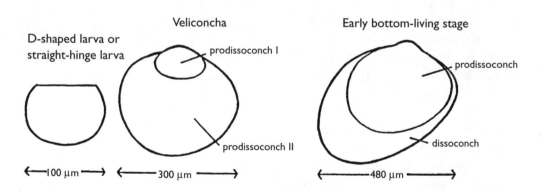

100 μm   300 μm   480 μm

Fig. 2. Terminology used to describe dimensions and shapes of bivalve larvae. The posterior end of the larval shell is typically blunter and shorter than the anterior end and has a higher shoulder. (Adapted from Chanley and Andrews, 1971; Rees, 1950)

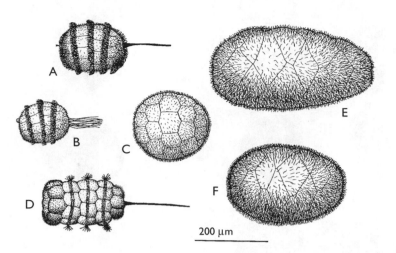

Fig. 3. Periclymma larvae.
(A) **Acila castrensis**.
(B) **Nucula** proxima.
(C) **Nucula**
delphinodonta. (D) **Yoldia**
limatula. (E) **Solemya**
reidi. (F) **Solemya** velum.
Local genera and species
are in bold. (From
Zardus and Morse, 1998,
Fig. 54)

200 μm

Key to bivalve periclymma larvae

1a. Test with 3 transverse bands of cilia and apical tuft of accessory
cilia. Late-stage larvae laterally compressed with shell
development. Test shed at metamorphosis (Fig. 3A) ....
.... Acila castrensis
1b. Test entirely covered with cilia. Length varies, 360–440 μm (Fig.
3E) .... Solemya reidi

Veliger Larvae

The subtle differences in larval shell shape between bivalve
species do not lend themselves to the normal dichotomous key.
What works better is a pictorial guide. We have modeled this
identification guide after Chanley and Andrews's (1971) guide
to the bivalve larvae of the Virginia coast. Because of the
difficulty in distinguishing straight-hinge or D-larvae, this
section describes only larvae in which the umbo has become
rounded. Use this section as follows: (1) Use the ocular
micrometer on the dissection microscope to determine the
length of the larva in question. (2) Find this length in Fig. 4. (3)
Match the shape of the specimen's larval shell. (4) Find details
of the selected species in the following descriptions.

The correct identification of larval bivalves takes considerable time and patience. Differences in shell shape among
species are in most cases extremely subtle. It is particularly
important to remember that few of the bivalve species common
to the Pacific Northwest coast have published descriptions of
their larvae (see Table 1).

Bankia setacea, **Feathery Shipworm** (Order Myoida, Family
Teredinidae). Characteristic dark rim around the margin of the

text continues on page 142

Fig. 4 (overleaf).
Comparative drawings of
15 species of larval
bivalves.

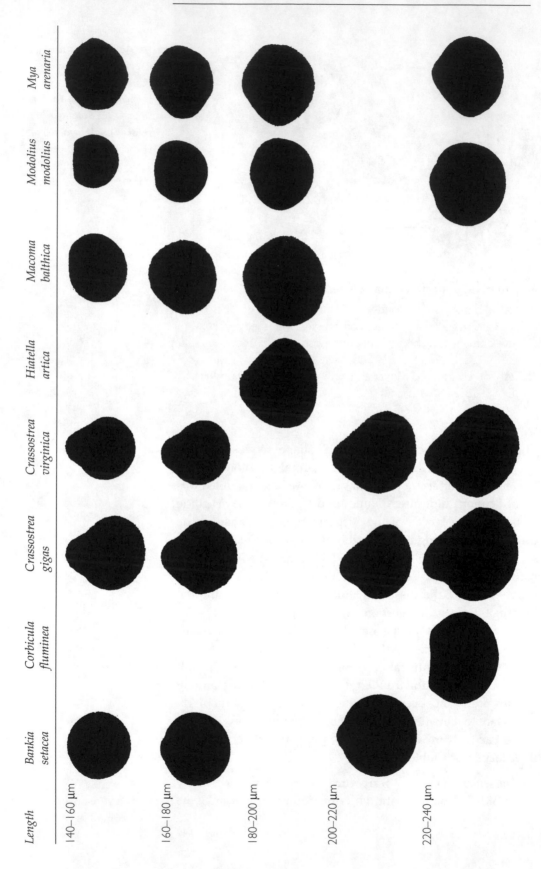

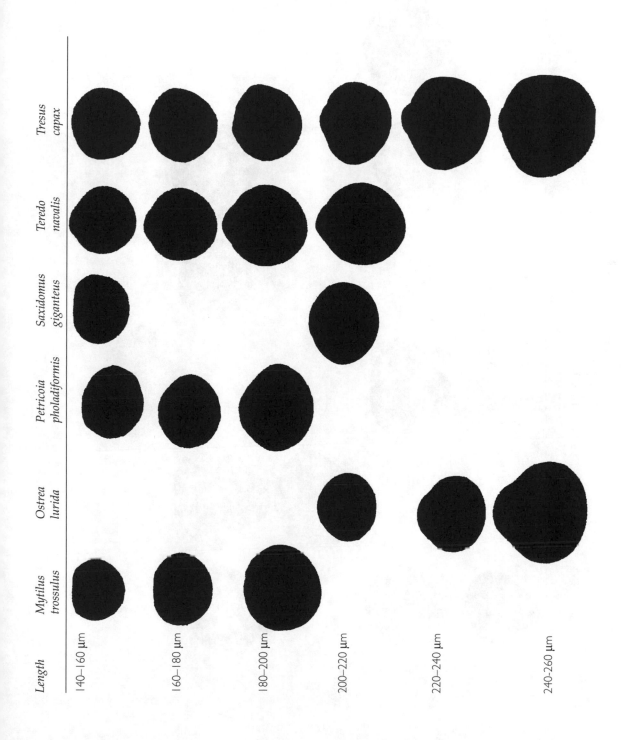

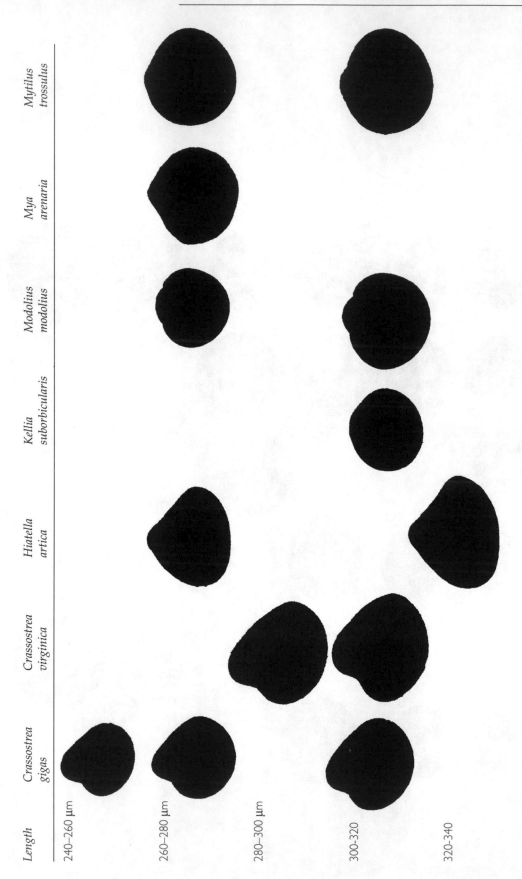

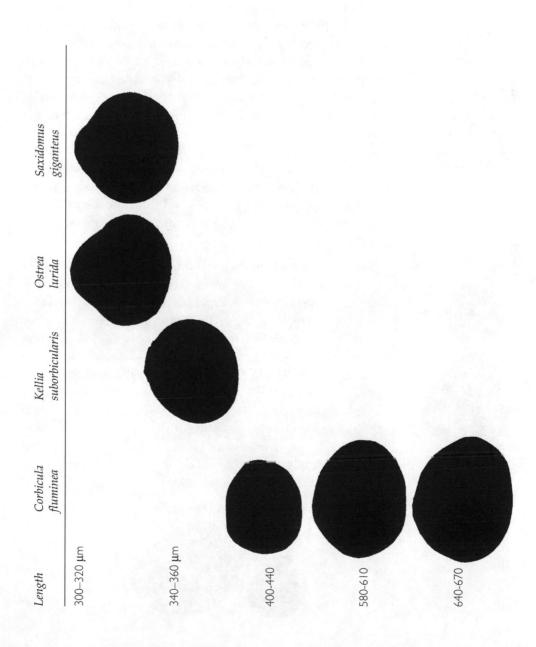

larval shell; at a length of ca 140 μm, a dark line appears just inside the edge of the shell and runs parallel to the rim of the shell. Before the appearance of the knobby umbo at ca 150 μm, the shell is nearly round and the shoulders are gradually sloping. The shell gets progressively yellower in color throughout the larval period. The shell becomes as tall if not taller than it is long; the shell depth becomes nearly as great as the length. Size at metamorphosis, ca 245 μm (Quayle, 1953).

Corbicula fluminea, **Asiatic Freshwater Clam** (Order Veneroida, Family Corbiculidae). Larvae are brooded within the adult clam until ca 210 μm length. Upon release the spat have a long, straight umbo, a steeply sloping posterior shoulder, and rounded posterior end; the anterior end is slightly longer and more pointed; the anterior shoulder is steeply sloping but slightly longer than the posterior end. Throughout development, the length of the umbo shortens and remains indistinct. The posterior end remains broadly rounded while the anterior end continues to lengthen and become more pointed; the anterior shoulder becomes more gradually sloping while the posterior end remains steeply sloping. Because larvae are brooded and released as spat, those caught in plankton tows are probably byssus thread drifters or individuals stirred up from the bottom by the currents and therefore are no longer considered true larvae (Kennedy et al., 1991; V. Kennedy, pers. comm.)

Crassostrea gigas, **Giant Pacific Oyster** (Order Osteoida, Family Ostreidae). These larvae are nearly indistinguishable from larvae of *Crassostrea virginica*. From early on, larvae are taller than they are long, and the umbo becomes extremely prominent and knobby in even small larvae (~140 μm); the umbo becomes progressively skewed throughout the planktonic life. The main body of the shell is initially round until ca 200 μm, at which time the anterior shoulder and end lengthen and become steeply sloping, being slightly pointed; the posterior shoulder is shorter and gradually sloping, eventually coming off the umbo at nearly a right angle. The shell is typically rather dark in color. Metamorphosis occurs at 275–330 μm (Loosanoff et al., 1966).

Crassostrea virginica, **Eastern Oyster.** See *Crassostrea gigas*.

Hiatella arctica, **Arctic Saxicave, Little Gaper, Red Nose** (Order Myoida, Family Hiatellidae). Larvae of this species have a distinct shape: the umbo is angular and slightly knobby; the posterior shoulder is long and steeply sloping; the bottom half of the posterior end is rather squared-off. The anterior shoulder is also long and steeply sloping, but less so than the posterior end. The anterior end comes to a distinct point halfway down the height of the shell. Metamorphosis is believed to occur at ca 345 μm (Rees, 1950).

Kellia suborbicularis, **North Atlantic Lepton** (Order Veneroida, Family Kelliidae). Larvae of this species are large and easily recognized by their very short, straight umbo. The posterior end is broadly rounded with a long, steeply sloping shoulder. The anterior end is also broadly rounded but is slightly longer and skewed toward the anteroventral margin of the shell. Size at metamorphosis is believed to be ca 370 μm (Rees, 1950; Strathmann, 1987).

Macoma balthica, **Balthica Macoma** (Order Veneroida, Family Tellinidae). These larvae retain a rather indistinct shape throughout the larval period. Young larvae (<160 μm) have a straight hinge and the body is broadly rounded with the anterior end only slightly longer than the posterior end. As the larvae develop, the anterior end lengthens and becomes more steeply sloped while the posterior end remains much shorter and more rounded. The result is a larval shell slightly skewed in the direction of the anterior end. The umbo remains broadly rounded and indistinct throughout development. Average size at metamorphosis is 255 μm (Sullivan, 1948).

Modiolus modiolus, **Northern Horse Mussel** (Order Mytiloida, Family Mytilidae). Straight-hinge larvae of *M. modiolus* are indistinguishable from *Mytilus trossulus* (*edulis*) straight-hinge larvae. At ca 170 μm, the umbo of *M. modiolus* becomes broadly rounded. At this size, the posterior end is short and also broadly rounded, while the anterior end is longer and slightly pointed. As size increases, the umbo becomes more prominent and knobby. The posterior shoulder is short and the posterior end is rather squared off; the anterior shoulder is longer and gradually sloping. The bottom half of the anterior end slopes sharply toward the ventral margin. This species is best distinguished from *M. trossulus* by the size of the umbo, which is longer and wider than in *M. trossulus*. Size at metamorphosis is ca 300 μm. (De Schweinitz and Lutz, 1976).

Mya arenaria, **Soft-shell Clam** (Order Myoida, Family Myidae). Young larvae (<150 μm) have a broadly rounded umbo with a

short, gradually sloping posterior shoulder and a longer, more steeply sloping anterior shoulder and end. As development proceeds, the umbo becomes angled and the shoulders become straighter and more steeply sloping. Both ends are pointed but the anterior end is slightly longer. Larvae retain this general shape throughout development. Metamorphosis occurs at 170–230 μm (Chanley and Andrews, 1971).

Mytilus trossulus (edulis), **Blue Mussel** (Order Mytiloida, Family Mytilidae). This larva retains its straight hinge much longer than other larvae, until almost 220 μm. At this point, the umbo becomes broadly rounded and the anterior end begins to lengthen and become more pointed. Eventually a small, knobby umbo develops. The posterior end is short but steeply sloping; the anterior end is longer but slopes more gradually down to a point. The bottom half of the anterior end slopes steeply toward the ventral margin. Compare *Modiolus modiolus*. Metamorphosis occurs at 215–305 μm (Chanley and Andrews, 1971).

Martel et al. (2000) provide a description of characteristics that can be used to differentiate between settling and early postlarval stages (e.g., prior to dissoconch secretion) of *Mytilus trossulus* and *M. californianus*. In addition, they suggest that the following characteristics can also be used to separate *M. galloprovincialis* (a species found in southern California) from *M. californianus*. *M. californianus* displayed 1) a shallower, flatter umbo (i.e., the PI curve was more pronounced, see Fig. 5), 2) the umbo was less conspicuous, barely extending above the larval hinge and only weakly curved (Fig. 5), and 3) wider separation between the provincular lateral teeth (Fig. 5).

Fig 5. Settling and early postlarval stages of (A) *Mytilus californianus* and (B) *M. trossulus* (Martel at al., 2000). The curved line above the upper figures indicates the length of the prodissoconch curve (the PI curve). Note the higher and more pronounced umbo displayed by *M. trossulus* and the closer spacing between the provincular lateral teeth (arrows) relative to *M. californianus*.

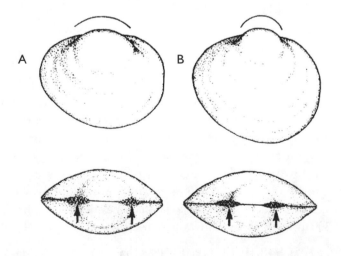

Ostrea lurida, **Native Pacific Oyster** (Order Ostreoida, Family Ostreidae). Larvae are brooded until 165–189 μm (Strathmann, 1987). Larvae less than ca 200 μm have a straight hinge, with the main body of the shell broadly rounded with ends of roughly equal length. At 200 μm, the umbo first becomes apparent and is initially just a rounded hump atop the valves. The anterior end begins to lengthen, and the anterior shoulder gets progressively longer and more steeply sloping while the posterior shoulder becomes shorter and less steeply sloping. The anterior end is longer than the posterior end. The umbo, like that of *Crassostrea,* becomes prominent and knobby in shape throughout development, although it does not become skewed. Metamorphosis occurs at ca 320 μm (Hori, 1933).

Petricola pholadiformis, **False Angle Wing** (Order Veneroida, Family Petricolidae). Larvae are free-swimming. Total length is 60–185 μm. Straight-hinge stage ends at ca 105 μm length, at which time a broadly rounded umbo develops. The anterior end is slightly longer than posterior. The ends of the shell are nearly equally rounded. The shoulders are straight and slope steeply. There is no distinctive color, though the margin is dark. The eye spot is not pigmented. The shell is heavier than in most clams. Metamorphosis occurs at ca 175 μm (Chanley and Andrews, 1971).

Saxidomus giganteus, **Butter Clam** (Order Veneroida, Family Veneridae). Rather indistinct larvae with a straight hinge until ca 160 μm. At this point, the anterior end is slightly longer and more pointed than the broadly rounded posterior end. As development proceeds, the umbo becomes broadly rounded as do both ends, with the anterior end being only slightly longer than the posterior end. The larvae are noticeably longer than they are tall and as a result have a rather squat appearance. Metamorphosis occurs at ca 230 μm (Breese and Phibbs, 1970).

Teredo navalis, **Common Shipworm** (Order Myoida, Family Teredinidae). Larvae have a distinct walnut-shaped (oval) shell, being as tall if not taller than long and with considerable depth. The larvae have a knobby umbo and virtually no shoulders. The ends are of equal length and rounded. The shell is usually a dark brown, golden color. Metamorphosis is at 190–200 μm (Chanley and Andrews, 1971).

Tresus capax, **Alaskan Gaper** (Order Veneroida, Family Mactridae). In small larvae with straight umbos (<140 μm), the anterior end is slightly longer and pointed, and the posterior end is short and more broadly rounded. The bottom half of the anterior end slopes sharply toward the ventral margin. As size increases, the umbo becomes progressively more angled but never becomes conspicuous. The posterior end becomes rather squared off, while the anterior end becomes more pointed. The anterior shoulder is longer than the posterior shoulder. Metamorphosis occurs at 270 μm (Bourne and Smith, 1972a,b).

References

Bayne, B. L. (1965). Growth and delay of metamorphosis of the larvae of *Mytilus edulis*. Ophelia. 2:1–47.

———— (1971). Some morphological changes that occur at the metamorphosis of the larvae of *Mytilus edulis*. In: Fourth European Marine Biology Symposium, pp. 259–80. Cambridge University Press.

Blacknell, W. M. and A. D. Ansell (1974). The direct development of the bivalve *Thyasira gouldi* (Philippi). Thalassia Jugosl. 10 (1/2):23–43.

———— (1975). Features of the reproductive cycle of an Arctic bivalve from a Scottish sea loch. Pubbl. Staz. Zool. Napoli. 39 (1):26–52.

Bourne, N. (1982). Distribution, reproduction, and growth of Manila Clam, *Tapes philippinarum* (Adams and Reeves), in British Columbia. J. Shell. Res. 2 (1):47–54.

Bourne, N. and D. W. Smith (1972a). The effect of temperature on the larval development of the Horse Clam, *Tresus capax* (Gould). Proc. Nat. Shell. Assoc. 62:35–37.

———— (1972b). Breeding and growth of the Horse Clam, *Tresus capax* (Gould), in southern British Columbia. Proc. Nat. Shell. Assoc. 62:38–46.

Breese, W. P. and F. D. Phibbs (1970). Some observations on the spawning and early development of the Butter Clam, *Saxidomus giganteus* (Deshayes). Proc. Nat. Shell. Assoc. 60:95–98.

Breese, W. P. and A. Robinson (1981). Razor Clams, *Siliqua patula* (Dixon): Gonadal development, induced spawning and larval rearing. Aquaculture. 22:27–33.

Bronson, J., T. Bettinger, L. Goodwin, and D. Burge (1984). Investigations of spat collection on artificial substrates for the Weathervane Scallop (*Patinopecten caurinus*) and the Rock Scallop (*Hinnites multirugosus*) in Puget Sound, Washington. State of Washington, Department of Fisheries.

Chanley, P. and J. D. Andrews (1971). Aids for identification of bivalve larvae of Virginia. Malacologia 11 (1):45–119.

Chew, K. K. and A. P. Ma (1987). Species Profiles: Life Histories and Environmental Requirements of Coastal Fishes and Invertebrates (Pacific Northwest)—Common Littleneck Clam. Army Corps of Engineers.

Culliney, J. L. (1975). Comparative larval development of the shipworms *Bankia gouldi* and *Teredo navalis*. Mar. Biol. 29:245–51.

Cummings, V. J., R. D. Pridmore, S. F. Thrush, and J. E. Hewitt (1993). Emergence and floating behaviors of post-settlement juveniles of *Macomona liliana* (Bivalvia: Tellinacea). Mar. Behav. Physiol. 24:25–32.

De Schweinitz, E. H. and R. A. Lutz (1976). Larval development of the Northern Horse Mussel, *Modiolus modiolus* (L.), including a comparison with the larvae of *Mytilus edulis* L. as an aid in planktonic identification. Biol. Bull. 150:348–60.

Epifanio, C. E., C. M. Logan, and C. Turk (1975). Culture of six species of bivalves in a recirculating seawater system. In: Proceedings of the 10th European Symposium on Marine Biology, pp. 97–108. Universa Press, Ostend, Belgium.

Fraser, C. M. (1929). The spawning and free swimming larval periods of *Saxidomus* and *Paphia*. R. Soc. Can. Proc. Trans. 4 (23):195–98.

Gallager, S. M., J. P. Bidwell, and A. M. Kuzirian (1989). Strontium is required in artificial seawater for embryonic shell formation in two species of bivalve molluscs. In: Origin, Evolution and Modern Aspects of Biomineralization in Plants and Animals. Plenum Press, New York.

Gallucci, V. F. and B. B. Gallucci (1982). Reproduction and ecology of the Hermaphroditic Cockle *Clinocardium nuttallii* (Bivalvia: Cardiidae) in Garrison Bay. Mar. Ecol. Prog. Ser. 7:137–45.

Goodwin, C. L. (1973). Subtidal Geoducks of Puget Sound, Washington. Washington Department of Fisheries.

Gray, S. (1982). Morphology and taxonomy of two species of the genus *Transennella* (Bivalvia: Veneridae) from western North America and a description of *T. confusa* Sp. Malacol. Rev. 15:107–17.

Gustafson, R. G. and R. G. B. Reid (1986). Development of the pericalymma larva of *Solemya reidi* (Bivalvia: Cryptodonta: Solemyidae) as revealed by light and electron microscopy. Mar. Biol. 93:411–27.

Gustafson, R. G., D. O'Foighil, and R. G. B. Reid (1986). Early ontogeny of the septibranch bivalve *Cardiomya pectinata* (Carpenter, 1865). J. Mar. Biol. Assoc. (UK) 66:943–50.

Haderlie, E. C. (1983). Depth distribution and settlement times of the molluscan wood borers *Bankia setacea* (Tryon, 1863) and *Xylophaga washingtona* Bartsch, 1921, in Monterey Bay. Veliger. 25 (4):339–42.

Hodgson, C. A. and N. Bourne (1988). Effect of temperature on larval development of the Spiny Scallop, *Chlamys hastata* Sowerby, with a note on metamorphosis. J. Shell. Res. 7 (3):349–57.

Hopkins, A. E. (1936). Ecological observations on spawning and early larval development in the Olympia Oyster (*Ostrea lurida*). Ecology. 17:551–56

Hori, J. (1933). On the development of the Olympia Oyster, *Ostrea lurida* Carpenter, transplanted from United States to Japan. Bull. Soc. Jpn. Sci. Fish. 1 (6):269–76.

Hutchings, J. A. and R. L. Haedrich (1984). Growth and population structure in two species of bivalves (Nuculanidae) from the deep sea. Mar. Ecol. Prog. Ser. 17:135–42.

Kennedy, V. and L.Van Heukelem (1985). Gametogenesis and larval production in a population of the introduced Asiatic Clam, *Corbicula* sp. (Bivalvia: Corbiculidae), in Maryland. Biol. Bull. 168:50–60.

Kennedy, V., S. C. Fuller, and R. A. Lutz (1991). Shell and hinge development of young *Corbicula fluminea* (Muller) (Bivalvia: Corbiculoidea). Amer. Malacal. Bull. 8 (2):107–11.

Kozloff, E. N. (1996). Marine Invertebrates of the Pacific Northwest. University of Washington Press, Seattle.

Lane, D. J. W., A. R. Beaumont, and J. R. Hunter (1985). Byssus drifting and the drifting threads of the young post-larval mussel *Mytilus edulis*. Mar. Biol. 84:301–8.

Lassuy, D. R. and D. Simons. (1989). Species Profiles: Life Histories and Environmental Requirements of Coastal Fishes and Invertebrates (Pacific Northwest)—Pacific Razor Clam. Army Corps of Engineers.

Lebour, M. V. (1938a). The life history of *Kellia Suborbicularis*. J. Mar. Biol. Assoc. (UK) 22:447–51.

—— (1938b). Notes on the breeding of some lamellibranchs from Plymouth and their larvae. J. Mar. Biol. Assoc. (UK) 23:119–44.

Leonard, V. K., Jr. (1969). Seasonal gonadal changes in two bivalve molluscs in Tomales Bay, California. Veliger 11:382–90.

Le Pennec, M. (1980). The larval and post-larval hinge of some families of bivalve molluscs. J. Mar. Biol. Assoc. (UK) 60: 601–17.

Loosanoff, V. L., H. C. Davis, and P. E. Chanley (1966). Dimensions and shapes of larvae of some marine bivalve mollusks. Malacologia 4 (2):351–435.

Lough, R. G. and J. J. Gonor (1971). Early embryonic stages of *Adula californiensis* (Pelecypoda: Mytilidae) and the effect of temperature and salinity on developmental rate. Mar. Biol. 8:118–25.

Lutz, R., J. Goodsell, M. Castagna, S. Chapman, C. Newell, H. Hidu, R. Mann, D. Jablonski, V. Kennedy, S. Siddall, R. Goldberg, H. Beattie, C. Falmagne, and A. Chestnut (1982). Preliminary observations on the usefulness of hinge structures for identification of bivalve larvae. J. Shell. Res. 2 (1):65–70.

Marriage, L. D. (1954). The Bay Clams of Oregon. Fish Commission of Oregon. Contribution No. 20:1–47.

Martel, A. and F.-S. Chia (1991). Drifting and dispersal of small bivalves and gastropods with direct development. J. Exp. Mar. Biol. Ecol. 150:131–47.

Martel, A. L., L. M. Auffrey, C. D. Robles, and B. M. Honda. 2000. Identification of settling and early postlarval stages of mussels (*Mytilus* spp.) from the Pacific coast of North America, using prodissoconch morphology and genomic DNA. Mar. Biol. 137:811-18.

Matveeva, T. A. (1976). The biology of the bivalve mollusk *Turtonia minuta* in different parts of its geographic range. Soviet J. Mar. Biol. 2 (6):370–76.

Morris, R. H., D. P. Abbot, and E. C. Haderlie (1980). Intertidal Invertebrates of California. Stanford University Press, Stanford.

Nickerson, R. B. (1977). A Study of the Littleneck Clam (*Protothaca staminea* Conrad) and the Butter Clam (*Saxidomus giganteus* Deshayes) in a Habitat Permitting Coexistence, Prince William Sound, Alaska. Proc. Nat. Shell. Assoc. 67:85–102.

O'Foighil, D. O. (1985). Sperm transfer and storage in the brooding bivalve *Mysella tumida*.Biol. Bull. 169:602–14.

Pauley, G. B., B. Van Der Raay, and D. Troutt (1988). Species Profiles: Life Histories and Environmental Requirements of Coastal Fishes and Invertebrates (Pacific Northwest)—Pacific Oyster. Army Corps of Engineers.

Quayle, D. B. (1953). The Larva of *Bankia setacea* Tryon. British Columbia Fisheries Department.

——— (1959). The early development of *Bankia setacea* Tryon. In: Marine Boring and Fouling Organisms, pp. 157–74. University of Washington Press, Seattle.

——— (1992). Marine wood borers in British Columbia. Can. Spec. Pub. Fish. Aquat. Sci. 115:1–34.

Rae, J. G. I. (1978). Reproduction in two sympatric species of *Macoma* (Bivalvia). Biol. Bull. 155:207–19.

——— (1979). The population dynamics of two sympatric species of *Macoma* (Mollusca: Bivalvia). Veliger 21 (3):384–99.

Rees, C. B. (1950). The identification and classification of Lamellibranch larvae. Hull Bull. Mar. Ecol. 3 (19):73–104.

Savage, N. B. and R. Goldberg (1976). Investigation of practical means of distinguishing *Mya arenaria* and *Hiatella* sp. larvae in plankton samples. Proc. Nat. Shell. Assoc. 66:42–53.

Shaw, W. N., T. J. Hassler, and D. P. Moran (1988). Species Profiles: Life Histories and Environmental Requirements of Coastal Fishes and Invertebrates (Pacific Southwest)—California Sea Mussel and Bay Mussel. Army Corps of Engineers.

Stafford, J. (1912). On the Recognition of Bivalve Larvae in Plankton Collections. In: Contributions to Canadian Biology, pp. 221–42. C. H. Parmelee, Ottawa.

Strathmann, M. F. (1987). Phylum Mollusca, Class Bivalvia. In: Reproduction and Development of Marine Invertebrates of the Northern Pacific Coast, pp. 309–353. University of Washington Press, Seattle.

Sullivan, C. M. (1948). Bivalve larvae of Malpeque Bay, P.E.I. Fish. Res. Bd. Can. LXXVII:1–36.

Thomas, K. A. (1994). The functional morphology and biology of *Pandora filosa* (Carpenter, 1864) (Bivalvia: Anomalodesmata: Pandoracea). Veliger 37 (1):23–29.

Townsley, P. M., R. A. Richy, and P. C. Trussell (1966). The laboratory rearing of the shipworm, *Bankia setacea* (Tryon). Proc. Nat. Shell. Assoc. 56:49–52.

Wilson, E. C. and G. L. Kennedy (1984). The Boring Clam, *Penitella conradi* (Bivalvia: Pholadidae) in nephrite from Monterey County, California. The Nautilus. 98 (4):159–62.

Yonge, C. M. (1952). Studies on Pacific Coast mollusks. V. Structure and adaptation in *Entodesma saxicola* (Baird) and *Mytilimeria nuttallii* (Conrad): with a discussion on evolution within the family Lyonsiidae (Eulamellibranchia). Univ. Calif. Publ. Zool. 55:439–50.

Zardus, J. D. and M. P. Morse. (1998). Embryogenesis, morphology and ultrastructure of the pericalymma larva of *Acila castrensis* (Bivalvia:Protobrachia:Nuculoida). Invertebr Biol. 117:221–44.

11

Mollusca: The Smaller Groups Polyplacophora, Scaphopoda, and Cephalopoda

Alan L. Shanks

Class Polyplacophora

The chitons are one of the more primitive molluscan groups. They are characterized by a shell composed of eight plates. Worldwide, there are around 800 living species. Locally, chitons are represented by one order (Neoloricata), three suborders (Lepidopleurina, Chitonia, and Acanthochitonina), and around 42 species (Table 1). Chitons are a common component of rocky intertidal and subtidal communities.

Chitons are dioecious. Nearly all of the local species are broadcast spawners. Brooding species tend to be found in the high latitudes in Arctic and Antarctic waters (Pearse, 1979). Locally, most species spawn in the late winter and spring (Strathmann, 1987). Some chiton species spawn repeatedly, with spawning events associated with the spring/neap tidal cycle. Experiments suggest that the local species, *Katharina tunicata,* is induced to spawn by substances in the waters associated with the spring phytoplankton bloom (reviewed in

Fig. 1. Dorsal side of larvae and juveniles of several chiton species, left to right: *Lepidopleurus asellus, Mopalia lignosa, Lepidochitona cinereus, Chaetopleura apiculata, Cryptochiton stelleri.* Upper row: Newly hatched larvae. Middle row: Larvae competent to settle. Bottom row: Juveniles shortly after settlement. Scales = 100 μm. Note that *Mopalia lignosa* and *Cryptochiton stelleri* are local species and *Lepidochitona* is a local genus. (From Pearse, 1979, Fig. 16)

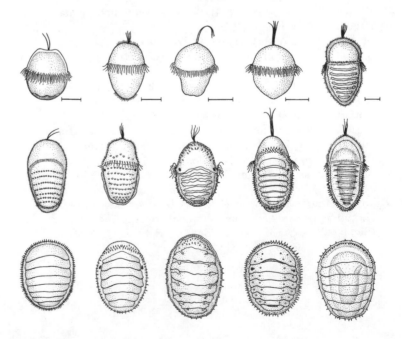

Table 1. Species in the class Polyplacophora from the Pacific Northwest (from Kozloff, 1996)

Suborder Lepidopleurina
Family Leptochitonidae
Leptochiton nexus
Leptochiton rugatus

Family Hanleyidae
Henleya oldroydi

Suborder Chitonina
Family Ischnochitonidae
Ischnochiton abyssicola
Ischnochiton interstinctus
Ischnochiton trifidus
Stenoplax fallaz
Stenoplax heathiana
Lepidozona cooperid
Lepidozona mertensii
Lepidozona retiporosa
Lepidozona scabricostata
Lepidozona willetti

Family Callistochitonidae
Callistochiton crassicostatus

Family Chaetopleuridae
Chaetopleura gemma

Family Lepidochitonidae
Dendrochiton flectens
Dendrochiton semiliratus
Lepidochitona dentiens
*Lepidochitona fernaldi**
Lepidochitona hartwegii
Nuttallina californica
Tonicella insignis
Tonicella lineata
Schizoplax brandtii

Family Mopaliidae
Mopalia ciliata
Mopalia cirrata
Mopalia cithara

*Brooding species (Strathmann, 1987).

Mopalia egretta
Mopalia hindsii
Mopalia imporcata
Mopalia laevior
Mopalia lignosa
Mopalia muscosa
Mopalia phorminx
Mopalia porifera
Mopalia sinuata
Mopalia spectabilis
Mopalia swanii
Katharina tunicata
Placiphorella rufa
Placiphorella verlata

Suborder Acanthochitonina
Family Acanthochitonidae
Cryptochiton steller

Pearse, 1979). Several researches have reported spawning to occur during periods of calm water.

Chiton larvae are non-feeding and have relatively short pelagic durations; most are in the plankton for one to two weeks (0.5 to 19 days, Pearse, 1979; Strathmann, 1987). If an appropriate settlement surface is not available, however, larvae can delay metamorphosis for up to a month (Pearse, 1979). The larvae of most free-spawning chiton species hatch as trochophores (Fig. 1). Chiton trochophores appear to be typical trochophores with no unique characteristics that would allow one to differentiate them easily from trochophores of other mollusks. Later in larval development, however, chiton larvae do develop unique characteristics that allow easy recognition (Fig. 2). Prior to settlement they first develop indentations on their dorsal side that separate the future shell valves. Seven calcified shell valves subsequently develop (the eighth valve

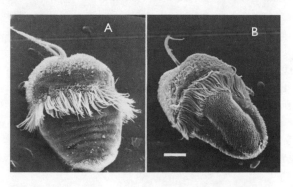

Fig. 2. Scanning electron micrographs of *Katharine tunicata*, a local species. (A) Dorsal side view of a three-day-old larva, showing the indentations separating the future shell valves. (B) Ventral side view of an eight-day-old larva, showing the developing foot. Scale = 50 μm, for both SEMs. (From Pearse, 1979, Fig. 17)

forms after settlement and metamorphosis). On the side of the larvae opposite the developing shell valves (the ventral surface), a well-developed foot forms. The distinctive shell valves of these larvae should allow one to identify a larva as that of a chiton. Few local chiton larvae have been described, however, so it is not possible at this time to develop a key to their identification.

Class Scaphopoda

Table 2. Species in the class Scaphopoda from the Pacific Northwest (from Kozloff, 1996)

Order Dentaliida
Family Dentaliidae
Dentalium agassizi
Dentalium pretiosum
Dentalium rectius

Order Gadilida
Family Gadilidae
Cadulus aberrans
Cadulus californicus
Cadulus tolmiei

Family Pulsellidae
Pulsellum salishorum

Scaphopoda (tusk snails) are benthic, subtidal burrowing mollusks with tusk-shaped shells. Locally, we have seven species grouped into three families and two orders (Table 2). This group has received little attention, and most of that occurred in the nineteen century. Anatomical and reproductive studies have concentrated on species in the genus *Dentalium*, in which there are three local species. The following description is from the review by McFadien-Carter (1979).

Tusk snails are broadcast spawners. A trochophore develops between 3 and 36 hours after the beginning of cleavage. The early trochophore in *Dentalium* is characterized by four bands of cilia encircling the middle of the larvae, an apical tuff, and the rudiments of the mantle (Fig. 3A). A disk-shaped ciliated velum at the center of which is an apical tuff characterizes the veliger stage. By this stage in development a shell has developed that envelops the larvae like a blanket (Fig. 3B). The shell has anterior and posterior apertures. Later in the veliger stage, a trilobed foot develops and the posterior shell aperture shrinks (Fig. 3C). At the prodissoconch stage, the posterior shell aperture closes so that the shell takes on the shape of a cylinder or goblet (Fig. 3D). The apical tuff is reduced, the foot increases in size, and the tentacular apparatus forms. Prior to settlement most of the velar lobe is lost, but enough remains that, on settlement, the newly settled larva crawls across the bottom by means of its foot and velum. The free-swimming stage of development lasts from two to six days. The larvae of no local tusk snails appear to have been described.

Class Cephalopoda, Order Octopoda

Table 3. Species in the class Cephalopoda from the Pacific Northwest (from Kozloff, 1996)

Order Octopoda
Family Octopodidae
Octopus dofleini
Octopus leioderma
Octopus rubescens

Three species of octopus are found in the Pacific Northwest (Table 3). Sexes are separate and dimorphic, with males possessing arms modified for the transfer of spermatophores during mating. Female octopuses deposit eggs in stalked chorions and then care for the brood during the several months needed for the maturation of the larvae. Octopus species that produce relatively small eggs generally have hatchlings that

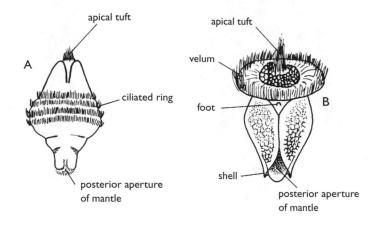

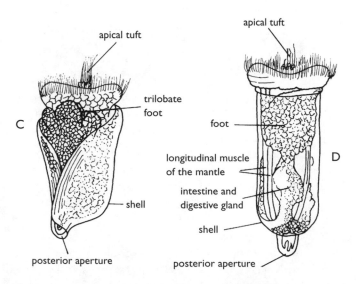

Fig. 3. Larval stages of *Dentalium*. (A) Early trochophore. (B) Veliger showing mantle development. (C) Later veliger; note the shrinking of the posterior shell aperture and the development of the trilobed foot. (D) Prodissoconch stage; note the cylindrical or goblet-shaped shell. (From McFadien-Carter, 1979)

spend time in the plankton prior to settling. Species producing large eggs tend to have hatchlings that settle to the bottom immediately (Wells and Wells, 1977). *Octopus dofleini* and *O. rubescens* produce pelagic juveniles; *O. leioderma* apparently produces benthic juveniles (Strathmann, 1987). The young of the former two species have been described, but those of *O. leioderma* have not. Figure 4 provides a general idea of the appearance of the pelagic juvenile stage of an octopus.

Key to the pelagic juvenile stage of local octopus (adapted from Green, 1973)

1a. Juvenile possess 10 arms .... Orders Sepioidea or Teuthoidea
1b. Juvenile possess 8 arms 2

2a. Juveniles with jellylike consistency; length of longest arm less than one-third dorsal mantle length; some lateral teeth in radula with many cusps (Fig. 4C) ....*Bolitaena diaphana*

Fig. 4. Pelagic juveniles of (A) *Octopus dofleini*, (B) *Octopus rubescens*, and (C) *Bolitanea diaphana*. Ventral views to the right of the scale bar; dorsal views to the left. (From Green, 1973, Figs. 2, 8, 12).

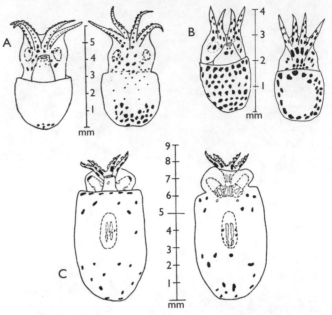

2b. Juveniles without jellylike consistency; length of longest arm more than one-third dorsal mantle length; lateral teeth in radula not with many cusps 3

3a. 2 rows of large chromatophores on each arm; generally only 4 chromatophores on funnel, ventral mantle completely covered with even pattern of chromatophores; juvenile may be 2 mm long and have 4 suckers arranged in single row. When preserved, arms likely straight (Fig. 4B) .... *Octopus rubescens*

3b. 1 row of large chromatophores on each arm; generally 7 or more chromatophores on funnel; anterior part of ventral mantle bare of chromatophores; larvae >2 mm and always with >4 suckers in double row. When preserved, arms likely curled (Fig. 4A) .... *Octopus dofleini*

References

Green, M. G. (1973). Taxonomy and distribution of planktonic octopods in the northeastern Pacific. M.S. Thesis, University of Washington. 98 pp.

Kozloff, E. N. (1996). Marine Invertebrates of the Pacific Northwest. University of Washington Press, Seattle.

McFadien-Carter, M. (1979). Scaphopoda. In: Reproduction of Marine Invertebrates Vol. V., A. C. Giese and J. S. Pearse (eds.), pp. 95–112.

Pearse, J. S. (1979). Polyplacophora. . In: Reproduction of Marine Invertebrates Vol. V., A. C. Giese and J. S . Pearse (eds.), pp. 27–86.

Strathmann, M. F. (1987). Reproduction and development of marine invertebrates of the northern Pacific coast. University of Washington Press, Seattle.

Wells, M. J. and J. Wells (1977). Cephalopoda: Octopoda. In: Reproduction of Marine Invertebrates Vol. IV., A. C. Giese and J. S. Pearse (eds.), pp. 291–336.

12

Arthropoda, Cirripedia: The Barnacles

Andrew J. Arnsberg

The Cirripedia are the familiar stalked and acorn barnacles found on hard surfaces in the marine environment. Adults of these specialized crustaceans are sessile. They are usually found in dense aggregations among conspecifics and other fouling organisms. For the most part, sexually mature Cirripedia are hermaphroditic. Cross-fertilization is the dominant method of reproduction. Embryos are held in ovisacs within the mantle cavity (Strathmann, 1987). Breeding season varies with species as well as with local conditions (e.g., water temperature or food availability). The completion of embryonic development culminates in the hatching of hundreds to tens of thousands of nauplii. There are approximately 29 species of intertidal and shallow subtidal barnacles found in the Pacific Northwest, of which 12 have descriptions of the larval stages (Table 1). Most of the species without larval descriptions (11 species) are parasitic barnacles, order Rhizocephala; a brief general review of this group is provided at the end of the chapter.

Development and Morphology

The pelagic phase of the barnacle life cycle consists of two larval forms. The first form, the nauplius, undergoes a series of molts producing four to six planktotrophic or lecithotrophic naupliar stages (Strathmann, 1987). Each naupliar stage is successively larger in size and its appendages more setose than the previous. The final nauplius stage molts into the non-feeding cyprid

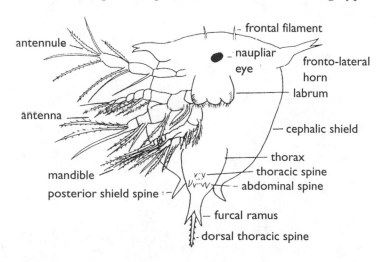

antennule — frontal filament

naupliar eye

fronto-lateral horn

labrum

antenna

cephalic shield

thorax

mandible — thoracic spine

posterior shield spine — abdominal spine

furcal ramus

dorsal thoracic spine

Fig. 1. Ventral view of a stage V nauplius larva. (From Miller and Roughgarden, 1994, Fig. 1)

Table 1. Species in the crustacean subclass Cirripedia from the Pacific Northwest

Taxa	Larval Description	Reference
Suborder Lepadomorpha		
Mitella polymerus	No	
Scalpellum columbianum	No	
Lepas anatifera	Yes	Moyse 1987
Lepas fascicularis	Yes	Bainbridge and Roskell, 1966
Lepas hilli	No	
Lepas pacifica	No	
Pollicipes polymerus	Yes	Lewis, 1975; Standing, 1980
Suborder Balanomorpha		
Chthamalus dalli	Yes	Korn and Ovsyannikova, 1979; Standing, 1980; Miller et al., 1989
Semibalanus balanoides	Yes	Pyefinch, 1949; Lang, 1980
Semibalanus cariosus	Yes	Standing, 1980; Branscomb and Vedder, 1982; Korn, 1989
Solidobalanus engbergi	No	
Solidobalanus hesperius	Yes	Barnes and Barnes 1959a
Balanus balanus	Yes	Barnes and Costlow, 1961; Lang, 1980
Balanus crenatus	Yes	Herz, 1933; Standing, 1980; Branscomb and Vedder, 1982; Ovsyannidova and Korn, 1984
Balanus glandula	Yes	Standing, 1980; Brown and Roughgarden, 1985
Balanus improvisus	Yes	Jones and Crisp, 1954; Standing, 1980
Balanus nubilus	Yes	Barnes and Barnes, 1959b; Standing, 1980; Miller and Roughgarden, 1994
Balanus rostratus	No	
Order Rhizocephala		
Angulosasscus tenuis	No	
Clistosaccus paguri	No	
Brarosaccus callosus	No	
Peltogaster boschmae	No	
Peltogaster paguri	No	
Peltogaster gracilis	No	
Loxothylacus panopaei	No	
Sylon hippolytes	No	
Thompsonia sp.	No	
Trypetesa sp.	No	
Dendrogaster sp.	No	

stage. The cyprid is responsible for finding a suitable habitat where it can settle and undergo a final metamorphosis into a juvenile sessile barnacle.

Both the nauplius and cyprid larval forms can be easily distinguished from other common zooplankters; there are, however, larval and adult forms of zooplankton that may create potential confusion for an inexperienced observer. Copepoda, another subclass of the subphylum Crustacea, also has a nauplius larval form. Figure 1 presents a generalized cirripede nauplius body form with distinguishing features noted. Figure 2 provides a comparison of the similar-looking copepod nauplius. The dorsal surface of the cirripede nauplius is covered with a cephalic shield (carapace) that is generally triangular or shield-shaped. Cirripede nauplii can be distinguished from copepod nauplii by the pair of conspicuous frontolateral horns on either side of the anterior end of the cephalic shield. A prominent naupliar eye present at the center of the anterior end may or may not be seen in other crustacean nauplii. Other cirripede naupliar characteristics include three pairs of biramous appendages and two prominent posterior structures, the dorsal thoracic spine and the ventral furcal ramus.

The cyprid larval form is unique to cirripedes. These are oblong, bivalved larvae with six pairs of thoracic appendages. A pair of compound eyes are present; both can be seen from the dorsal view, but only one is visible in the side view. This is a non-feeding larval form. Lipid reserves in the form of oil droplets are visible at the anterior end of the cyprids of some species. Ostracods are the organisms most easily confused with barnacle cyprids (Fig. 3). The easiest way to distinguish cyprids from ostracods is to note the compound eyes in the former.

Identification and Description of Local Taxa

A majority of this chapter deals with species of the order Thoracica. In addition there are two suborders in our estuaries and coastal waters, Lepadomorpha and Balanomorpha, with characteristics for the most part unique to each. Lepadomorphs have a unilobed labrum, whereas balanomorphs possess a trilobed labrum (see Fig. 1). The exception to this rule is found

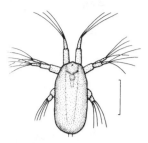

Fig. 2. Copepod nauplius. Scale = 100 µm. (From *A Guide to Marine Coastal Plankton and Marine Invertebrate Larvae*, Second Edition, by DeBoyd L. Smith and Kevin B. Johnson. Copyright 1996 by DeBoyd L. Smith and Kevin B. Johnson. Reprinted by permission of Kendall/Hunt Publishing Company.)

Fig. 3. (A) Generalized barnacle cyprid. (B) Ostracod. (A from Brown and Roughgarden, 1985; B from *A Guide to Marine Coastal Plankton and Marine Invertebrate Larvae, Second Edition*, by DeBoyd L. Smith and Kevin B. Johnson. Copyright 1996 by DeBoyd L. Smith and Kevin B. Johnson. Reprinted by permission of Kendall/Hunt Publishing Company)

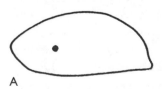

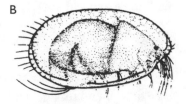

A

B

in *Chthamalus dalli*, a balanomorph with a unilobed labrum. Lepadomorph nauplii of the genus *Lepas* have distinct projections on their cephalic shield margin and a dorsal thoracic spine that is about twice as long as the carapace length (see Fig. 18). The cephalic shield of balanomorph nauplii, and those of the lepadomorph *Pollicipes polymerus*, is nearly as wide as long. Generally, lepadomorphs cyprids are the largest in Pacific Northwest waters. *Pollicipes polymerus*, however, is an exception; its cyprids are small.

The identification of barnacle larvae is fairly difficult. A nauplius must be identified to stage before it can be differentiated to species. Species can be separated by examining the appendage setation, which requires a compound micro-scope (see Miller and Roughgarden, 1994, for species-specific setation patterns). The key to nauplii presented here is based on characteristics that can usually be distinguished through a good dissecting microscope, but it is usually much easier to stage and identify nauplii under a compound microscope.

Cyprid larvae are also difficult to identify to species. The separation of species is dependent on morphological differences such as pigmentation and variations in carapace margin outline. Overall cyprid size is useful in limiting possible species identification. Standing (1980) provides a compre-hensive comparison of the cyprids of seven of the more common barnacles. Identification can be done under a dissecting microscope.

Nauplius Larvae

Balanomorph nauplii can be identified to stage based on the basis of the setation of their antennule and the presence of abdominal spines (Fig. 4) (Lang, 1979; Miller and Roughgarden, 1994). Lepadomorph nauplii may be staged similarly, as some characteristics used to stage balanomorph nauplii are conservative in lepadomorph nauplii. For example, in all naupliar stages of *Pollicpes polymerus*, a lepadomorph, antennule setation is identical to that of the balanomorphs (Lewis, 1975). Additionally, series two abdominal spine configurations for stage IV and VI *P. polymerus* nauplii are like that of the balanomorphs.

All stage I nauplii have folded frontolateral horns and are small in size. Because of the short duration of this stage, they are rarely present in plankton tows. The frontolateral horns of a stage II larva point away from the body approximately perpendicular to the long axis, and the tip of each horn is unsplit. There is one pair of abdominal spines and no preaxial setae on the antennule. Stage III nauplii are characterized by

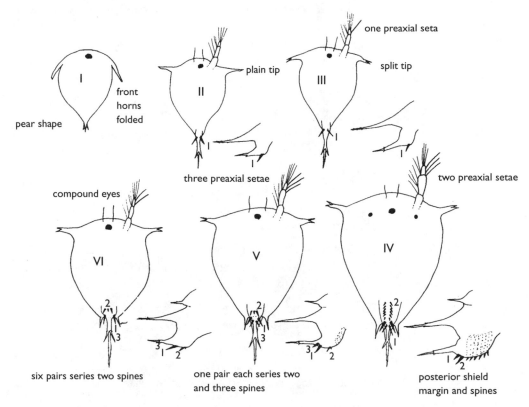

one preaxial seta

plain tip

split tip

front horns folded

pear shape

three preaxial setae

two preaxial setae

compound eyes

six pairs series two spines

one pair each series two and three spines

posterior shield margin and spines

Fig. 4. Diagnostic morphological features, including lateral views of furcal ramus, for each of the six stages of balanomorph nauplii. Abdominal spines are grouped in three series, as noted by arabic numerals (Modified from Lang, 1979)

split frontolateral horn tips and one preaxial seta on the antennule. An additional preaxial seta and additional pair of abdominal spines are gained on the antennule and abdominal process, respectively, in stage IV nauplii. Stage V nauplii are distinguished by three preaxial setae on the antennule and three pairs of abdominal spines. Compound eyes are rarely present in stage V nauplii. Compound eyes are, however, a diagnostic characteristic of stage VI nauplii. Additionally, there are six pairs of abdominal spines on stage VI nauplii.

Key to cirriped naupliar stages (modified from Lang, 1980)

1a. Frontolateral horns extended; furcal ramus well developed 2
1b. Frontolateral horns folded against body; furcal ramus
    rudimentary spine .... Stage I

2a. Posterior shield spine absent 3
2b. Posterior shield spine present 4

3a. Frontolateral horns with plane tips; antennule without preaxial
    setae .... Stage II
3b. Frontolateral horns with split tips; antennule with 1 preaxial
    seta .... Stage III

4a. Furcal ramus with 1–2 pairs of large spines and up to 3 small spines   5

4b. Furcal ramus with 2 pairs of large spines, 6 pairs of small spines, antennule with 3 preaxial setae; compound eyes present .... Stage VI

5a. Furcal ramus with 1 pair of large spines, antennule with 2 preaxial setae .... Stage IV

5b. Furcal ramus with two pairs of large spines, antennule with three preaxial setae, compound eyes not present .... Stage V

Key to cirriped nauplii

1a. Labrum unilobed 2

1b. Labrum trilobed 4

2a. Cephalic shield outline smooth 3

2b. Cephalic shield outline with projections, frontal horns long in stages I and II (page 171) .... *Lepas* spp.

3a. Dorsal outline round (page 170) .... *Chthamalus dalli*

3b. Dorsal outline goblet-shaped (page 172) ...... *Pollicipes polymerus*

4a. Anterior margin flat between frontolateral horns (stages III–VI) 5

4b. Anterior margin rounded between frontolateral horns 7

5a. Bulky outline, wide triangular or goblet-shaped 6

5a. Slender outline, long and narrow, triangular; relatively long frontolateral horns pointed anteriorly (page 169) .... *Solidobalanus hesperius*

6a. Frontolateral horns curved laterally; cephalic shield triangular (page 164) .... *Balanus glandula*

6b. Frontolateral horns straight; cephalic shield square or goblet-shaped (page 166) .... *Balanus nubilus*

7a. Larger nauplii (all stages ≥450 μm), width less than cephalic shield length 8

7b. Smaller nauplii (all stages ≥500 μm), width nearly equal to length; anterior margin between frontolateral horns slightly curved (page 163) .... *Balanus improvisus*

8a. Anterior margin between frontolateral horns prominently curved 9

8b. Anterior margin between frontolateral horns slightly curved, smaller, narrower, more triangular shape, frontolateral horns anteriorly directed in all stages (page 165) .... *Balanus crenatus*

9a. Nauplius with rudimentary furcal ramus and small abdominal spines, frontolateral horns perpendicular to shield axis in stages I–III and swept back in stages IV–VI 10

9b. Nauplius with distinct well-formed furcal ramus and abdominal spines, frontolateral horns swept back in all stages (page 167) .... *Balanus balanus*

10a. Large, bulky, slightly rounded outline, stages II and III with long dorsal thoracic spine, frontolateral horns perpendicular to long axis in stages II and III and anteriorly directed in stages IV–VI; maximum size of stage II–VI, respectively, ca 350, 350, 450, 550, and 650 μm (page 165) .... *Semibalanus cariosus*

10b. Same general characters as above; maximum size of stage II–VI, respectively, 400, 500, 600, >700, and >800 μm (page 169) .. .. *Semibalanus balanoides*

Cyprid Larvae

Cyprid larvae can be identified by carefully comparing shape, size, carapace texture, and carapace pigmentation. The location of the collection site may be of some help in excluding possible species; for example, *Balanus improvisus* cyprids are not likely to be collected outside an estuary. Any cyprid belonging to the genus *Lepas* should rarely be found within an estuary, since they are strictly oceanic species; adults are only found attached to floating objects.

In the following key, size ranges are from animals collected by Standing (1980) at Bodega Harbor, California, except where noted. Although size ranges may be different for specimens collected farther north, the relative size differences should be conservative. Length refers to the long axis (anterior to posterior), depth refers to dorsal–ventral distance, and breadth refers to length from left to right side at the widest point of the cyprid. Note that the shape of the carapace margin is visible only in a side view.

Key to cirriped cyprids (adapted from Standing, 1908)

1a. Small, length <625 μm 2
1b. Medium to large, length >625 μm 4

2a. Anterior end broadly rounded in side view; posterodorsal margin with break in curve 3
2b. Anterior end angular in side view (arrow in figure 2b); posterodorsal margin evenly curved, relatively translucent when preserved, no special pigmentation when fresh, carapace surface shiny under low magnification, smooth under high magnification, length 440–580 μm (page 170) ....*Chthamalus dalli*

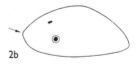

3a. Carapace depth greatest about one-third of the way from anterior end; carapace breadth narrow in dorsal view, surface of carapace sculptured under high magnification, posterior margin triangular; mainly found on outer coast; length from 420–540 μm (page 172) .... *Pollicipes polymerus*

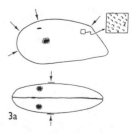

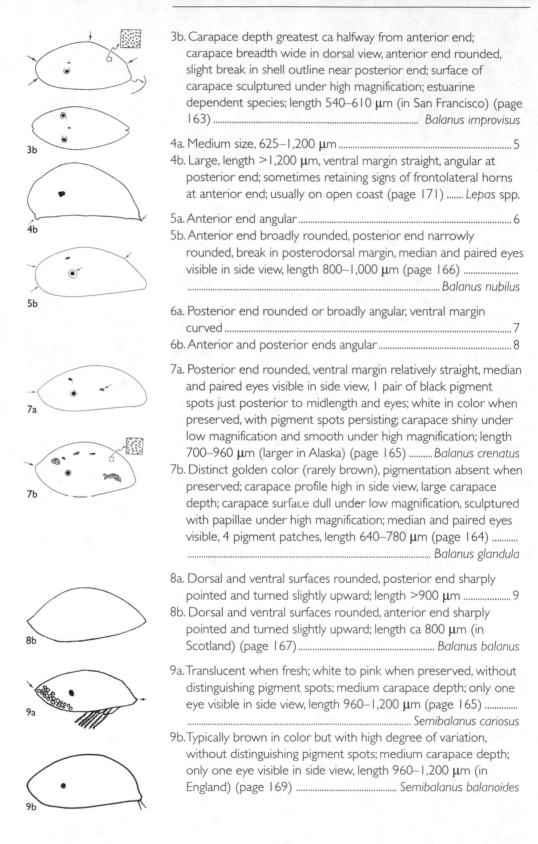

3b. Carapace depth greatest ca halfway from anterior end; carapace breadth wide in dorsal view, anterior end rounded, slight break in shell outline near posterior end; surface of carapace sculptured under high magnification; estuarine dependent species; length 540–610 μm (in San Francisco) (page 163) .... *Balanus improvisus*

4a. Medium size, 625–1,200 μm 5

4b. Large, length >1,200 μm, ventral margin straight, angular at posterior end; sometimes retaining signs of frontolateral horns at anterior end; usually on open coast (page 171) .... *Lepas* spp.

5a. Anterior end angular 6

5b. Anterior end broadly rounded, posterior end narrowly rounded, break in posterodorsal margin, median and paired eyes visible in side view, length 800–1,000 μm (page 166) ....
.... *Balanus nubilus*

6a. Posterior end rounded or broadly angular, ventral margin curved 7

6b. Anterior and posterior ends angular 8

7a. Posterior end rounded, ventral margin relatively straight, median and paired eyes visible in side view, 1 pair of black pigment spots just posterior to midlength and eyes; white in color when preserved, with pigment spots persisting; carapace shiny under low magnification and smooth under high magnification; length 700–960 μm (larger in Alaska) (page 165) .... *Balanus crenatus*

7b. Distinct golden color (rarely brown), pigmentation absent when preserved; carapace profile high in side view, large carapace depth; carapace surface dull under low magnification, sculptured with papillae under high magnification; median and paired eyes visible, 4 pigment patches, length 640–780 μm (page 164) ....
.... *Balanus glandula*

8a. Dorsal and ventral surfaces rounded, posterior end sharply pointed and turned slightly upward; length >900 μm 9

8b. Dorsal and ventral surfaces rounded, anterior end sharply pointed and turned slightly upward; length ca 800 μm (in Scotland) (page 167) .... *Balanus balanus*

9a. Translucent when fresh; white to pink when preserved, without distinguishing pigment spots; medium carapace depth; only one eye visible in side view, length 960–1,200 μm (page 165) ....
.... *Semibalanus cariosus*

9b. Typically brown in color but with high degree of variation, without distinguishing pigment spots; medium carapace depth; only one eye visible in side view, length 960–1,200 μm (in England) (page 169) .... *Semibalanus balanoides*

Order Thoracica, Suborder Balanomorpha

Balanus improvisus. Adults live in low intertidal and subtidal areas within estuaries and some enclosed bays on rocks, pilings, and hard-shelled animals. This cirripede is particularly tolerant of brackish waters; high numbers of cyprids and adults have been found in low salinity (3.4 PSU) water in the San Francisco Bay estuarine system (Standing, 1980). Adults and larvae of this species are found in the upper portions of Coos River and South Sough estuary, Oregon (R. Emlet, pers. comm.). *B. improvisus* ranges from the Columbia River, Washington, to the Salinas River, California, and occasionally farther south. This species was introduced, probably from the North Atlantic.

The nauplius is similar in appearance to other balanoids. Figure 5 shows outlines for the six stages of *B. improvisus* collected in British estuaries. Jones and Crisp (1954) reported that the paired eyes are present in the stage V nauplius, although unpigmented. This is earlier than in other cirripede species. A table of larval dimensions and additional comparative drawings can be found in Jones and Crisp (1954).

Cyprids of this species can usually be found low in the water column in estuaries and some enclosed bays. Standing (1980) found a mean length and width of *B. improvisus* cyprids collected in San Francisco Bay of 584 µm (range, 540–610 µm) and 285 µm, respectively (Fig. 6). Cyprids collected by Jones and Crisp (1954) from the east coast of England were smaller, with a mean length of 523 µm. This cyprid is most similar to *Chthamalus dalli*, *Balanus glandula*, and *Pollicipes polymerus*, but it can be differentiated by its size and carapace texture. *Chthamalus dalli* has a smooth carapace, whereas *B. improvisus* possesses a carapace sculptured with rounded pits. *Balanus improvisus* cyprids have a more rounded anterior end (in side

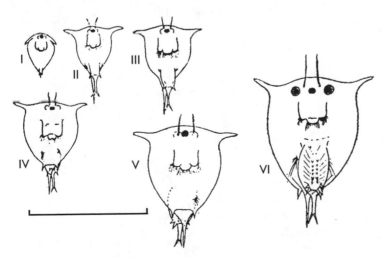

Fig. 5. *Balanus improvisus* nauplii. Stages I, II, III, and VI in ventral view; stages IV and V in dorsal view. Scale = 500 µm (From Jones and Crisp, 1954, Fig. 1)

Fig. 6. *Balanus improvisus* cyprid, (A) lateral and (B) dorsal views; carapace sculpturing inset. Scale = 500 µm. (From Standing, 1980, Fig. 3)

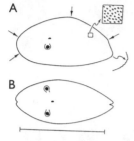

view) than do the cyprids of *C. dalli*. Further, the posterodorsal margin of *B. improvisus* cyprids breaks just in front of the posterior end (see Fig. 6). *Balanus improvisus* cyprids are usually smaller than *B. glandula* cyprids but larger than those of *P. polymerus*. For additional distinctions between *B. improvisus* and other cyprids, see Standing (1980).

Balanus glandula. Adults of this species live in the upper and occasionally lower intertidal zone on both exposed and protected shores (Standing, 1980). *Balanus glandula* is found from the Aleutian Islands south to Bahía de San Quintín, Baja California. Like all other local balanomorph nauplii (except *Chthamalus dalli*), this one has a three-lobed labrum. The first naupliar stage is ephemeral; 99% molt to stage II within one hour (Brown and Roughgarden, 1985). A table of naupliar and cyprid sizes of specimens collected from Monterey Bay, California, is provided in Brown and Roughgarden (1985). In addition to their larger size, stage IV, V, and VI nauplii can be distinguished from earlier stages by the presence of a pair of well-developed posterior shield spines at the posterior margin of the dorsal shield (Brown and Roughgarden, 1985). Outline drawings of stage I–VI nauplii are shown in Fig. 7 (see also Fig. 9).

Live *B. glandula* cyprids are translucent or brown (Brown and Roughgarden, 1985) and are usually found high in the

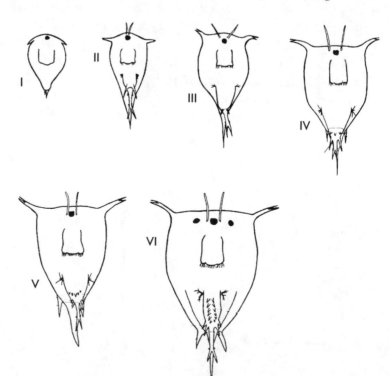

Fig. 7. Stage I–VI *Balanus glandula* nauplii as seen from ventral aspect. (From Brown and Roughgarden, 1985, Fig. 3)

water column and mainly during the non-winter months (Standing, 1980). They range in size from 640 to 780 mm (Fig. 8). Chartreuse pigmentation on live specimens is distinctive. See Standing (1980) for a detailed description of pigmentation and other unique characteristics. Preserved specimens are pale brown or, more commonly, golden. Similar to the smaller *P. polymerus* cyprid, the carapace of *B. glandula* is sculptured with papillae (Standing, 1980).

Balanus crenatus (Bruguiere). Adults are found in the low intertidal and subtidal to depths of 182 m. They live on pier pilings, hard-shelled animals, and occasionally seaweed. They range from Alaska to Santa Barbara, California (Morris et al., 1980; Standing, 1980). Herz (1933) was one of the first to describe *B. crenatus* development, though he incorrectly described eight naupliar stages. Ovsyannidova and Korn (1984) provide a detailed description of differences between the six naupliar stages in specimens collected from the Sea of Japan. Branscomb and Vedder (1982) describe the naupliar stages of *B. crenatus* as well as those for *B. glandula* and *Semibalanus cariosus* (Fig. 9). Early stage *B. crenatus* nauplii are roughly triangular (Pyefinch, 1948). The last three naupliar stages are more shield-shaped. *B. crenatus* nauplii have a shape similar to that of *S. cariousus*, but they tend to be smaller. *Balanus crenatus* nauplii are similar in shape and size to those of *B. glandula*; comparison of the frontolateral horns can be used to differentiate between the species. At each naupliar stage, the frontolateral horns of *B. glandula* are angled more strongly to the anterior than they are in *B. crenatus*.

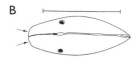

Fig. 8. *Balanus glandula* cyprid, (A) lateral and (B) dorsal views; carapace sculpturing inset. Scale = 500 μm. (From Standing, 1980, Fig. 4)

A

B

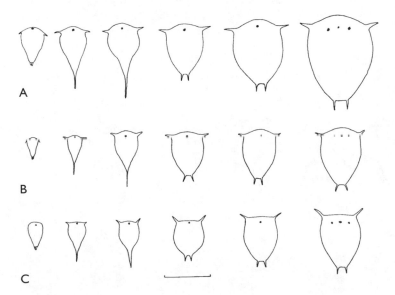

A

B

C

Fig. 9. Cephalic shield outlines of the six naupliar stages of (A) *Semibalanus cariosus*, (B) *Balanus crenatus*, and (C) *Balanus glandula*. Scale = 500 μm. (From Branscomb and Vedder, 1982, Fig. 1)

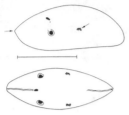

Fig. 10. *Balanus crenatus* cyprid, (A) lateral and (B) dorsal views. Scale = 500 µm. (From Standing, 1980, Fig. 5)

Standing (1980) provides a description of *B. crenatus* cyprids—their size, vertical distribution in the water column, and characteristics that can be used to differentiate them from other cyprids. This species produces a medium-sized cyprid with a mean length of 852 µm (range, 700–960 µm). Standing (1980) found that on the west coast of North America cyprid size increases with latitude. These cyprids are generally found low in the water column. *Balanus crenatus* cyprids are most similar to cyprids of *B. nubilus* but differ in having a narrow, angular anterior end and an evenly curved posterodorsal margin (Fig. 10). An important distinguishing characteristic is the presence of a single pair of black pigment spots, somewhat smaller than the compound eyes, just posterior to midline of the cyprid. These spots persist after preservation. See Standing (1980) for additional details on differentiating balanomorph cyprids.

Balanus nubilus. Adults occur in the lower intertidal and subtidal zones of exposed coasts from southern Alaska to Bahía de San Quentín, Baja California (Barnes and Barnes, 1959b; Standing, 1980). There are six naupliar stages (Fig. 11). From stage II onward, the cephalic shield is funnel-shaped. Stage III–VI nauplii are recognizable by the nearly flat, noncurved anterior margin between the relatively large and anteriorly oriented frontal horns (Miller and Roughgarden, 1994). A median naupliar eye is present in all six stages, and compound eyes are apparent at stage VI although they are not always deeply pigmented (Barnes and Barnes, 1959b). The boxy angular outline of these nauplii distinguishes them from all other Pacific Northwest barnacle nauplii.

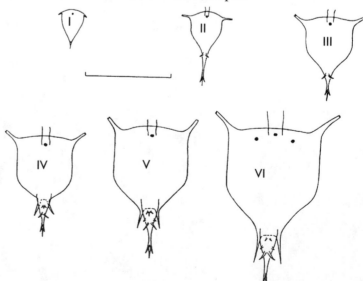

Fig. 11. Stage I–VI *Balanus nubilus* nauplii. Scale = 500 µm. (From Barnes and Barnes, 1959b, Fig. 1)

Standing (1980) provides the best description of *B. nubilus* cyprids. Cyprids are medium-sized (mean length, 932 µm; range 800–1,000 µm). Like the nauplii of this species, the cyprid poses a distinctive profile. Viewed from the side, there is a break in the posterodorsal margin and a sharply curved posterior end (Fig. 12). The anterior end is rounded. This cyprid is most similar to that of *B. crenatus* but can be differentiated from it by the break in the posterodorsal margin and the lack of a posterieor pigment spot. The rounded anterior end separates this cyprid from *Semibalanus cariosus*. All of the above-mentioned characteristics as well as a smooth carapace separate this cyprid from that of the smaller *B. glandula*. An additional description is provided by Miller and Roughgarden (1994).

Balanus balanus. This subtidal species is rare in the San Juan Archipelago (Strathmann, 1987). The carapace of the first three naupliar stages is triangular (Fig. 13), and the frontolateral horns are turned slightly to the posterior. The carapace of nauplier stages IV–VI is roughly goblet-shaped, with the frontolateral horns perpendicular to the long axis of the carapace. The cyprid is ca 800 µm in total length, with sharply pointed ends (Fig. 13). The anterior end turns slightly upward (Barnes and Costlow, 1961; Lang, 1980).

Semibalanus cariosus (= *Balanus cariosus*). *Semibalanus cariosus* adults are found in the lower midtidal and sometimes low intertidal zones of wave-exposed shores on rocks and hard-shelled animals; they range from Alaska south to Morro Bay, California (Standing, 1980). Nauplii of this species are shield-shaped like those of *Balanus glandula* and *B. crenatus*, but they

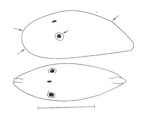

Fig. 12. *Balanus nubilus* cyprid, (A) lateral and (B) dorsal views. Scale = 500 µm. (From Standing, 1980, Fig. 6)

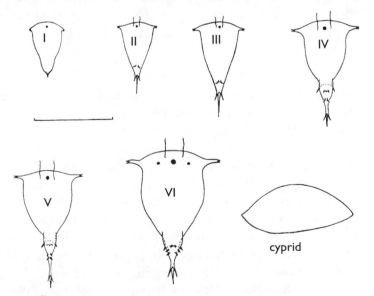

Fig. 13. *Balanus balanus* nauplii stages I–VI and cyprid. Scale = 500 µm. (From Barnes and Costlow, 1961).

Fig. 14. Ventral view of stage I–VI nauplii of *Semibalanus cariosus.* The abdominal process is slightly displaced to show the dorsal spine. Scale = 100 µm. (From Korn, 1989, Fig. 4)

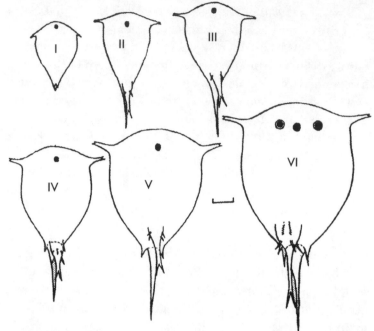

are wider, larger, and possess a more pronounced anterior curvature between the frontolateral horns (Fig. 14; see also Fig. 9). The naupliar stages are similar in shape to those of *Semibalanus balanoides,* but they are smaller in size. Approximate maximum size of stage II –VI, respectively, is 350, 350, 450, 550, and 650 µm (Korn, 1989).

Standing (1980) reports that *S. cariosus* cyprids tend to be found mainly in the spring and summer and low in the water column. These cyprids tend to be large, with a mean length of 1,111 µm (range, 960–1,200 µm). They are opaque when preserved except for a translucent area along the dorsum, and they lack unique pigmentation. The carapace is smooth, angular at both anterior and posterior ends, and has no pigment spots (Fig. 15).

Semibalanus balanoides (= Balanus balanoides). Adults are intertidal with a large vertical range and do not occur south of Washington state. The anterior of the nauplii is strongly curved. The frontolateral horns are roughly perpendicular to the long axis of the body in the first three stages and are tilted anteriorly in the last three stages (Fig. 16). The carapace is shield-shaped. The nauplii are similar in shape to those of *S. cariosus* but larger at each stage. Approximate maximum size of stage II–VI, respectively, is 400, 500, 600, >700, and >800 µm(Lang, 1980). The cyprid stage is typically brown but with a high degree of variation. There is no distinguishing pigmentation. In a side view, only one eye is visible. In samples collected off England, sizes range from 960 to 1,200 µm (Pyefinch, 1948; Lang, 1980).

Fig. 15. *Semibalanus cariosus* cyprid. (Drawn from color slide provided by R. Emlet)

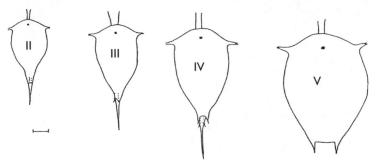

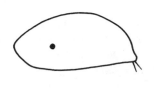

Fig. 16. *Semibalanus balanoides* naupliar stages II–V and cyprid. Scale = 0.1 mm and applies to the nauplii; the cyprid is ca 0.9 mm total length. (Nauplii from Lang, 1980; cyprid from Pyefinch, 1948)

Solidobalanus hesperius (Pilsbry). This species is not well studied. Adults of what Pilsbry (1916) called the *laevidomus* form of *Solidobalanus hesperius* are found from Alaska south to Monterey, California. Pilsbry separated a unique form of this animal after thorough examination of the cross sections of calcareous plates of adults from different eastern Pacific locations; species variation based on these findings is, however, seldom considered in recent literature.

Barnes and Barnes (1959a) described the six naupliar stages but did not provide a description of the cyprid (Fig. 17). Stage I nauplii are pear-shaped. The later stages are characteristically triangular; their length is about two and a half times greater than their width. Stage III nauplii and beyond posses a prominently flat anterior margin between the frontolateral horns. The frontolateral horns are long. In nauplier stages V and VI the posterior shield spines are long, extending more than half the length of the furcal ramus. There is no published description of the cyprid form of this species.

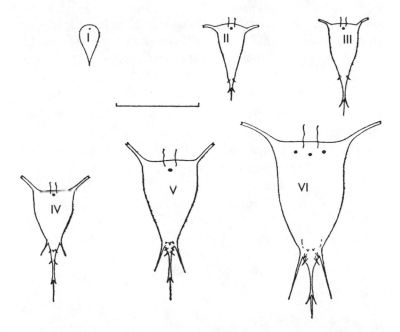

Fig. 17. *Solidobalanus hesperius* naupliar stages I–VI. Scale = 500 µm. (From Barnes and Barnes, 1959a, Fig. 1)

Fig. 18. *Chthamalus dalli* naupliar stages I–VI and cyprid. Note that the anterior end of the cyprid is on the left. Scale = 100 μm. (From Miller et al., 1989, Fig. 5)

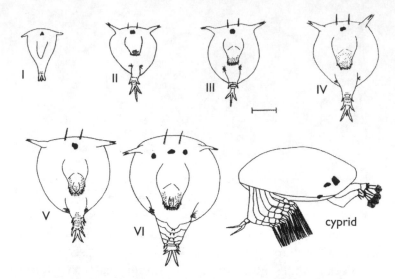

Solidobalanus engbergi. There are no descriptions of the larvae of this species. Adults live in the littoral zone from Alaska to Oregon (Austin, 1985).

Chthamalus dalli. Adults of this common northern barnacle inhabit the high intertidal zone from Alaska to San Diego, California (Standing, 1980; Miller et al., 1989). Miller et al. (1989) provide a complete description of distinctive characteristics of *C. dalli* nauplii. Unique features include a rounded dorsal shield outline and a unilobate labrum (Fig. 18). In addition, *C. dalli* nauplii are small; the only similarly sized nauplii are those of *Pollicipes polymerus*. The nauplii of the two species can be differentiated by noting the following differences: the cephalic shield of *C. dalli* nauplii is round, whereas that of *P. polymerus* nauplii is goblet-shaped; *C. dalli* nauplii have large spines on the furcal ramus which are not present in *P. polymerus* nauplii. See Miller et al. (1989) for further discussion of the differentiation of *C. dalli*.

Standing (1980) states that *C. dalli* cyprids are found at the surface and mid-depths of the water column. Like the nauplius stage, the cyprid is small, with a mean length of 529 μm (Figs. 18, 19). Live specimens are translucent, and when preserved they become a bit more opaque at the anterior end. They lack unique pigmentation. The carapace is completely smooth, distinguishing this cyprid from that of the similarly sized *P. polymerus*. Both the nauplii and cyprid of this species are identical in appearance to the more southern *C. fissus* (Miller et al., 1989). Korn and Ovsyannikova (1979) provide an additional resource with detailed descriptions of naupliar and cyprid features of *C. dalli*.

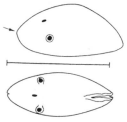

Fig. 19.*Chthamalus dalli* cyprid, lateral (a) and dorsal (b) views. Scale = 500 μm. (From Standing, 1980, Fig. 2)

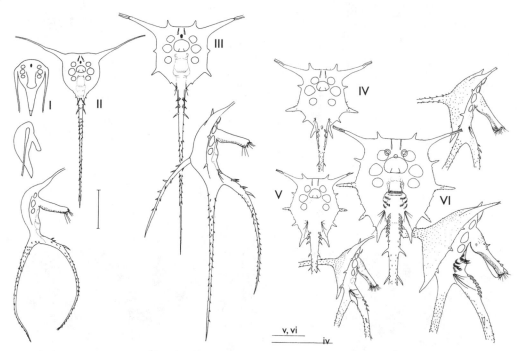

Fig. 20. *Lepas anatifera* naupliar stages I–VI. Scale = 200 μm for stages I–III and 500 μm for stages IV–VI. (Modified from Moyse, 1987, Figs. 4, 5)

Order Thoracica, Suborder Lepadomorpha

Lepas anatifera. Like all adults of the genus *Lepas*, these are found on open-ocean flotsam such as logs and glass fishing floats. Moyse (1987) provides detailed illustrations of *L. anatifera*. The unique shape of the nauplii of these lepadomorph barnacle larvae easily distinguishes them from other Pacific Northwest cirripedes (Fig. 20). The stage I nauplii have frontolateral horns folded back against the body as in the balanomorph stage I nauplii, but these horns are nearly as long as the entire nauplius. In stage II and all later stages, the dorsal thoracic spine is more than twice the body length. With each stage, the cephalic shield gains projections around its edges. As in all lepadomorphs, the nauplii have a single-lobed labrum.

Moyse (1987) gives little information on the cyprid. *Lepas* cyprids are large and translucent. In a side view the ventral margin is relatively straight, with a uniformly rounded dorsal margin. The anterior end is broadly rounded, sometimes showing signs of retaining the naupliar frontolateral horns (anterior arrow, Fig. 21). The posterior ventral end is angular (posterior arrow, Fig. 21). *Lepas pacifica* and *L. fascicularis*, two other members of the genus, may be present off our coastline; little has been published on the appearance of the larvae of these two species. Additional information can be found in Bainbridge and Roskell (1966) and Darwin (1851).

Fig. 21. Outline of an unidentified *Lepas* cyprid. Overall length is 2,200 μm. (Drawn from color slide provided by R. Emlet)

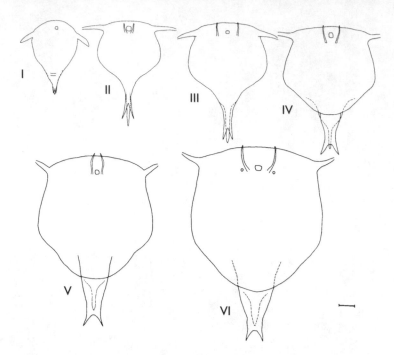

Fig. 22. *Pollicipes polymerus* naupliar stages I–VI, dorsal view. Scale = 50 μm. (From Lewis, 1975, Fig. 2)

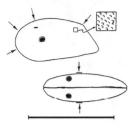

Fig. 23. *Pollicipes polymerus* cyprid, lateral (a) and dorsal (b) views; carapace sculpturing inset. Scale = 500 μm. (From Standing, 1980, Fig. 1)

Pollicipes polymerus (= Mitella polymerus). Aggregations of adults occur in the upper rocky intertidal zone in areas of surge or wave action from British Columbia to Cabo San Lucas, Baja California (Lewis, 1975; Strathmann, 1987). The naupliar carapace is a rounded and shaped like a goblet (Fig. 22). The frontal horns move progressively anterior with each successive stage. The furcal ramus exceeds the length of the dorsal thoracic spine in stage V and VI nauplii, unlike that in all other Pacific Northwest cirripedes (see Lewis, 1975, for illustrations of this feature).

The cyprid larvae do not posses the characteristic oil cells of other cirripedes, suggesting that they either feed in the plankton or settle out of the water column rapidly (Lewis, 1975). The entire carapace is sculptured with uniform, rounded, regularly spaced papillae (ca 4.5 μm diam. by 1.5 μm high) (Fig. 23). Transparent when fresh, this cyprid differs from other species in having a broadly rounded anterior end in the side view, a break in the posterodorsal margin, a narrow dorsal carapace profile and small papillae (Standing, 1980).

Lewis (1975) provides a description of *P. polymerus* embryological development and larval growth. Standing (1980) provides a detailed comparison of *P. polymerus* larvae to other common Pacific Northwest cirriped larvae.

Order Rhizocephala

The Rhizocephalan barnacles are parasites on decapod crustaceans particularly hermit crabs and other anomurans.

There are two suborders, the Kentrogonida and the Akentrogonida, and at least eleven local species (Table 2). A detailed description of the amazing life cycle of the rhizocephalan barnacles can be found in Høeg (1992).

Most species in the Kentrogonidae release propagules at the naupliar stage. Several Kentrogonidae species and all Akentrogonidae species release propagules that develop to the cyprid stage.

The nauplii are small lecithotrophic, and there are only four stages. Depending on the species and ambient water temperature, development takes from four to twenty days. Consistent with their being non-feeding, the nauplii lack gnathobases (paired endites, inwardly directed lobes, used to manipulate or move food) on their limbs (Høeg 1992). Most nauplii possess pigmented eyes and all contain numerous large yolky cells (Fig. 24). In some species in the family Peltogastridae nauplii stages II – IV are enclosed in a hollow annulus consisting of very thin cuticle reinforced by a mesh of thicker ribs (Høeg, 1992). This hollow annulus may aid in floatation of the larvae.

Rhizocephalan cyprids are small, ranging in size from 60 μm to as much as 400 μm. In the Kentrogonidae, the male cyprid tends to be smaller and morphologically different from the female (Høeg 1992); after the molt from the naupliar to the

Table 2. Species in the order Rhizocephala from the Pacific Northwest (from Kozloff, 1996)

Suborder Kentrogonida
Family Clistrosaccidae
Angulosaccus tenuis
Clistrosaccus paguri

Family Peltogastridae
Briarosaccus callosus
Peltogaster boschmae
P. paguri
P. gracilis

Family Sacculinidae
Loxothylacus panopaei

Family Sylonidae
Sylon hippolytes

Suborder Akentrogonida
Family Akentrogonidae
Thompsonia **sp.**

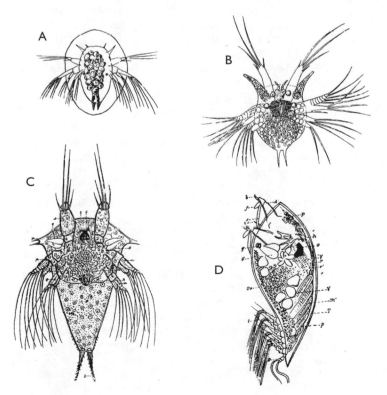

Fig. 24. Larvae of rhizocephalan barnacles. (A, B) *Peltogaster* sp., total length about 0.3 mm. (C, D) *Sacculina carcini*, total length about 0.2 mm. The numerous small circles in the nauplii represent yolk cells. (From Hoeck, 1964, Figs. 48–50, 52)

Fig. 25. *Peltogaster paguri* cyprid. Note the numerous setae covering the carapace, a characteristic of rhizocephalan cyprids. Scale = 25 µm (From Høeg 1992, Fig. 13A)

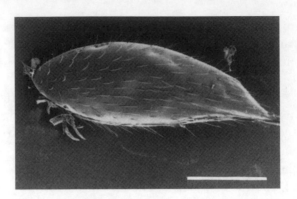

cyprid stage, the cyprid must go through several days of development before it is competent to settle. Rhizocephalan cyprids may be distinguished by their small size, most lack a compound eye associated with the frontal filament (except the genus *Thompsonia*), most species have a naupliar eye (except some species in the genus *Peltogaster*), and the carapace is covered with setae (Fig. 25).

References

Austin, W. C. (1985). An annotated checklist of marine invertebrates in the cold temperate northeast Pacific., vol. 3, pp. 682. Khoyatan Marine Laboratory, Cowichan Bay, British Columbia.

Bainbridge, V. and J. Roskell (1966). A re-description of the larvae of *Lepas fascicularis* Ellis and Solander with observations on the distributions of *Lepas* nauplii in the north eastern Atlantic. In: Some Contemporary Studies in Marine Science, H. Barnes (ed.), pp. 67–81. George Allen & Unwin Ltd., London.

Barnes, H. and M. Barnes (1959a). The naupliar stages of *Balanus hesperius* Pilsbry. Can. J. Zool 37, 237–44.

——— (1959b). The naupliar stages of *Balanus nubilis* Darwin. Can. J. Zool. 37, 15–23.

Barnes, H. and J. D. Costlow (1961). The larval stages of *Balanus balanus* (L.) DaCosta. J. Mar. Biol. Ass. U.K. 41:5968.

Branscomb, E. S. and K. Vedder (1982). A description of the naupliar stages of the barnacles *Balanus glandula* Darwin, *Balanus cariosus* Pallas, and *Balanus crenatus* Bruguière (Cirripedia, Thoracica). Crustaceana. 42, 83–95.

Brown, S. K. and J. Roughgarden (1985). Growth, morphology, and laboratory culture of larvae of *Balanus glandula* (Cirripedia: Thoracica). J. Crust. Bio. 5, 574–90.

Darwin, C. 1851. A monograph on the subclass Cirripedia, with figures of all species. The Lepadidae; or, pedunculated cirripedes. Roy. Soc. London 1-400.

Herz, L. E. (1933). The morphology of the later stages of *Balanus creatus* Breguiere. Biol. Bull. 64:432–42.

Hoeck, P. P. C. (1964). Die Cirripedien. In: Nordisches Plankton, Entomostraca. 265–332. Verlag von Lipsius & Tischer, Kiel.

Høeg, J. T. (1992). Rhizocephala. In: Microscopic Anatomy of Invertebrates Vol. 6: Crustacea, F. W. Harrison and A. G. Humes (eds.), pp. 313–45.

Høeg, J. and J. Lutzen 1985. Crustacean Rhizocephala. In: Marine Invertebrates of Scandinavia. No. 6. Norwegian University Press.

Jones, L. W. G. and D. J. Crisp (1954). The larval stages of the barnacle *Balanus improvisus* Darwin. Proc. Zool. Soc. London 123, 765–80.

Korn, O. M. (1989). Reproduction of the barnacle *Semibalanus cariosus* in the Sea of Japan. Biol. Morya 5, 40–48.

Korn, O. M. and L. L. Ovsyannikova (1979). Development of larvae of the barnacle *Chthamlus dalli*. Biol. Morya. 5:60–69.

Kozloff, E. N. 1996. Marine Invertebrates of the Pacific Northwest. University of Washington Press, Seattle.

Lang, W. H. (1979). Larval development of shallow water barnacles of the Carolinas (Cirripedia: Thoracica) with keys to naupliar stages. National Oceanographic and Atmospheric Administration, Technical Report National Marine Fisheries Service Circular, 1–23.

———— (1980). Cirripedia: Balanomorph nauplii of the NW Atlantic shores. Fich. Ident. Zoopancton 163:6 pp.

Lewis, C. A. (1975). Development of the Gooseneck Barnacle *Pollicipes polymerus* (Cirripedia: Lepadomorpha): Fertilization through Settlement. Mar. Biol. 32, 141–53.

Miller, D. M., S. M. Blower, D. Hedgecock, and J. Roughgarden (1989). Comparison of larval and adult stages of *Chthamalus dalli* and *Chthamalus fissus* (Cirripedia: Thoracica. J. Crust. Biol. 9, 242–56.

Miller, K. M. and J. Roughgarden (1994). Descriptions of the larvae of *Tetraclita rubescens* and *Megabalanus californicus* with a comparison of the common barnacle larvae of the central California coast. J. Crust. Biol. 14, 579–600.

Morris, R. H., D. P. Abbott, and E. C. Haderlie (1980). Intertidal Invertebrates of California. 690 pp. Stanford University Press, Stanford.

Moyse, J. (1987). Larvae of lepadomorph barnacles. In: Crustacean 5: Barnacle Biology, vol. 5, A. J. Southward (ed.). A. A. Balkema, Rotterdam.

Ovsyannikova, I. I. and O.M. Korn. 1984. Naupliar development of the barnacle *Balanus crenatus* in Peter the Great Bay (Sea of Japan). Biol. Morya 5:34–40.

Philsbry, H. A. 1916. The sessile barnacles (Cirripedia) contained in the collections of the United States National Museum; including a monograph on the American species. U. S. Natl. Mus. Bull. 93:1-366.

Pyefinch, K. A. (1948). Methods of identification of the larvae of *Balanus balanoides* (L.), *B. crenatus* Brug. and *Verruca stroemia* O. F. Müller. J. Mar. Bio. Assoc. (UK) 27, 451–63.

Standing, J. D. (1980). Common inshore barnacle cyprids of the Oregonian Faunal Province (Crustacea: Cirripedia). Proc. Biol. Soc. Wash. 93, 1184–1203.

Strathmann, M. F. (1987). Reproduction and Development of Marine Invertebrates of the Northern Pacific Coast. University of Washington Press, Seattle.

Strathman, R. R., and E. S. B. 1979. Adequacy of cues to favorable sites used by settling larvae of two intertidal barnacles. In: Reproductive Ecology of Marine Invertebrates, vol. 9, S. E. Stancyk (ed.), pp. 77–89. University of South Carolina, Columbia.

13

Arthropoda: Isopoda

Steve Sadro

Isopoda is the second-largest order of crustaceans. It comprises eight suborders, seven of which are known from Pacific Northwest waters. Although most of the 4,000 species are marine (of which nearly 90 are local), some species are found in fresh or brackish water and some terrestrial species are found in damp habitats. There is considerable morphological plasticity in this order, yet most isopods are characterized by ovoid or elongate bodies that are frequently dorsoventrally compressed.

Sexual dimorphism in isopods is rare, but where it occurs (e.g., gnathiids, some epicarideans and bopyrids) it can be quite pronounced. Sexual reproduction is facilitated through copulation, resulting in internal fertilization. Copulation must occur between molts during the biphasic molt cycle of the female. The molting process of isopods is unique: the posterior portion of the body is molted first, allowing copulation, followed by the anterior portion, which creates a brood pouch, or marsupium. Egg laying may begin within a few hours of copulation, and eggs are typically brooded in the marsupium. Free-living forms typically have direct development, with juvenile development largely complete by the time of hatching. The emerging young are essentially miniatures of the adults. Juveniles are characterized by a lack of the last pair of thoracic legs, those of the eighth thoracomere (Lee and Miller, 1980; Schram, 1986).

Some parasitic isopods have larval stages markedly different in morphology from the adult forms, but few of the species, and none locally, have been described in any detail (Bonnier, 1900; Shino, 1942a,b; Markham, 1974; Bourdon and Bruce, 1979; Sassaman, 1985). Members of the suborder Epicaridea (Table 1), whose adult forms are ectoparasitic on other crustaceans, have larvae that are also parasitic. Eggs, which are brooded as in the free living forms, hatch into free-swimming 12-legged epicarid larvae (or epicaridium) with seven distinct thoracic and six abdominal segments. This free-swimming larva then parasitizes a zooplankter, usually a copepod. While on the zooplanktonic host, it transforms into a 14-legged microniscus (or microniscium) larva. The third and final larval molt yields the cryptoniscus (or cryptoniscium) larva (Fig. 1), which is

Table 1. Species in the order Isopoda from the Pacific Northwest (from Kozloff, 1996) that produce planktonic larvae

Suborder Epicaridea
Family Bopyridae
Argeia pugettensis
Bopyroides hippolytes
Hemiarthrus abdominalis
Ione cornuta
Munidion parvum
Phyllodurus abdominalis
Pseudione galacanthae
Pseudione giardi

Family Dajidae
Arthophryxus beringanus
Holophryxus alaskensis
Prophryxus alascensis

Family Entoniscidae
Portunion conformis

Family Hemioniscidae
Hemioniscus balani

Family Liriopsidae
Liriopsis pygmaea

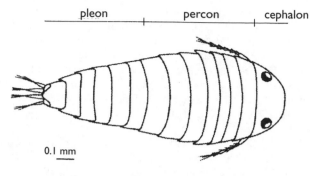

pleon | percon | cephalon

0.1 mm

Fig. 1. Cryptoniscus stage larva, dorsal view. (Modified from Sassaman, 1985)

longer, up to 7 mm, and has a seventh pereonite with another pair of legs. The cryptoniscus larva detaches from its zooplanktonic host and enters a benthic phase as it searches for its ultimate crustacean host. Once a suitable host is found, the cryptoniscus larva either metamorphoses into a female or undergoes lesser modification to become a male. Keying the larvae beyond the suborder is difficult since the larval stages in the four epicaridean families are superficially indistinguishable. Further investigation into the fine morphological structure of epicaridean larvae will most likely open the way for more precise identification (Hatch, 1947; Schultz, 1969; Miller, 1975).

The likelihood of encountering isopod larvae in the plankton is low. Of the three larval stages discussed above, the cryptoniscium larvae are perhaps those most likely to be encountered. Special care should be given in examining specimens, since some adult isopod species are pelagic, and benthic species can get suspended in the plankton as well. It should be possible to distinguish between cryptoniscium larvae and other superficially similar adult isopods on the basis of size and number of pereonites and legs.

References

Bonnier, J. (1900). Contribution à l'étude des Epicarides, les Bopyridae. Trav. Station Zool. Wimereux 8:1–146.

Bourdon, R. and A. J. Bruce (1979). *Bopyrella saronae* sp. nov., a new boyprid parasite (Isopoda) of the shrimp *Saron marmoratus* (Olivier). Crustaceana 37:191–97.

Hatch, M. H. (1947). The Chelifera and Isopoda of Washington and adjacent regions. Univ. Wash. Publ. Biol. 10:155–274.

Kozloff, E. N. (1996). Marine Invertebrates of the Pacific Northwest. University of Washington Press, Seattle.

Lee, W. L. and M. A. Miller (1980). Isopoda and Tanaidacea: The Isopods and allies. In: Intertidal Invertebrates of California, R. H. Morris, D. P. Abbott, and E. C. Haderlie (eds.), pp. 536–558. Stanford University Press, Stanford.

Markham, J. (1974). A Systematic Study of Parasitic Bopyrid Isopods in the West Indian Faunal Region. University of Miami, Miami. 344 pp.

Miller, M. A. (1975). Phylum Arthropoda: Crustacea, Tanaidacea and Isopoda. In: Light's Manual: Intertidal Invertebrates of the Central California Coast, R. I. Smith and J. T. Carlton (eds.), pp. 281–312. University of California Press, Los Angeles.

Sassaman, C. (1985). *Cabirops Montereyensis*, a new species of hyperparasitic isopod from Monterey Bay, California (Epicaridea: Cabiropsidae). Proc. Biol. Soc. Wash. 98:778–89.

Schram, F. R. (1986). Crustacea. Oxford University Press, New York and Oxford.

Schultz, G. A. (1969). How to Know the Marine Isopod Crustaceans. W. C. Brown Co., Dubuque.

Shiino, S. M. (1942a). On the parasitic isopods of the family Entoniscidae, esp. those found in the vicinity of Seto. Mem. Coll. Sci. Kyoto Ser. B 17:37–76.

——— (1942b). Bopyrids from the South Sea Islands with description of a hyperparasitic cryptoniscid. Palao Tropical Biological Station Studies 2:437–58.

14

Arthropoda: Decapoda

Amy L. Puls

The order Decapoda includes crayfish, lobsters, shrimp, and crabs. It is the largest order in the subphylum Crustacea, encompassing almost one-fourth of the known species of crustaceans. Like all crustaceans, decapods have a chitinous exoskeleton, a segmented body, two pairs of antennae, biramous appendages, and a unique naupliar larval stage. What distinguishes them from other crustaceans is five pairs of jointed thoracic appendages (legs), hence the name Decapoda.

In Pacific Northwest waters, there are approximately 87 species of shrimp-like (so-called "Natantia") decapods, of which the larvae of only 12 have been described (Table 1); most of these are in the genus Pandalus. There are approximately 83 non-shrimp-like decapods (the "Reptantia," including the crabs, hermit crabs, thalassinids, and other non-shrimp-like decapods), the larvae of 47 of which have been described (Table 1). Larval descriptions tend to be available for economically important groups, (e.g., the Pandalid shrimp and cancrid crabs) and some ecologically important intertidal groups (e.g., hermit crabs—Paguridae—and Grapsid and Xanthid crabs), but overall we have larval descriptions for only approximately a third of the decapod species.

Within the Decapoda, species display a vast diversity in habitat, feeding, and morphology. Adults can be found in the open ocean, on wave-swept sandy beaches, in the intertidal and subtidal of rocky shores, and in mudflats in estuaries and bays. They are carnivores, herbivores, scavengers, filter feeders, and detritivores. Morphologically they can be cylindrical, dorsoventrally or laterally flattened; they can be as small as half a centimeter, as in the case of commensal crabs, or have a carapace length of 45 cm and chelipeds that span 4 m, as in the case of the Japanese spider crab.

Reproduction and Development
The majority of decapods reproduce by copulating shortly after the female molts. Fertilization of the eggs is external for the majority of decapods but internal for brachyuran crabs. With the exception of the penaeids and sergestids, females attach the eggs to their pleopods, where they are brooded until

text continues on page 184

Table 1. Species in the order Decapoda from the Pacific Northwest, with references to larval description

Taxa	References	Reference Contents
Order Decapoda	Gurney 1942	larval development of decapods
	Hart 1971	larval key to British Columbia decapod families
"Natantia"—Shrimp-like Decapods		
Suborder Dendrobranchiata		
Family Penaeidae		
Bentheogennema borealis		
Bentheogennema burkenroadi		
Gennadas incertas		
Gennadas propinquus		
Gennadas tinayrei		
Hemipenaeus spinidorsalis		
Family Sergestidae		
Eusergestes similis		
Petalidium subspinosum		
Sergia tenuiremis		
Suborder Pleocyemata		
Infraorder Caridea	Strathmann 1987	reproduction and development of Caridea
Family Alpheidae		
Betaeus harrimani		
Betaeus setosus		
Family Crangonidae		
Argis alaskensis		
Argis levior		
Crangon alaskensis		
Crangon alba		
Crangon franciscorum	Israel 1936	all larval stages
Crangon handi		
Crangon nigricauda		
Crangon stylirostris		
Lissocrangon stylirostris		
Mesocrangon munitella		
Metacrangon acclivis		
Metacrangon munita		
Metacrangon spinosissima		
Metacrangon variabilis		
Neocrangon abyssorum		
Neocrangon communis		
Neocrangon resima		
Paracrangon echinata	Kurata 1964	all larval stages
Rhynocragon alata		
Family Hippolytidae	Pike and Williamson 1960	larvae of Spirontocaris and related genera
Eualus avinus		
Eualus barbatus		
Eualus berkeleyorum		
Eualus biunguis		
Eualus fabricii		
Eualus lineatus		
Eualus macrophthalmus		

Taxa	References	Reference Contents
Eualus suckleyi	Haynes 1981b	first two zoeal stages, with illustrations
Heptacarpus brevirostris		
Heptacarpus camtschaticus		
Heptacarpus carinatus		
Heptacarpus decorus		
Heptacarpus flexus		
Heptacarpus herdmani		
Heptacarpus kincaidi		
Heptacarpus littoralis		
Heptacarpus moseri		
Heptacarpus paludicola		
Heptacarpus pictus		
Heptacarpus pugettensis		
Heptacarpus sitchensis		
Heptacarpus stimpsoni		
Heptacarpus taylori		
Heptacarpus tenuissimus		
Hippolyte clarki		
Lebbeus groenlandicus	Haynes 1978a	all larval stages, with illustrations
Lebbeus schrencki		
Lebbeus washingtonianus		
Spirontocaris arcuata		
Spirontocaris holmesi		
Spirontocaris lamellicornis		
Spirontocaris prionota		
Spirontocaris sica		
Spirontocaris spina		
Spirontocaris synderi		
Spirontocaris truncata		
Family Oplophoridae		
Acanthinephyra curtirostris		
Hymenodora acanthitelsonis		
Hymenodora frontalis		
Hymenodora glacialis		
Hymenodora gracilis		
Notostomus japonicus		
Systellaspis braueri		
Systellaspis cristata		
Family Pandalidae		
Pandalopsis ampala		
Pandalopsis dispar	Berkeley 1930	four larval stages, with illustrations
Pandalus eous (borealis)	Berkeley 1930	six larval stages, with illustrations
Pandalus danae	Berkeley 1930	all larval stages, with illustrations
Pandalus gurneyi		
Pandalus jordani	Modin and Cox 1967	eleven larval stages, with illustrations
Pandalus platyceros	Price and Chew 1972	all larval stages, with illustrations
Pandalus stenolepis	Needler 1938	all larval stages, with illustrations
Pandalus tridens	Haynes 1980	six larval stages, with illustrations
Family Pasiphaeidae		
Parapasiphae sulcatifrons		
Pasiphaea pacifica		
Pasiphaea tarda		

table continues

Taxa	References	Reference Contents

"Reptantia"—Crabs, Hermit Crabs, Thalassinids, and Other Non-shrimp-like Decapods
Suborder Pleocyemata
Infraorder Thalassinidae
Family Axiidae
Axiopsis spinulicauda
Calastacus stilirostris
Calocaris investigatoris
Calocaris quinqueseriatus

Family Callianassidae

Neotrypaea (Callianassa) californiensis	McCrow 1972	all zoeal stages, with illustrations
Neotrypaea (Callianassa) gigas		
Callianopsis goniophthalma		

Family Upogebiidae

Upogebia pugettensis	Hart 1937	all larval stages, with illustrations
Infraorder Anomura	Haynes 1984	larval morphology and development in the Anomura
	Lough 1975	key to 5 anomuran families, also keys for zoeae and megalopae for some Oregon species
	MacDonald et al. 1957	larvae of the British species of Diogenes, Pagurus, Anapagurus, and Lithodes
	Strathmann 1987	reproduction and development

Family Diogenidae

Paguristes turgidus	Hart 1937	all larval stages, with illustrations
Paguristes ulreyi		

Family Galatheidae

	Gore 1979	larval development
Munida quadrispina		
Munidopsis quadrata		

Family Hippidae

Emerita analoga	Johnson and Lewis 1942	all larval stages, with illustrations
Family Lithodidae	Haynes 1984	lithodid larval morphology, key to N. Pacific zoeae
Acantholithodes hispidus		
Cryptolithodes sitchensis		
Cryptolithodes typicus	Hart 1965	all larval stages, with illustrations
Haplogaster grebnitzkii		
Lithodes couesi		
Lopholithodes foraminatus		
Lopholithodes mandtii		
Oedignathus inermis		
Paralomis multispina		
Paralomis verrilli		
Phyllolithodes papillosus		
Rhinolithodes wossnessenskii	Haynes 1984	first two zoeal stages, first zoea illustrated
Family Paguridae	Nyblade 1974	larval morphology and development of 18 species, photographs and illustrations of final-stage zoea
Orthopagurus minimus		
Pagurus aleuticus		
Pagurus armatus		

Taxa	References	Reference Contents
Pagurus beringanus	Hart 1937	all larval stages, with illustrations
Pagurus capillatus		
Pagurus caurinus		
Pagurus confragosus		
Pagurus cornutus		
Pagurus dalli		
Pagurus granosimanus		
Pagurus hemphilli		
Pagurus hirsutiusculus	Fitch and Lindgren 1979	all larval stages, with illustrations
Pagurus ochotensis		
Pagurus quaylei		
Pagurus samuelis	MacMillan 1971	all larval stages, with illustrations
	Coffin 1960	all larval stages, with illustrations
Pagurus setosus		
Pagurus tanneri		
Family Parapaguridae		
Parapagurus pilosimanus		
Family Porcellanidae	Gonor 1970, Gonor and Gonor 1973a	four species' (below) larval stages, with illustrations
Pachycheles pubescens	see above	
Pachycheles rudis	see above	
Petrolisthes cinctipes	see above	
Petrolisthes eriomerus	see above	
Infraorder Brachyura	Lough 1975	key to five families, keys to some Oregon zoeae and megalopae
	Strathmann 1987	reproduction and development
Family Atelecyclidae		
Telmessus cheiragonus		
Family Calappidae		
Mursia gaudichaudi		
Family Cancridae	Iwata and Konishi 1981	larval development of eight *Cancer* species
Cancer antennarius	Roesijadi 1976	all larval stages, with illustrations
Cancer branneri (gibbosulus)		
Cancer gracilis	Ally 1975	all larval stages, with illustrations
Cancer jordani		
Cancer magister	Poole 1966	all larval stages, with illustrations
Cancer oregonensis	Lough 1975	all larval stages, with illustrations
Cancer productus	Trask 1970	all larval stages, with illustrations
Family Grapsidae		
Hemigrapsus nudus	Hart 1935	all larval stages, with illustrations
Hemigrapsus oregonensis	Hart 1935	all larval stages, with illustrations
Pachygrapsus crassipes	Schlotterbeck 1976	zoeae, with illustrations
Planes cyaneus		
Planes marinus		
Family Majidae		
Chionoecetes angulatus		
Chionoecetes bairdi	Haynes 1973, 1981a, Jewett and Haight 1977	prezoea, zoea I, II, megalopa, with illustrations
Chorilia longipes		
Mimulus foliatus		
Oregonia bifurca		

table continues

Taxa	References	Reference Contents
Oregonia gracilis	Hart 1960	all larval stages, with illustrations
Pugettia gracilis		
Pugettia producta		
Pugettia richii		
Scyra acutifrons		
Family Pinnotheridae		
Fabia subquadrata	Lough 1975	all larval stages, with illustrations
Pinnixa eburna		
Pinnixa faba		
Pinnixa littoralis		
Pinnixa occidentalis		
Pinnixa schmitti		
Pinnixa tubicola		
Pinnotheres pugettensis		
Pinnotheres taylori	Hart 1935	all larval stages, with illustrations
Scleroplax granulata		
Family Portunidae		
Carcinus maenas	Rice and Ingle 1975	all larval stages, with illustrations
Family Xanthidae		
Lophopanopeus bellus bellus	Hart 1935	all larval stages, with illustrations
Lophopanopeus bellus diegensis	Knudsen 1959	all larval stages
Rithropanopeus harrisii	Connolly 1925	all larval stages, with illustrations

hatching. Decapod larvae can undergo two forms of development. In the first, direct development, larval development is completed within the egg; at hatching a first instar juvenile is released. In the second, indirect development, development within the egg is through the embryonic stage; at hatching the larva released must molt through one or more stages before becoming a first instar juvenile. Indirect development is the more common pattern among the decapods. A typical life cycle for a species with indirect development is depicted in Fig. 1. For additional general information on decapods, see Brusca and Brusca (1990) or any invertebrate zoology textbook. For more information on reproduction and development in crustaceans, see Bliss (1982). For more specific information on the reproduction of local decapod families, an excellent starting place is Strathmann (1987).

In penaeids and sergestids the first free-swimming larval stage is the nauplius (Fig. 2A). It is characterized by three pairs of cephalic (head) appendages, with the second and third having longer setae for locomotion. In all other decapod families the naupliar stage is passed while in the egg, and the first free-swimming larval stage is the zoea.

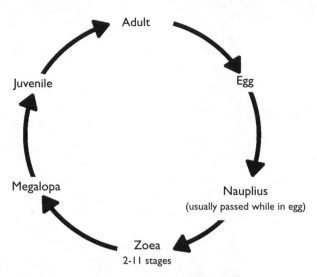

Fig. 1. Generalized
decapod life cycle.

Zoeae (Fig. 2B, C) are distinguished by a rostral spine and plumose setae on the thoracic appendages, which are used for locomotion. Growth is accomplished by a succession of molts; each intermolt period is called a stage. Each species molts through a set number of stages; however, the number of zoeal stages varies greatly within the Decapoda and is only somewhat consistent at the family level. After spending weeks to months in the plankton as a zoea, the larva molts to the last stage.

This last stage has several different names including postlarva, decapodid, megalopa, and glaucothoe. Williamson (1957) suggested that megalopa, the last-stage Brachyura larva, be applied to all larvae that can locomote by use of pleopods, and this suggestion is adopted here. The megalopa (Fig. 2D–F) resembles the adult form and is the stage during which settlement to the juvenile habitat occurs. It is able to swim using the setose pleopods on its abdomen or walk on the benthos with its thoracic appendages. After settling, the megalopa molts to the first instar juvenile.

Morphology

Although the external morphologies of the various decapod families may look extremely different, they all have the same basic body plan. The body is divided into three main regions: the head or cephalic region, the thorax, and the abdomen (Figs. 3A, 4A, 5A). In decapods, the head and thorax have fused; this is often called the cephalothorax. Each region is further divided into somites, or segments. Each segment has a pair of biramous appendages—appendages that are split in two near the base; the inner branch is the endopod, and the outer branch is the

Fig. 2. Decapod larvae. (A) Naupliar larva. (B) Anomuran zoea. (C) Brachyuran zoea. (D) Anomuran megalopa. (E) Brachyuran megalopa. (F) Caridean megalopa. Illustration (A from Cook and Murphy, 1971; B, D from Hart, 1937; C from Rice and Ingle, 1975); E adapted from Lough, 1975; F from Berkeley, 1930)

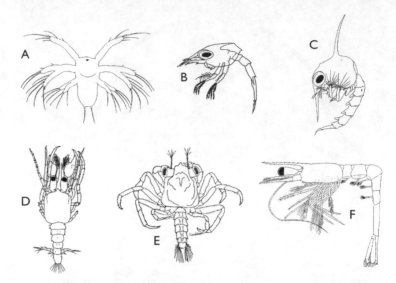

exopod. Some groups have lost the exopod from their pereopods (precursors of the walking legs) (Figs. 3A, 5A).

Covering the cephalothorax dorsally is the carapace (Figs. 3A, 4A, 5A). The anterior-most projection of the carapace is the rostral spine, or rostrum (Figs. 3A, 4A, 5A). Other spines and "teeth" of varying sizes may be present and are typically defined by their location on the carapace.

The head is composed of five indistinguishable fused segments. There are five corresponding pairs of appendages. From anterior to posterior these are the antennules, antennae, mandibles, maxillules, and maxillae (Figs. 3B–E, 4B–E, and 5B–E). In the adult the antennules and antennae become the first and second antennae, and the mandibles, maxillules, and maxillae all become incorporated into mouth parts.

The thorax is composed of eight segments that are dorsally fused. There are eight pairs of corresponding appendages. The first three pairs are maxillipeds (Figs. 3A, F, 4A, F, 5A, F), which become incorporated into or are associated with the mouth in the adult. The last five pairs are pereopods (Figs. 3A, 5A), which become the adult walking legs, some of which may be chelate.

The abdomen is composed of six segments and a telson (Figs. 3A, H, 4A, H, 5A, H), although sometimes the sixth segment remains fused or partially fused to the telson. In late-stage zoea, all or most of the abdominal segments have a pair of pleopods (Figs. 3A, 4A, 5A) that remain or are reduced in the adult. The last abdominal segment may or may not have a pair of uropods (Figs. 3A, 4A) which, when present, form the tail fan in conjunction with the telson.

text continues on page 192

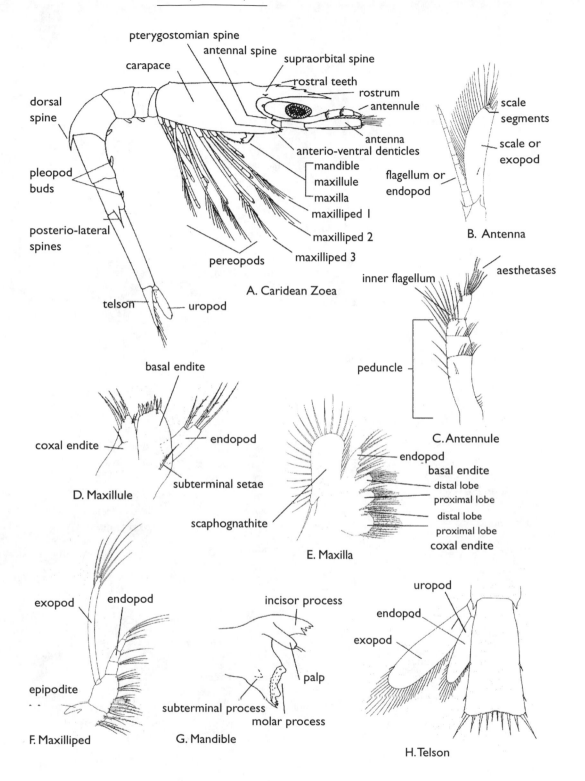

Fig. 3. External morphology and terminology describing caridean zoea.
(From Haynes, 1985)

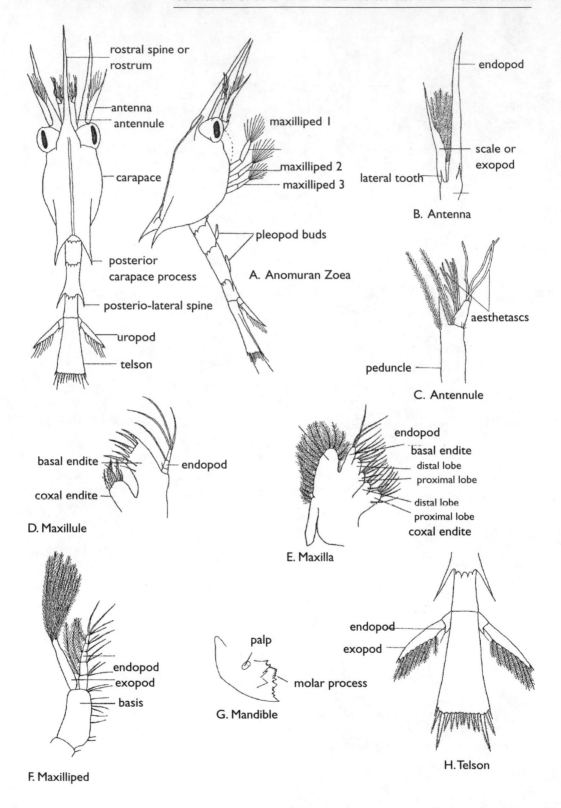

Fig. 4 External morphology and terminology describing anomuran zoea. (From Nyblade, 1974)

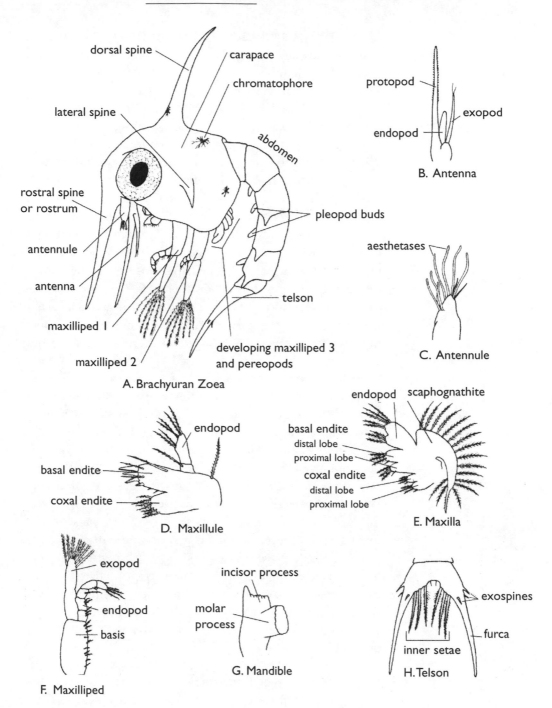

Fig. 5. External morphology and terminology describing brachyuran zoea.
(from Martin, 1984)

Description and Identification of Local Taxa

The following keys and species descriptions cover only pelagic decapod larvae found off the coast of the Pacific Northwest. The keys have been adapted from Hart (1971), Lough (1975), Gonor (1970), and Haynes (1985). The first key is used to distinguish zoea from megalopa stages. Subsequent keys assist in species identification and direct readers to the following species descriptions and, for some families, additional keys. The key for zoeae enables identification of a larva to the family level; only in cases where all species of a family are described is a key to the species given. For families with a limited number of described species, additional descriptive information is provided for those that have been described. Because megalopae are further developed, they possess unique features that make identification easier; therefore, the megalopa key can enable identification of a megalopa to the genus level and in most cases to the species level. The number of described megalopae is, however, limited, so there is a chance that specimens in hand are not included in the key.

In the descriptions of the known species, if an appendage is described in an early stage but fails to be mentioned in later stages, it has remained unchanged. When possible, the most readily apparent characteristics are used in both the species descriptions and the keys. Refer to Figs. 3, 4, and 5 for the terminology of the morphological characters used in the keys. Note that size approximations in the keys are often from laboratory-reared larvae; larvae that developed in the plankton may have different characteristic sizes. The easiest way to distinguish a species is by chromatophore pattern, since it remains constant regardless of stage. Such patterns are not, however, used in the keys and descriptions, with a few exceptions, because chromatophores are lost in preserved samples. More complete descriptions of all the larval appendages and chromatophore patterns can be found in the original literature.

Key to zoea and megalopa stages

1a. Long setae on thoracic appendages; when present, pleopod
     buds without setae .... zoea, Key A
1b. Larval morphology resembles adult form; pleopods with long
     setae .... megalopa, Key B

A. Key to decapod zoeae

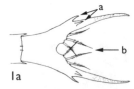

1a. Telson forked and may be armed with exospines (a) and inner
     setae (b); no uropod development in any zoeal stage ....
     .... (Brachyura), 2

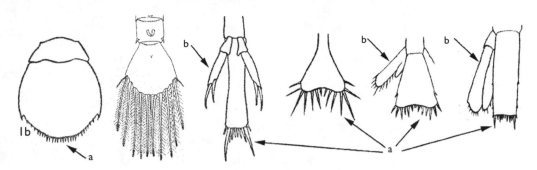

1b. Telson broader and flattened, may be armed with spines (a)
   along posterior margin; uropod (b) development in late-stage
   larvae (uropods may be rudimentary or absent in early stages
   and some species) 9
1c. Telson composed of 2 cylindrical rami (a) bearing long setae 17

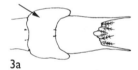

2a. Carapace with lateral spines 3
2b. Carapace without lateral spines 7

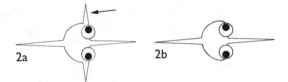

3a. Fifth abdominal segment expanded laterally ....
   .... Pinnotheridae (p. 242)
3b. Fifth abdominal segment not expanded laterally 4

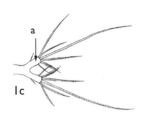

4a. Telson exospines present (1–3 small pairs; arrows) 5
4b. Telson exospines absent .... Grapsidae (p. 236)

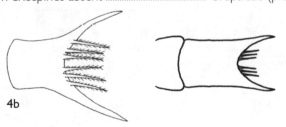

5a. Antenna protopod (a) spinulate with length ≤half rostrum
   length (b) .... Cancridae (p. 233)
5b. Antenna protopod (a) either spinulate or smooth with length
   ≥three-fourths rostrum length (b) 6

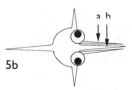

6a. Antenna protopod smooth; 1 pair of dorsal exospines on telson
   (sometimes with 2 very tiny inconspicuous lateral spines in early
   stages) .... Xanthidae (p. 246)

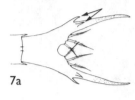

7a

6b. Antenna protopod spinulate; 1 pair of dorsal and 1 pair of lateral exospines on telson .... Majidae (p. 240)

7a. Telson exospines present (1–3 small pairs; arrows) 8
7b. Telson exospines absent ....
.... Pinnotheridae (p. 244) or Zoea 1 Pachygrapsus (p. 239)

8a. Antenna protopod (a) spinulate with length ≤half rostrum length (b) .... Portunidae, *Carcinus maenas* (p. 244)
8b. Antenna protopod (a) spinulate with length ≥three-fourths rostrum length (b) .... Majidae (p. 240)

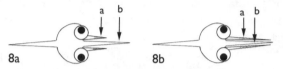

9a. Telson posterior margin (a) rounded with numerous minute spines; carapace with long rostrum (b) and pair of posteriolateral spines (c); lateral spines absent in stage 1 zoeae ..
.... Hippidae, *Emerita analoga* (p. 217)

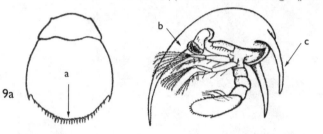

9b. Not as above 10

10a. Telson posterior margin (a) rounded with numerous long plumose setae; rostrum (b) measuring .twice length of carapace; carapace also with pair of posterior spines (c) ....
.... Porcellanidae (p. 230)

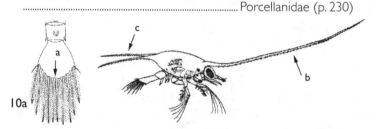

10b. Not as above 11

11a. Carapace with posterior processes (a); rostrum (b) commonly ≥length of antenna endopod (c); swimming setae on exopods of maxillipeds (d) not on pereopod (early and late stages); exopod of uropod (e) flat and bladelike (later stages) 12
11b. Carapace often with orbital (a), antennal (b), and/or pterygostomian spines (c); rostrum (d) rarely longer than antenna (e); swimming setae on exopods of maxillipeds and pereopod (f) (later stages); exopod and endopod of uropod (g) flat and broad (later stages) 14

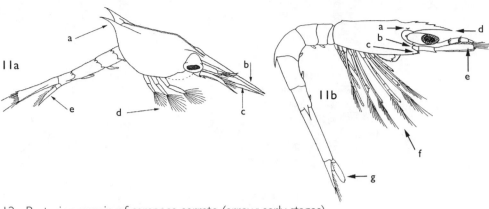

IIa

IIb

12a. Posterior margin of carapace serrate (arrow; early stages) ....
.... Galatheidae (no local larval descriptions)

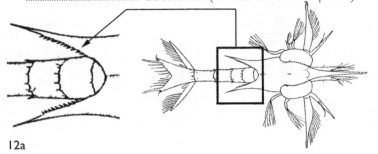

12a

12b. Posterior margin of carapace not serrate 13

13a. Posterior processes of carapace generally close together with
parallel edges (arrow); slender appearance ....
Diogenidae (p. 216); Paguridae (p. 221); or Parapaguridae (no
local larval descriptions)

13b. Posterior processes of carapace generally farther apart with
sloped diverging edges (arrow); stout appearance ....
.... Lithodidae (p. 219)

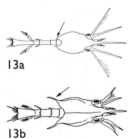

13a

13b

14a. Telson posterior margin with medial tooth (a) (not medial
setae) 15

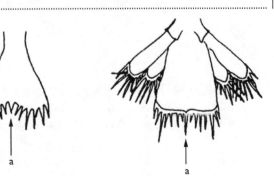

14a

14b. Telson posterior margin without medial tooth 16

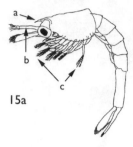

15a

15a. Rostrum (a) small, one-third length of antennules (b); carapace and abdomen without spines; functional exopods on maxillipeds 1–3 and pereopod 1–3(c) ....
.... Upogebiidae, *Upogebia pugettensis* (p. 214)

15b. Rostrum longer than antennules; abdominal segments with dorsal and/or lateral spines ....
. Axiidae (no local larval descriptions) or Callianassidae (p. 213)

16a. Telson deeply cleft; pereopod pairs 1–3 chelate (later stages); functional exopods on all pereopods 17

16b. Telson not deeply cleft; pereopod pair 3 never chelate; number of pereopods with functional exopods variable 18

17a. Carapace with many spines (a) (early stages); carapace with several paired spines and one medial-posterior spine (b) (later stages) .... Sergestidae (no local larval descriptions)

17a

17b. Carapace with long rostrum (a) with paired orbital spines (b) at base; abdominal segments with dorsal spines, segment 2 (c) bearing largest spine (later stages) ....
.... Penaeidae (no local larval descriptions)

17b

18a. Fifth pereopod developed early and much longer than others
.... Alpheidae (no local larval description)

18b. Fifth pereopod no longer than others 19

19a. Rostrum usually wide (a), eyestalks usually hemispherical (b), and bases of antennules touching (c); tip of antennal scale always unsegmented, inner flagellum of antennule a setose spine or oblong projection, functional exopod usually only on pereopod 1 and never on 3–5, in later stages pereopod 1 subchelate and telson widening posteriorly ....
.... Crangonidae (no descriptions included)

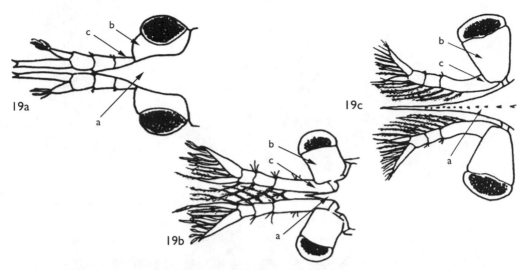

19b. Rostrum usually narrow (a), eyestalks usually cylindrical (b),
     and bases of antennules close together but not touching (c);
     rostrum without teeth in all stages, basal endite of maxillule
     lacking subterminal seta, functional exopods on pereopods 1–2,
     1–3, or 1–4, abdomen with posterio-ateral spines ....
     ....Hippolytidae (p. 202)
19c. Rostrum usually slender (a), eyestalks usually taper toward
     base (b), and bases of antennules relatively far apart (c); rostrum
     always ≥one-fourth length of carapace (later stages), basal
     endite of maxillule with subterminal seta, functional exopods on
     only pereopods 1–2 or 1–3, abdomen lacking posteriolateral
     spines .... Pandalidae (p. 203)

B. Key to decapod megalopae

1a. Shape shrimp-like; pereopods all long and thin resembling adult
    form; first and/or second pereopods either chelate or
    subchelate ....
    Caridea (pp. 204-15), or Thalassinidea* 13 (see also pp. 215-34)
1b. Shape crab-like; first pereopod pair chelate, second through fifth
    pereopods used as walking legs .... Brachyura, 2

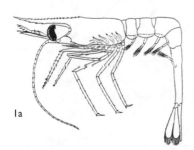
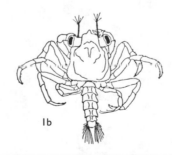

* Can be confused with euphausids and mysids; in euphausids the last two
pereopod pairs are reduced or vestigial, and rarely are any of the
pereopods chelate; the majority of mysids have small bubble-like statocysts
on the endopods of the uropods.

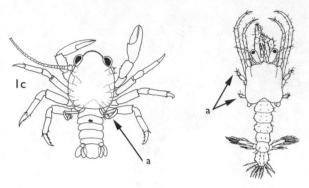

1c. Shape crab-like; first pereopod pair chelate, second and third pereopods used as walking legs, fourth and/or fifth pereopod pairs greatly reduced (a) .... Anomura, 13

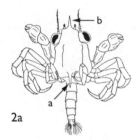

2a

2a. Carapace with posterior spine (a) and long pointed rostrum (b) .... Cancridae, 3
2b. Carapace with more than 1 spine 4
2c. Carapace without spines; rostrum blunt or flexed downward . 5

3a. Length of carapace from tip of rostrum to back of carapace, 5.3–6.6 mm; width of carapace across widest point, 3.5–4.6 mm .... *Cancer magister* (p. 235)
3b. Length of carapace from tip of rostrum to back of carapace, 3.4–3.6 mm; width of carapace across widest point, 2.0–2.1 mm .... *Cancer oregonensis* (p. 235) or *Cancer productus* (p. 236)
3c. Length of carapace from tip of rostrum to back of carapace, 2.0–3.3 mm; width of carapace across widest point, 1.2–2.4 mm .... *Cancer antennarius* (p. 234) or *Cancer gracilis* (p. 234)

4a. Carapace with elongated posterior spine (a) and 2 anteriolateral spines (b) .... Pinnotheridae, *Pinnotheres* sp.
4b. Carapace with 1 posterior spine (a) and 1 mid-dorsal spine (b) .... Majidae, *Pugettia* sp.

4a

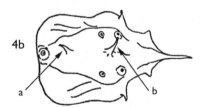

4b

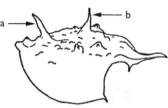

4c. Carapace with pair of anteriolateral spines (a), pair of mid-dorsal spines (b), small pair of lateral spines (c), and elongated posterior spine (d) .... Majidae, *Oregonia gracilis* (p. 241)

4c

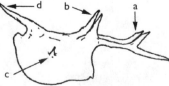

4d. Carapace with pair of anteriolateral spines (a), pair of mid-
dorsal spines (b), small pair of lateral spines (c), and set of
posterior spines (d) .... Majidae, *Chionoecetes bairdi* (p. 240)

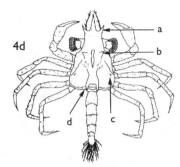
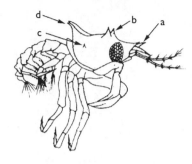

5a. Carapace oval, wider than long, sometimes with lateral teeth
and granular texture .... Pinnotheridae, 6

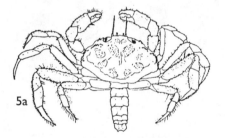

5b. Carapace roughly rectangular or square 7

6a. Largest pinnotherid, carapace ca 2.5 by 3.8 mm ....
.... Pinnotheridae, *Fabia subquadrata* (p. 242)

6b. Lateral edges of carapace with or without teeth; carapace
surface granular or smooth, ca 1.8 by 2.8 mm (a); 1.4 by 2 mm
(b); 1.4 by 2.4 mm (c); 1.5 by 2.4 mm (d); 1.4 by 1.8 mm ....
.... *Pinnixa* spp.

7a. Carapace with small teeth at anterior corners ....
.... Xanthidae, *Lophopanopeus bellus* (p. 246)
7b. Not as above 8

8a. Carapace with prominent cone-like projections 9

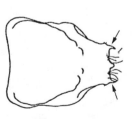

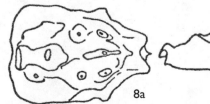

8b. Not as above 10

9a. Larger species; carapace ca 1.6 by 1.2 mm..
.... Majidae, *Pugettia productua*

9b. Smaller species; carapace ca 1.2 by 0.9 mm ....
.... Majidae, *Pugettia gracilis*

10a. Carapace roughly rectangular 11

10b. Carapace roughly square 12

11a. Telson posterior margin with setae (other than uropod setae; arrow); carapace ca 1.8 by 1.5 mm ....
.... Grapsidae, *Hemigrapsus nudus* (p. 238)

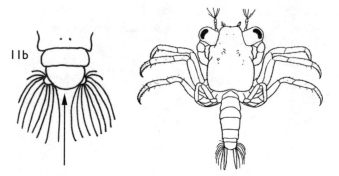

11b. Telson posterior margin without setae (arrow); carapace 1.4–1.7 by 1.1–1.3 mm..
.... Grapsidae, *Hemigrapsus oregonensis* (p. 238)

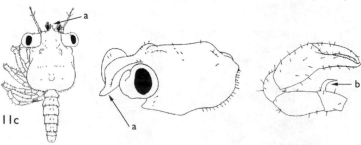

11c. Rostrum small and flexed downward (a); cheliped with prominent ischiobasal hook (b); carapace 1.3–1.4 by 1.0–1.2 mm .... Portunidae, *Carcinus maenas* (p. 244)

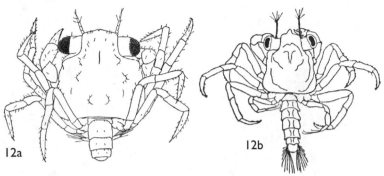

12a. Carapace ca 1.1 by 1.0 mm ....
.... Xanthidae, *Rhithropanopeus harrisii* (p. 247)
12b. Carapace ca 3.0 by 4.1 mm (average, Coos Bay, Oregon) ....
.... Grapsidae, *Pachygrapsus crassipes* (p. 239)

13a. Fifth pereopod pair greatly reduced; carapace rounded;
      abdomen wide .... Porcellanidae, 14

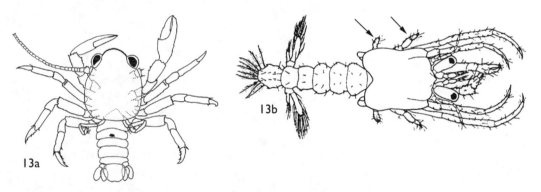

13b. Fourth and fifth pereopod pairs greatly reduced ....
.... Diogenidae, Paguridae, or Parapaguridae, 17

14a. Chelipeds long, slender, and dorsoventrally flattened ....
.... *Petrolisthe*, 15
14b. Chelipeds heavy, broad, and swollen, not dorsoventrally
      flattened .... *Pachycheles*, 16

15a. Carpus of cheliped with single small spine on inner margin (a);
      central notch in posterior margin of telson indistinct (b) ....
.... *Petrolisthes cinctipes* (p. 232)

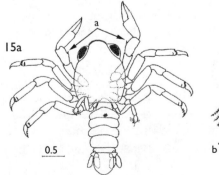
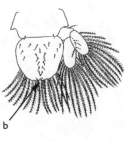

15b. Carpus of cheliped with 2–3 small spines on inner margin (a); central notch in posterior margin of telson distinct (b) ....
.... *Petrolisthes eriomerus* (p. 232)

16a. Carpus of cheliped with 2–3 prominent spines on inner margin ....*Pachycheles pubescens* (p. 232)

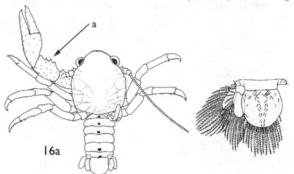

16b. Carpus of cheliped with single prominent spine on inner margin .... *Pachycheles rudis* (p. 232)

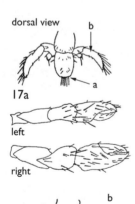

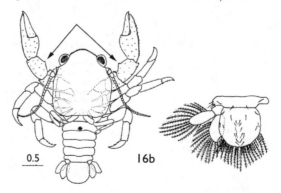

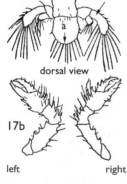

17a. Telson posterior margin (a) with 6 setae; right uropod (b) smaller than left; right cheliped larger than left, chelae smooth, moderately broad, hairy; carapace ca 1.6 mm long, with 1.4 mm wide; total length ca 3.4 mm...... Paguridae, *Pagurus granosimanus*
17b. Telson posterior margin (a) with 10 setae; right uropod (b) slightly larger than left; chelipeds of equal size, narrow, heavily toothed, hairy; antennal scale scimitar-shaped; carapace ca 1.2 mm long, with 1.0 mm wide; total length ca 3.4 mm ....
.... Diogenidae, *Paguristes turgidus*

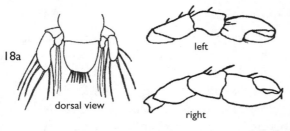

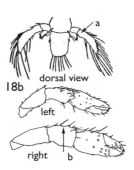

17c. Telson posterior margin with 8 setae 18

18a. Telson, uropods, and chelipeds comparatively small and reduced; chelae smooth with few hairs and no teeth; carapace ca 1.3 mm long, with 1.0 mm wide; total length ca 2.4 mm ....
.... Paguridae, *Pagurus hirsutiusculus*

18b. Uropods stout and unequal in size, with right slightly smaller (a); antennae long reaching beyond chelipeds; chelipeds stout with 4–5 spines on inner margin of right carpus (b) ....
.... Paguridae, *Pagurus ochotensis*

18c. Uropods nearly equal in size; chelae broad with rounded tips and serrated margin 19

19a. Chela and carpus margins heavily toothed (a); carapace ca 1.84 mm long, with 1.8 mm wide; total length ca 4.4 mm ....
.... Paguridae, *Orthopagurus schmitti*

19b. Chela and carpus margins not as heavily toothed as above(a); carapace ca 1.5 mm long, with 1.4 mm wide; total length ca 3.2 mm .... Paguridae, *Pagurus beringanus*

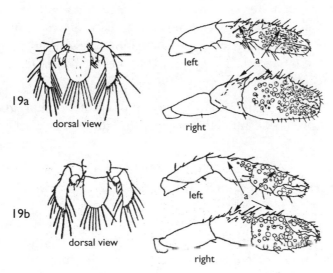

Family Hippolytidae

Eualus suckleyi, **Shortscale Eualid.** Of the local species of *Eualus*, only the larvae of *E. suckleyi* have been described. *Eualus suckleyi* can reach a length of 79 mm as an adult. Its range extends from the Chukchi and Bering Seas to Grays Harbor, Washington. Adults are found subtidally at 11–1,025 m (Jensen, 1995). Haynes collected females carrying eggs from late April to early May. Although Haynes describes only two larval stages, he suggests that they molt through five to nine zoeal stages. Information on the length of the larval period was not found. Sizes below are from laboratory-reared larvae. Larvae described in Haynes (1981b).

Fig. 6. *Eualus suckleyi.*
(From Haynes, 1981b)

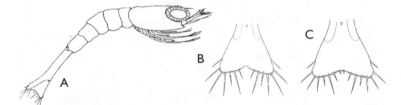

Zoea I: Length from tip of rostrum to posterior margin of telson, 3.0–3.5 mm (Fig. 6A). Rostrum thin and without teeth, about half length of antennae. First, second, and third maxilliped pairs with four, five, and five swimming setae on exopods, respectively. Abdomen with five segments and triangular telson. Abdominal segment 5 with pair of posteriolateral spines. Posterior margin of telson with seven setae on each half (Fig. 6B). Uropods visible under thin membrane.

Zoea II: Length from tip of rostrum to posterior margin of telson, 3.5–4.2 mm. Telson now with eight setae on each half (Fig. 6C).

Lebbeus groenlandicus, **Spiny Lebbeid.** Of the species in the genus *Lebbeus* found off the coast of the Pacific Northwest, only the larvae of *L. groenlandicus* have been described. *Lebbeus groenlandicus* grows to a length of 40 mm as an adult. Its range extends from the Bering Sea to the Puget Sound. Adults are found on shell mixed with sand or gravel from the low intertidal to 518 m (Jensen, 1995). They molt through two zoeal stages and one megalopa. Information on the timing and length of larval period was not found. Size approximations below are from larvae reared *in situ* in Kachemak Bay, Alaska. Larvae described in Haynes (1978a).

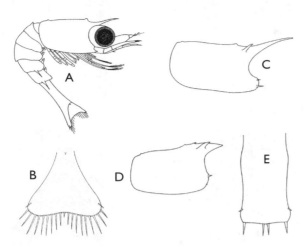

Fig. 7. *Lebbeus groenlandicus.* (From Haynes, 1978a)

Zoea I: Length from tip of rostrum to posterior margin of telson, 6.4–7.4 mm (Fig. 7A). Rostrum thin and pointed, ca half length of antenna. First, second, and third maxilliped pairs with four, five, and five swimming setae on exopods, respectively. Abdomen with five segments and triangular telson. Abdominal segments four and five with posterio-lateral spines. Posterior margin of telson with 19–21 setae (Fig. 7B). Uropods visible under thin membrane. Pleopod buds present on abdominal segments 1–5.

Zoea II: Length from tip of rostrum to posterior margin of telson, 8.1–8.7 mm. Carapace with pair of spines lateral to base of rostrum (Fig. 7C). First, second, and third maxilliped pairs with five, sixteen, and sixteen swimming setae on exopods, respectively. First and second pereopod pairs chelate. Pleopods slightly more developed.

Megalopa: Length from tip of rostrum to posterior margin of telson, 7.4–7.6 mm. Rostrum shorter, with single small medial spine at base (Fig. 7D). Telson rectangular with 1 pair of spines on posteriolateral margin and 2 pairs of spines on posterior margin (Fig. 7E).

Family Pandalidae

Pandalopsis dispar, **Sidestriped Shrimp.** *Pandalopsis dispar* can reach a length of 208 mm as an adult. Its range extends from the Pribilof Islands to Manhattan Beach state park, Oregon. Adults can be found on soft bottoms at 46–649 m (Jensen, 1995). In the descriptions below, stage I zoeae were from laboratory-reared larvae and stages II, III, and V(?) zoeae were collected from plankton samples. Larvae described in Berkeley (1930).

Zoea I: Length from tip of rostrum to posterior margin of telson, 10 mm (Fig. 8A). Rostrum slopes upward and has

Fig. 8. *Pandalopsis dispar.* (From Berkeley, 1930)

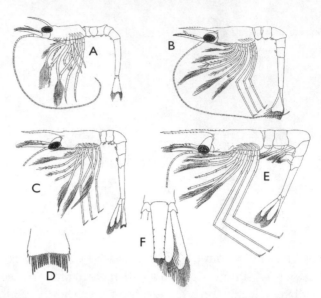

five dorsal teeth and three ventral teeth. Antennal flagellum longer than whole body. Abdomen composed of six unequal segments and triangular telson. Five pairs of pleopod buds present on ventral side of abdomen. Telson posterior margin has twelve pairs of setae. First, second, and third maxilliped pairs with setae. Third, fourth, and fifth pereopod pairs without exopods.

Zoea II: Length from tip of rostrum to posterior margin of telson, 13 mm (Fig. 8B). Rostrum now has six dorsal teeth (as well as three ventral teeth) on anterior half and additional three to four teeth at base. Carapace now has pair of supraorbital spines. All thoracic appendages with swimming setae on exopods, and second pereopod pair endopods chelate.

Zoea III: Length from tip of rostrum to posterior margin of telson, 16 mm (Fig. 8C). Rostrum has eight dorsal teeth between eyes and tip, one tooth near tip, and five to six small ventral teeth (in addition to three or four teeth near base). Pleopods distinctly biramous but lack setae. Uropods free, and telson more rectangular, with eight pairs of setae on posterior margin and two pairs on lateral margins (Fig. 8D).

Zoea V(?): There is at least one zoeal stage between this stage and the third zoea, but there may be more. Length from tip of rostrum to posterior margin of telson, 30 mm (Fig. 8E). Rostrum with total 15 dorsal teeth and 10 ventral teeth. Pleopods with long setae. Telson now narrower at base, with seven pairs of setae on posterior margin and six pairs of lateral spines (Fig. 8F).

Pandalus eous (borealis). **Alaskan Pink Shrimp.**
Larvae described in Berkeley (1930).

Zoea I: Length from tip of rostrum to posterior margin of telson, 5 mm (Fig. 9A, B). Rostrum slender, extending two-thirds way up antennal scale. Abdomen composed of five segments and triangular telson. Telson posterior margin with seven pairs of setae. Three maxilliped pairs have swimming setae on exopods, and pereopods poorly developed.

Zoea II: Length from tip of rostrum to posterior margin of telson, 7 mm (Fig. 9C). Carapace has pair of supraorbital spines. First, second, and third pereopod pairs with swimming setae on exopods. Small pleopod buds present on ventral side of abdomen.

Zoea III: Length from tip of rostrum to posterior margin of telson, 8–9 mm (Fig. 9D). Rostrum has two small teeth at base. Abdomen now composed of six segments and telson. Telson posterior margin now with eight pairs of setae, and uropods free (Fig. 9E).

Zoea IV: Length from tip of rostrum to posterior margin of telson, 9–10 mm (Fig. 9F). Rostrum now with four small teeth near base. Pleopods biramous. Telson rectangular, with five pairs of setae on posterior margin and three pairs of lateral spines (Fig. 9G).

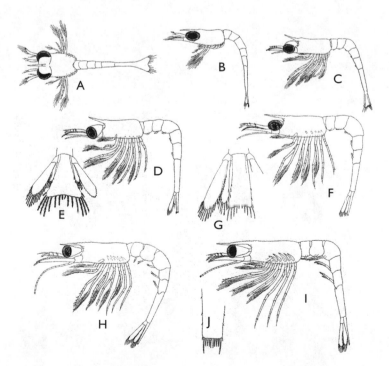

Fig. 9. *Pandalus eous (borealis)*. (From Berkeley, 1930)

Zoea V: Length from tip of rostrum to posterior margin of telson, 14 mm (Fig. 9H). Rostrum now with 11–12 dorsal teeth. second pereopod pair chelate.

Zoea VI: Rostrum with 14 teeth and one near tip (Fig 9I). Pleopods with setae. Telson (Fig. 9J) now with five pairs of setae on posterior margin and four pairs of lateral spines.

Pandalus danae, **Dock Shrimp**. *Pandalus danae* can reach a length of 140 mm as an adult. Its range extends from the Alaskan Peninsula to Bahía de San Quintín, Baja California. During the day, adults are most commonly found from the intertidal to 185 m on mixed-composition bottoms hidden under algae or in crevices. At night in marinas they come out of hiding and can be found on pilings (Jensen, 1995). Larvae described in Berkeley (1930).

Zoea I: Length from tip of rostrum to posterior margin of telson, 6 mm (Fig. 10A). Rostrum thin and extending two-thirds way up antennal scale. Abdomen composed of six segments and triangular telson. Telson posterior margin with seven pairs of setae (Fig. 10B). First, second, and third maxilliped pairs with swimming setae on exopods.

Zoea II: Length from tip of rostrum to posterior margin of telson, 8 mm (Fig. 10C). Carapace now with pair of supraorbital spines. Telson posterior margin now with eight pairs of setae (Fig. 10D). First and second pereopod pairs with swimming setae on exopods.

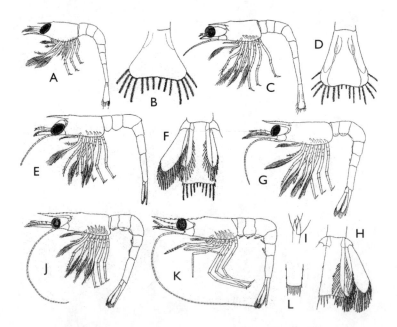

Fig. 10. *Pandalus danae.*
(From Berkeley, 1930)

Zoea III: Length from tip of rostrum to posterior margin of telson, 9 mm (Fig. 10E). Telson rectangular, with five pairs of setae on posterior margin and three pairs of lateral spines (Fig. 10F). Uropods free and setae bare.

Zoea IV: Length from tip of rostrum to posterior margin of telson, 12 mm (Fig. 10G, H). Rostrum has eight to ten dorsal teeth. Second pereopod pair chelate (Fig. 10I). Small pleopod buds present on ventral side of abdomen.

Zoea V: Length from tip of rostrum to posterior margin of telson, 14 mm (Fig. 10J). Rostrum with additional four or five ventral teeth.

Megalopa: Length from tip of rostrum to posterior margin of telson, 17 mm (Fig. 10K). Carapace has lost supraorbital spines. Telson narrows posteriorly (Fig. 10L). Telson posterior margin with spine, three fine hairs, and three pairs of small lateral spines. Pleopods biramous. Maxillipeds and pereopods lack exopods.

Pandalus jordani, **Pacific Ocean Shrimp.** *Pandalus jordan* averages about 100 mm in length but can reach 140 mm. Its range extends from Unalaska, Alaska, to at least as far south as San Diego, California. Adults are found mainly at 45–370 m depth (Morris et al., 1980). Off the coast of Crescent City,

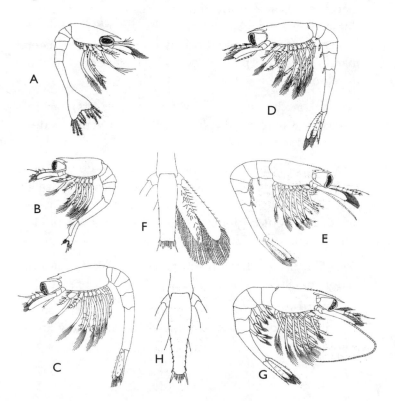

Fig. 11. *Pandalus jordani.* (From Modin and Cox, 1967)

California, spawning occurs in November and December. Hatching occurs from late February through mid-May. Modin and Cox (1967) report *P. jordani* molting through 11–13 stages in about 80 days. Sizes below are from laboratory-reared larvae. Larvae described in Modin and Cox (1967).

Zoea I: Length from tip of antennal scale to posterior margin of telson, 5 mm (Fig. 11A). Rostrum thin and pointed, ca one-third length of carapace. Abdomen composed of five segments and triangular telson. Posterior margin of telson with seven pairs of setae. First, second, and third maxilliped pairs with swimming setae on exopods. Pereopods poorly developed.

Zoea II: Length from tip of antennal scale to posterior margin of telson, 6.5 mm (Fig. 11B). Carapace with pair of supraorbital spines. First and second pereopod pairs now with swimming setae on exopods. Abdomen now composed of six segments and telson. Posterior margin of telson with additional pair of setae. Uropods free.

Zoea III: Length from tip of antennal scale to posterior margin of telson, 7 mm. Third pereopod pair now with swimming setae on exopods.

Zoea IV: Length from tip of antennal scale to posterior margin of telson, 7.5 mm (Fig. 11C). Rostrum with two small dorsal teeth near base. Telson becoming rectangular. Posterior margin of telson with five pairs of setae and lateral margins with three pairs of spines. Uropods larger and more developed.

Zoea V: Length from tip of antennal scale to posterior margin of telson, 8 mm. Small pleopod buds on ventral side of abdominal segments 2–5.

Zoea VI: Length from tip of antennal scale to posterior margin of telson, 9.5 mm (Fig. 11D). Rostrum now with eight or nine dorsal teeth. Antennal scale and flagellum equal in length. Pleopods beginning to show signs of segmentation. Endopod of second pereopod chelate.

Zoea VII: Length from tip of antennal scale to posterior margin of telson, 11 mm.

Zoea VIII: Length from tip of antennal scale to posterior margin of telson, 12 mm. Antennal flagellum now longer than scale.

Zoea IX: Length from tip of antennal scale to posterior margin of telson, 13 mm (Fig. 11E, F). Endopod of first pereopod now cheliform.

Zoea X: Length from tip of antennal scale to posterior margin of telson, 14.5 mm.

Zoea XI: Length from tip of antennal scale to posterior margin of telson, 17 mm (Fig. G). Pleopods with setae. Telson (Fig. H) narrows posteriorly. Telson with seven pairs of small spines on lateral margin and five pairs of setae on posterior margin.

Pandalus platyceros, **Spot Shrimp.** *Pandalus platyceros* can reach a length of 253 mm as an adult. Its range extends from Unalaska, Alaska, to San Diego, California. Adults can be found from the intertidal zone to 487 m on rocky bottoms and vertical rock faces (Jensen, 1995). Females carrying eggs were collected in Dabob Bay, Washington, in early January. *Pandalus platyceros* molts through four zoeal stages and one megalopa. In the laboratory it took ca 35 days from hatching through the molt to the megalopa. Sizes below from laboratory-reared larvae. Larvae described in Price and Chew (1972).

Zoea I: Length from tip of rostrum to posterior margin of telson, 8.1 mm (Fig. 12A, B). Rostrum thin, extending just beyond tips of antennal scale. Rostrum with 12–13 dorsal

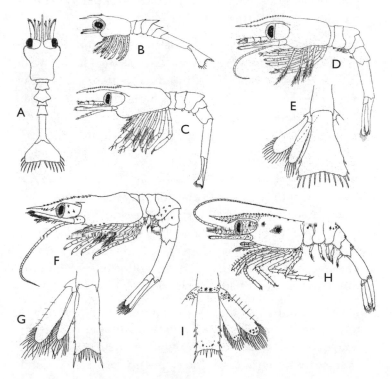

Fig. 12. *Pandalus platyceros.* (From Price and Chew, 1972)

teeth. Along outer edge of carapace, 23–25 fine denticles. Posterior margins of abdominal segments also with fine denticles. Abdomen composed of five segments unequal in size and triangular telson. Posterior margin of telson with eight pairs of setae. First, second, and third maxilliped pairs with swimming setae. Endopod of second pereopod pair chelate. Five pairs of small pleopod buds visible on ventral side of abdomen.

Zoea II: Length from tip of rostrum to posterior margin of telson, 10 mm (Fig. 12C). Denticles on carapace and abdominal segments fewer and finer. Carapace also with pair of antennal spines, pterygostomian spines, and supraorbital spines. Rostrum now with additional three to four ventral teeth. First, second, and third pereopod pairs with swimming setae on exopods. Fourth and fifth pereopod pairs lack exopods. Pleopods beginning to show signs of segmentation.

Zoea III: Length from tip of rostrum to posterior margin of telson, 11 mm (Fig. 12D). Fine denticles now gone from carapace and abdominal segments. Pleopods now biramous. Uropods free (Fig. 12E).

Zoea IV: Length from tip of rostrum to posterior margin of telson, 11.5–12 mm (Fig. 12F). Telson rectangular, with five pairs of setae on posterior margin and three pairs of spines on lateral margins (Fig. 12G).

Megalopa: Length from tip of rostrum to posterior margin of telson, 12–13 mm (Fig. 12H). Carapace no longer with supraorbital spines. Maxillipeds and pereopods now resemble adult form. Pleopods well developed with setae. Telson posterior margin with spine, two fine hairs, and another spine (Fig. 12I). Telson lateral margins with four pairs of small spines.

Pandalus stenolepis, **Rough Patch Shrimp.** *Pandalus stenolepis* can reach a length of 82 mm as an adult. Its range extends from Unalaska, Alaska, to Heceta Bank, Oregon. Adults can be found from 18 to 229 m on mud or cobble bottoms (Jensen, 1995). Descriptions of zoeae I and II are based on laboratory-reared larvae and zoeae III–VII on larvae collected in plankton samples. Larvae described in Needler (1938).

Zoea I: Length from tip of rostrum to posterior margin of telson, 5 mm (Fig 13A, B). Carapace margins and posterior margins of abdominal segments with fine denticles. Rostrum ca same length as carapace and slopes upward. First, second,

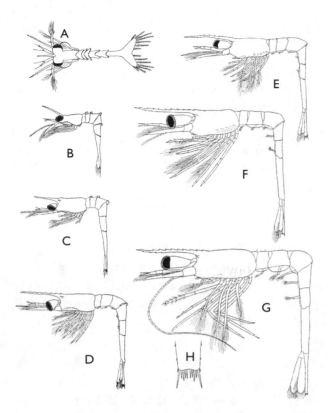

Fig. 13. *Pandalus stenolepis.* (From Needler, 1938)

and third maxilliped pairs with swimming setae on exopods. Pereopods poorly developed. Abdomen composed of five segments and triangular telson. Posterior margin of telson with seven pairs of setae.

Zoea II: Length from tip of rostrum to posterior margin of telson, 6 mm (Fig. 13C). Carapace now with pair of supraorbital spines. Rostrum with four to five dorsal teeth. Telson posterior margin with additional pair of setae.

Zoea III: Length from tip of rostrum to posterior margin of telson, 8 mm (Fig. 13D). Rostrum with eight to nine dorsal teeth and two ventral teeth. First, second, and third pereopod pairs with swimming setae on exopods. Small pleopod buds present on ventral side of abdomen. Uropods free, and telson margin with five posterior pairs of setae and three small pairs of lateral spines.

Zoea IV: Length from tip of rostrum to posterior margin of telson, 9 mm (Fig. 13E). Rostrum with 10–12 dorsal teeth and four ventral teeth. Pleopods beginning to show signs of segmentation.

Zoea V: Length from tip of rostrum to posterior margin of telson, 12 mm (Fig. 13F). Denticles on carapace and abdominal segments reduced considerably. Endopods of second pereopod pair chelate. Pleopods clearly biramous. Telson rectangular.

Zoea VI: Length from tip of rostrum to posterior margin of telson, 14 mm (Fig. 13G,H). Denticles on carapace and abdominal segments gone.

Megalopa: Length from tip of rostrum to posterior margin of telson not much longer than in zoea VI. Carapace no longer with pair of supraorbital spines. Rostrum with 11 dorsal teeth, six ventral teeth, and trifid tip (tip divided in three). Maxillipeds and pereopods resemble adult form. Pleopods now have setae.

Pandalus tridens, **Yellow Leg Pandalid.** *Pandalus tridens* can reach a length of 123 mm as an adult. Its range extends from the western Bering Sea to San Nicolas Island, California. Adults can be found at 5–1,984 m on rocky or muddy substrate (Jensen, 1995). Larvae described in Haynes (1980).

Zoea I: Length from tip of rostrum to posterior margin of telson, 3.1–3.5 mm (Fig. 14A). Rostrum thin, without teeth, and ca two-thirds length of carapace. Very small pterygostomian spines present but usually hidden by eyes. Abdomen composed of five segments and triangular telson.

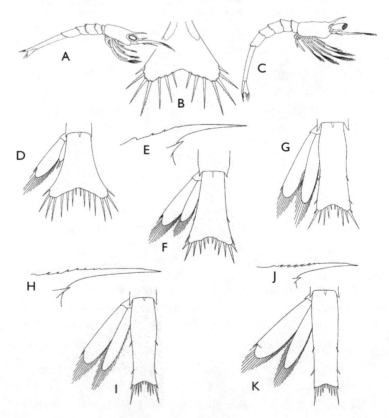

Fig. 14. *Pandalus tridens.*
(From Haynes, 1980)

Fine denticles present on posterior margins of abdominal segments. Telson posterior margin has seven pairs of setae (Fig. 14B). First, second, and third maxilliped pairs with swimming setae on exopods. Pereopods poorly developed and lack setae.

Zoea II: Length from tip of rostrum to posterior margin of telson, 3.9–4.6 mm (Fig. 14C). Rostrum now projects dorsally. Carapace with pair of supraorbital spines, antennal spines, and clearly visible pterygostomian spines. Abdomen now composed of six segments and telson. Telson posterior margin with eight pairs of setae. First and second pereopod pairs have swimming setae on exopods.

Zoea III: Length from tip of rostrum to posterior margin of telson, 5.6–6.3 mm. Denticles along posterior margin of abdominal segments reduced. Third pereopod pair now with swimming setae on exopods. Uropods free (Fig. 14D).

Zoea IV: Length from tip of rostrum to posterior margin of telson, 7.0–8.4 mm. Rostrum projects horizontally and has two teeth near base (Fig. 14E). Pleopods present as small bumps on ventral side of abdominal segments. Telson now with six pairs of setae on posterior margin and two pairs on lateral margins (Fig. 14F).

Zoea V: Pleopod buds longer. Telson narrower with well-developed uropods (Fig. 14G).

Zoea VI: Length from tip of rostrum to posterior margin of telson, 10.2–11.2 mm. Rostrum with six dorsal teeth and ca one-third length of carapace (Fig. 14H). Second pereopod pair chelate. Pleopods biramous. Telson nearly rectangular (Fig. 14I).

Zoea VII: Length from tip of rostrum to posterior margin of telson, 13.0 mm. Rostrum with seven dorsal teeth and almost half length of carapace (Fig. 14J). Telson lateral margins nearly parallel (Fig. 14K).

Family Callianassidae

Neotrypaea (Callianassa) californiensis, **Bay Ghost Shrimp.** *Neotrypaea californiensis* can reach a length of 120 mm as an adult. Its range extends from Mutiny Bay, Alaska, to Estero Punta Banda, Baja California. Adults can be found burrowing in bays and estuaries in the middle to low intertidal zone in sand and muddy sand, where they build multibranching tunnels (Jensen, 1995). In Yaquina Bay, Oregon, the main breeding season occurs in late spring and early summer.

Fig. 15. *Neotrypaea californiensis.*(From McCrow, 1972)

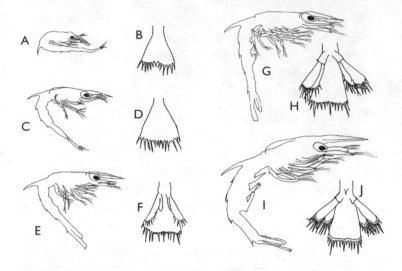

Neotrypaea californiensis molts through five zoeal stages. The length of the larval period is about six weeks. Sizes below are from larvae taken in plankton samples. Larvae described in McCrow (1972).

Zoea I: Length from tip of rostrum to posterior margin of telson, 2.8–3.6 mm (Fig. 15A). Rostrum long and cylindrical. Prominent spine on dorsal side of second abdominal segment. Telson triangular. Posterior margin of telson has five setae, one small fine hair, and small spine on each half extending from medial tooth outward (Fig. 15B).

Zoea II: Length from tip of rostrum to posterior margin of telson, 3.8–4.4 mm (Fig. 15C). Rostrum now dorsoventrally flattened with serrated margin. Posterior margin of telson with additional pair of spines added on each side of medial tooth (Fig. 15D).

Zoea III: Length from tip of rostrum to posterior margin of telson, 4.7–5.2 mm (Fig. 15E). Telson now with uropods (Fig. 15F).

Zoea IV: Length from tip of rostrum to posterior margin of telson, 5.5–6.3 mm (Fig. 15G,H). Three pairs of pleopod buds on ventral side of abdominal segments 3–5.

Zoea V: Length from tip of rostrum to posterior margin of telson, 6.8–7.5 mm (Fig. #15). Pleopods long and slender. Uropods well developed (Fig. 15J).

Family Upogebiidae

Upogebia pugettensis, **Blue Mud Shrimp.** *Upogebia pugettensis* can reach 150 mm in length as an adult. Its range extends from

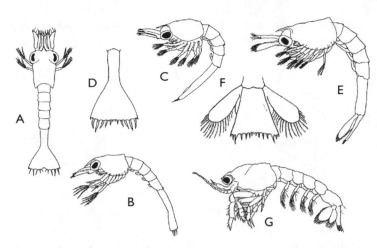

Fig. 16. *Upogebia pugettensis*. (From Hart, 1937)

Valdez Narrows, Alaska, to Morro Bay, California. Adults can be found burrowing in low intertidal mud or muddy sand, where they build Y-shaped burrows (Jensen, 1995). At Friday Harbor, Washington, females carrying eggs were found in December through February and in late spring/early summer. *Upogebia pugettensis* molts through three zoeal stages and one megalopa. Larvae are present in the plankton from February through June, but information on the length of their larval period was not found. Sizes below are from laboratory-reared larvae. Larvae described in Hart (1937).

Zoea I: Length from tip of rostrum to posterior margin of telson, 3.7 mm (Fig. 16A, B). Rostrum small and pointed, ca one-third length of antennules. Four swimming setae on exopods of first and second maxilliped pairs. Abdomen composed of five segments and triangular telson. Posterior margin of telson slightly indented in center. Telson posterior margin with five setae, one small, fine hair, and small spine on each half extending from center outward (Fig. 16A). Four pairs of small buds on ventral side of abdomen eventually develop into pleopods.

Zoea II: Length from tip of rostrum to posterior margin of telson, 4.4 mm (Fig. 16C). Six swimming setae on exopods of first, second, and third maxilliped pairs. First and second pereopod pairs also with six swimming setae on exopods, and third pereopod pair with five swimming setae on exopods. Telson now with additional pair of setae and small median tooth on posterior margin (Fig. 16D). Uropods of telson visible enclosed under membrane. Pleopods buds larger.

Zoea III: Length from tip of rostrum to posterior margin of telson, 5.4 mm (Fig. 16E). Six setae on exopods of first maxilliped pair and third pereopod pair, rest of appendages with seven. Abdomen now composed of six segments and telson. Telson has four setae, two spines, small fine hair, and another spine flanking each side of median tooth extending from center outward (Fig. 16F). Uropods now free, and pleopod buds long and slender.

Megalopa: Length from tip of rostrum to posterior margin of telson, 4.0 mm (Fig. 16G). Rostrum blunt in appearance. Maxillipeds and pereopods now resemble adult forms. Propodus (second-to-last segment) of first and fifth pairs of pereopods with small tooth on distal, dorsal side. Telson more rectangular, with large rounded uropods on each side, with long plumose setae on distal edges. Pleopods, now with long setae, present on abdomen segments 2–5.

Family Diogenidae

Paguristes turgidus, **Hermit Crab.** *Paguristes turgidus* can grow to a carapace length of 32 mm and is usually found occupying shells of *Fusitriton oregonensis*, the Oregon triton. Its range extends from the Chukchi Sea to San Diego, California. Adults are found subtidal on muddy sand at 5–465 m (Jensen, 1995). *Paguristes turgidus* molts through three non-feeding zoeal stages and one megalopa. Females are ovigerous November through September, and hatching occurs August through September. The length of the larval period is reported as being three to four times faster than with pagurid zoeae (Strathmann 1987). Sizes below are from laboratory-reared larvae. Larvae described in Hart (1937).

Fig. 17. *Paguristes turgidus.* (From Hart, 1937)

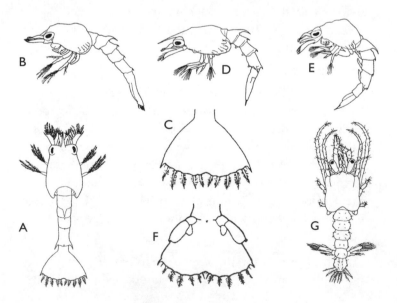

Zoea I: Length from tip of rostrum to posterior margin of telson, 4.5 mm (Fig. 17A, B). Rostrum narrow, with length almost equal to antennules. Exopods of first and second maxilliped pairs have four swimming setae. Abdomen composed of five segments and triangular telson. Second abdominal segment with large spine on dorsal, distal margin. Abdominal segments 3–5 with similar smaller spines. Fifth abdominal segment with additional pair of posteriolateral spines. Posterior margin of telson has five setae, one small fine hair, and small spine on each half extending from center outward (Fig. 17C).

Zoea II: Length from tip of rostrum to posterior margin of telson, 4.5 mm (Fig. 17D). Exopods of first and third maxilliped pairs have seven swimming setae, and exopods of second maxilliped pair with eight swimming setae. Abdomen now composed of six segments and telson. Telson has additional pair of medial setae on posterior margin. Uropods visible enclosed under thin membrane on ventral side of telson. Four pairs of small buds on ventral side of abdomen eventually develop into pleopods.

Zoea III: Length from tip of rostrum to posterior margin of telson, 4.4 mm (Fig. 17E). Exopods of first and second maxilliped pairs have eight swimming setae, and third maxilliped pair with seven setae on exopods. Telson with additional pair of medial setae on posterior margin (Fig. 17F). Uropods now free, and pleopod buds long and slender.

Megalopa: Length from tip of rostrum to posterior margin of telson, 3.0 mm (Fig. 17G). Rostrum small, almost blunt. Maxillipeds and pereopods now resemble adult forms. First pair of pereopods chelate and unequal in size. Second and third pereopods long and thin. Fourth and fifth pereopods small and armed with suctorial setae, fifth terminally chelate. Telson hexagonal, bearing two lateral spines and 10 setae on posterior margin. Pleopods with long setae present on abdominal segments 2–5.

Family Hippidae

Emerita analoga, **Pacific Sand Crab.** *Emerita analoga* can reach 40 mm in length. Its range extends from Kodiak, Alaska, to Chile. Adults are found on sandy beaches in the surf zone, moving up and down with the tide (Jensen, 1995). According to MacGinitie (1938), peak mating season is in May and June, which agrees with Johnson and Lewis's (1942) observation that the majority of the larvae are hatched in July and August.

Fig. 18. *Emerita analoga.*
(From Johnson and
Lewis, 1942)

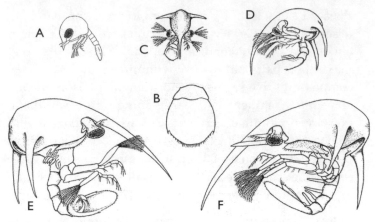

Emerita analoga molts through five zoeal stages and one megalopa. Information on the length of the larval period was not found. Sizes below are from larvae taken in plankton samples. Larval description in Johnson and Lewis (1942).

Zoea I: Carapace ca 0.53 mm wide by 0.70 mm long (Fig 18A). Rostrum short, stout, and pointed. No lateral spines present exopods of first and second maxilliped pairs have four swimming setae. Abdomen composed of four segments. Telson with rounded posterior margin having 26 spines (Fig. 18B); this number remains constant throughout remainder of zoeal stages.

Zoea II: Carapace ca 0.8 mm wide by 1.0 mm long (Fig. 18C). Rostrum longer and thinner. Two posteriolateral spines now present, almost as long as rostrum. First and second maxilliped pairs now have six swimming setae on exopods.

Zoea III: Carapace ca 1.3 mm wide by 1.6 mm long (Fig. 18D). Rostrum and posteriolateral spines longer. First and second maxilliped pairs with eight swimming setae on exopods. Telson now with uropod rudiments on ventral side.

Zoea IV: Carapace ca 2.0 mm wide by 2.4 mm long (Fig. 18E). First and second maxilliped pairs with 16 swimming setae on exopods. Abdomen now composed of five segments. Pereopod buds visible posterior to maxillipeds. Uropods on telson more developed.

Zoea V: Carapace ca 2.6 mm wide by 3.5 mm long (Fig. 18F). Number of swimming setae on exopods of first and second maxilliped pairs still 16. Pereopods longer and more slender with signs of segmentation. Uropods well developed, with endopod three-fourths length of exopod. Pleopod buds on ventral side of abdomen long and slender.

Family Lithodidae

Cryptolithodes typicus, **Butterfly Crab, Turtle Crab.** *Cryptolithodes typicus* can reach a carapace width of 80 mm as an adult. Its range extends from Amchitka Island, Alaska, to Santa Rosa Island, California. Adults are found from the low intertidal to 45 m (Jensen, 1995). *Cryptolithodes typicus* molts through four zoeal stages and one megalopa. The larval period from hatching to (but not including) the first juvenile zoea was 24 days in the laboratory. Sizes below are from laboratory-reared larvae. Larvae described in Hart (1965).

Zoea I: Length from tip of rostrum to posterior margin of telson, 3.0 mm (Fig. 19A, B). Rostrum wide at base, narrowing to point, extending almost to tips of antenna. First and second maxilliped pairs with four swimming setae on exopods. Abdomen composed of five segments and long rectangular telson. Abdominal segments 2–5 with posterio-lateral spines and two small "teeth" on the posterior margin. Telson posterior margin with five setae, one small, fine hair, and another setae on each half extending from center outward (Fig. 19C).

Zoea II: Length from tip of rostrum to posterior margin of telson, 3.25 mm (Fig. 19D,E). First, second, and third maxilliped pairs with eight swimming setae on exopods. Pleopod buds may be present as small bumps.

Zoea III: Length from tip of rostrum to posterior margin of telson, 3.4 mm (Fig. 19F, G). Pleopods present as small buds. Otherwise resembles previous zoeal stages.

Zoea IV: Length from tip of rostrum to posterior margin of telson, 3.6 mm (Fig. 19H, I). Pleopod buds longer and larger. Otherwise resembles previous zoeal stages.

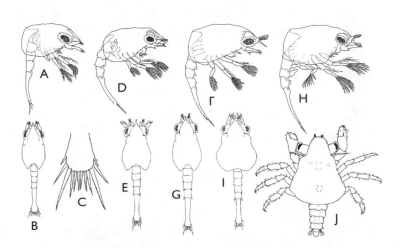

Fig. 19. *Cryptolithodes typicus.* (From Hart, 1965)

Megalopa: Length from tip of rostrum to posterior margin of telson, 2.8 mm (Fig. 19J). Rostrum blunt; carapace triangular. Maxillipeds and pereopods now resemble adult forms. First pereopod pair chelate, with right larger than left. Second, third, and fourth pereopod pairs long and stout. Fifth pereopod pair smaller and hidden under carapace. Posterior margin of telson bare of setae. Uropods and pleopods well developed, with long plumose setae.

Rhinolithodes wossnessenskii, **Rhinoceros Crab.** *Rhinolithodes wossnessenskii* can reach a carapace width of 64 mm as an adult. Its range extends from Kodiak, Alaska, to Crescent City, California. Adults are often found in crevices on rock or gravel bottoms at 6 –73 m (Jensen, 1995). Haynes collected females bearing eggs in March near Auke Bay, Alaska. Although he was able to describe only two zoeal stages, it is likely that *R. wossnessenskii* molts through four zoeal stages and one megalopa, as do morphologically similar lithodid species. Sizes below are from laboratory-reared larvae. Larvae described in Haynes (1984).

Zoea I: Size from tip of rostrum to posterior margin of telson, ca 4.45 mm (Fig. 20A, B). Rostrum long and pointed, with length almost equal to antennae. Carapace with spine at mid-dorsal posterior margin and two posteriolateral spines. First and second maxilliped pairs with four swimming setae on exopods. Abdomen composed of five segments and rectangular telson (Fig. 20C). Abdominal segments 2–5 with small teeth on dorsal posterior margin. Abdominal segment 5 with large pair of posteriolateral spines. Telson posterior margin with small spine, five setae, one small fine hair, and small spine on each half, extending from center outward.

Zoea II: Size from tip of rostrum to posterior margin of telson, ca 4.81 mm. First and second maxilliped pairs with seven swimming setae on exopods, and third maxilliped pair with six swimming setae on exopods. Other characteristics nearly identical to zoea I.

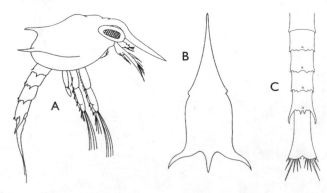

Fig. 20. *Rhinolithodes wossnessenskii.* (From Haynes, 1984)

Family Paguridae

The Paguridae are hermit crabs. Adults can range in size from as small as 6 mm carapace length up to 46 mm carapace length. Most are found intertidally, although some species are found at depths of 600 m or deeper.

Pagurid zoeae are morphologically very similar. The best way to distinguish them is by chromatophore pattern. Chromatophore patterns for several types of pagurid zoea are presented in Fig. 21. Chromatophores are, however, lost during preservation. For those trying to identify preserved pagurids, less extensive keys to four zoeal stages and megalopae based on preserved samples are presented here. There are no additional species descriptions, but the pagurid keys are illustrated by Figs. 22–30.

Key to pagurid zoeal stages (from Lough, 1975)

1a. Uropods absent; telson and sixth abdominal segment fused ... 2
1b. Uropods present; telson and sixth abdominal segment
    articulated 3

2a. Telson spines usually 6 + 6; 4 swimming setae on maxillipeds;
    eyes fixed .... Zoea I, Key A
2b. Telson spines usually 7 + 7; .4 swimming setae on maxillipeds;
    eyes movable .... Zoea II, Key B

3a. Pleopods absent, or small buds .... Zoea III, Key C
3b. Pleopods present .... Zoea IV, Key D

A. Key to pagurid zoeae I

1a. Carapace posterior processes short, end in small point or hook
    2
1b. Carapace posterior processes more elongate, end in distinct
    point 5

2a. Antennal exopodite usually not as broad and shorter than
    rostrum; telson shape more triangular 3
2b. Antennal exopodite markedly broad but more tapered, equal
    or slightly less in length than rostrum; telson shape more
    rectangular 4

3a. Telson with broad base; total length, 2.68 mm ....
    .... (Fig. 25) *Orthopagurus schmitti*
3b. Telson base not as broad; total size, 2.20 mm; ....
    .... (Fig. 26) *Pagurus beringanus*

4a. Telson base squarer, with small notch; total size, 2.28 mm ....
    .... (Fig. 22) *Pagurus granosimanus*
4b. Telson base with distinct notch separating convex halves; total
    size, 2.20 mm .... (Fig. 29) *Pagurus hirsutiusculus*

5a. Antennal endopodite with distinct double-pronged spine at apical end; antennal exopodite with distinct short spines; carapace with prominent mid-ridge; total size, 5.36 mm .... (Fig. 27) *Pagurus tanneri*

5b. Antennal endopodite and exopodite not as above 6

6a. Telson base narrow with notch separating convex halves; antennal exopodite slender; carapace posterior processes comparatively short; total size, 2.96 mm .... (Fig. 23) *Pagurus ochotensis*

6b. Telson broader at base with distinct notch separating more straight-edged base; antennal exopodite broader, carapace posterior processes more elongate; total size, 3.36 mm .... (Fig. 24) *Pagurus* sp. C

6c. Carapace posterior processes elongate with widely diverging edges, almost lithodid-like appearance; largest pagurid, total size, 6.7 mm .... (Fig. 28) *Pagurus* sp. I

Fig 21. Chromatophore patterns in pagurid zoea. Hatched circles, red chromatophores; Empty circles, yellow chromatophores; Stippled circles, orange chromatophores; Open hatching, diffuse blue pigment. (From Nyblade, 1974)

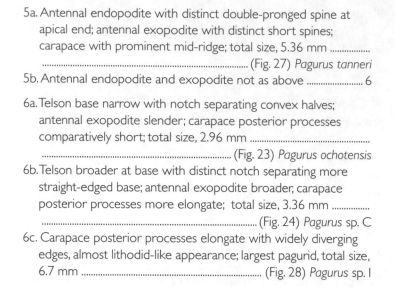

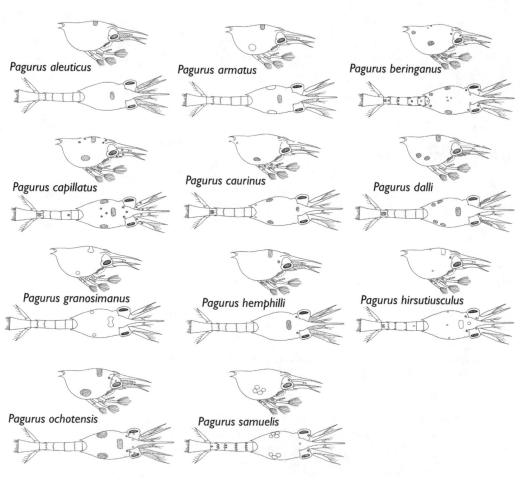

Pagurus aleuticus

Pagurus armatus

Pagurus beringanus

Pagurus capillatus

Pagurus caurinus

Pagurus dalli

Pagurus granosimanus

Pagurus hemphilli

Pagurus hirsutiusculus

Pagurus ochotensis

Pagurus samuelis

B. Key to pagurid zoeae II

1a. Antennal exopodite broad, shorter than rostrum 2
1b. Antennal exopodite tapered, more elongate, length as long as
 rostrum 3

2a. Telson with broad base; total length, 3.64 mm ....
 .... (Fig. 25) *Pagurus schmitti*
2b. Telson base not as broad; total size, 2.64 mm ....
 .... (Fig. 26) *Pagurus beringanus*

3a. Carapace posterior processes short, end in small point or hook
 4
3b. Carapace posterior processes more elongate, end in distinct
 point 5

4a. Telson base square, no notch, with 6 + 6 spines, fourth spine
 longest; total size, 2.6 mm .... (Fig. 22) *Pagurus granosimanus*
4b. Telson base with notch separating convex halves, usual 7 + 7
 spines with fifth spine longest; total size, 2.6 mm ....
 .... (Fig. 29) *Pagurus hirsutiusculus*

5a. Antennal endopodite with distinct double-pronged spine at
 apical end; antennal exopodite with distinct short spines;
 carapace with prominent mid-ridge; total size, 7.2 mm ....
 .... (Fig. 28) *Pagurus tanneri*
5b. Antennal exopodite and endopodite not as above 6

6a. Telson base with notch separating convex halves ....
 .... (Fig. 23) *Pagurus ochotensis*
6b. Telson base broader and concave, no notch 7

7a. Carapace posterior processes elongate, with widely diverging
 edges; almost lithodid-like appearance; total size, 7.6 mm ....
 .... (Fig. 28) *Pagurus* sp. I
7b. Carapace posterior processes not as elongate, edges parallel;
 total size, 4.8 mm .... (Fig. 24) *Pagurus* sp. C

C. Key to pagurid zoeae III

1a. 2 primary setae on uropod inner margin; uropods narrow 2
1b. 2 primary setae on uropod inner margin; uropod broader 3

2a. Telson posterior margin with 5 + 5 spines, fourth spine longest;
 no additional secondary setae on uropod inner margin; total
 length, 1.12 mm .... (Fig. 22) *Pagurus granosimanus*
2b. Telson posterior margin with usual 7 + 7 spines, fifth spine
 longest; 1 additional small secondary seta on inner margin of
 uropods; total size, 3.68 .... (Fig. 29) *Pagurus hirsutiusculus*

3a. Telson posterior margin square; antennal exopodite broad;
 carapace posterior processes short; stout appearance 4
3b. Telson posterior margin concave; antennal exopodite elongate;
 carapace posterior processes elongate; slender appearance 5

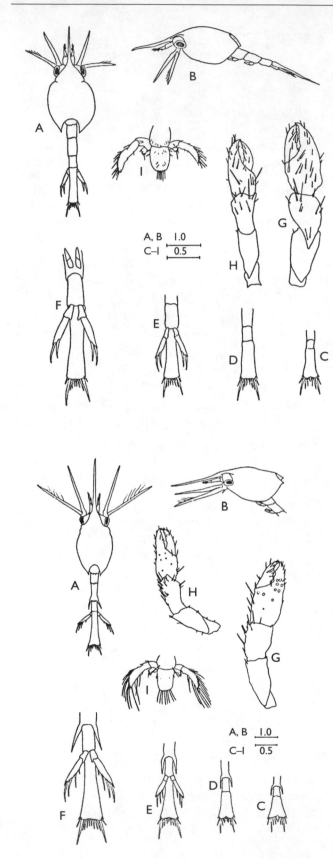

Fig 22. *Pagurus granosimanus*. (A, B) Zoea IV, generalized. (C–F) Telson (ventral view), zoea I–IV, respectively. (G–I) Right and left cheliped, telson of megalopa (dorsal view). Scales in millimeters. (From Lough, 1975, Fig. 4)

Fig 23. *Pagurus ochotensis*. (A, B) Zoea IV, generalized. (C–F) Telson (ventral view), zoea I–IV, respectively. (G–I) Right and left cheliped, telson of megalopa (dorsal view). Scales in millimeters. (From Lough, 1975, Fig. 5)

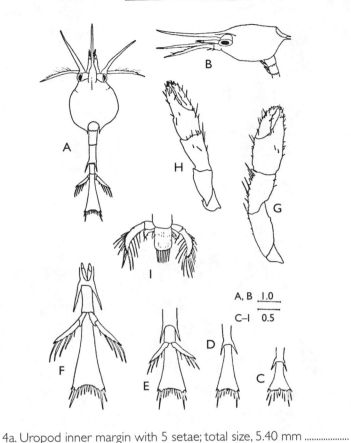

Fig 24. *Pagurus* sp. C. (A, B) Zoea IV, generalized. (C–F) Telson (ventral view), zoea I–IV, respectively. (G–I) Right and left cheliped, telson of megalopa (dorsal view). Scales in millimeters. (From Lough, 1975, Fig. 6)

4a. Uropod inner margin with 5 setae; total size, 5.40 mm ....
.... (Fig. 24) *Orthopagurus schmitti*
4b. Uropod inner margin with 3 setae; total size, 3.28 mm ....
.... (Fig. 26) *Pagurus beringanus*

5a. Antennal endopodite with distinct double-pronged spine at apical end; antennal exopodite with distinct short spines; uropod inner margin with 5 setae; carapace with prominent mid-ridges; total size, 8.96 mm .... (Fig. 27) *Pagurus tanneri*
5b. Antennal exopodite and endopodite not as above 6

6a. Uropod inner margin with 5 long setae; carapace posterior processes elongate, with widely diverging edges; almost lithodid-like in appearance; largest species, total size, 9.60 mm ....
.... (Fig. 28) *Pagurus* sp. I
6b. Uropod inner margin with 3 setae; carapace posterior processes comparatively short with small central notch; total size, 5.60 mm .... (Fig. 23) *Pagurus ochotensis*
6c. Uropod inner margin with 4 setae, fourth seta weakly developed; carapace posterior processes intermediate in length; total size, 6.08 mm.. (Fig. 24) *Pagurus* sp. C

D. Key to pagurid zoeae IV

1a. 2 primary setae on uropod inner margin, uropods narrow 2
1b. 2 primary setae on uropod inner margin, uropods broader 3

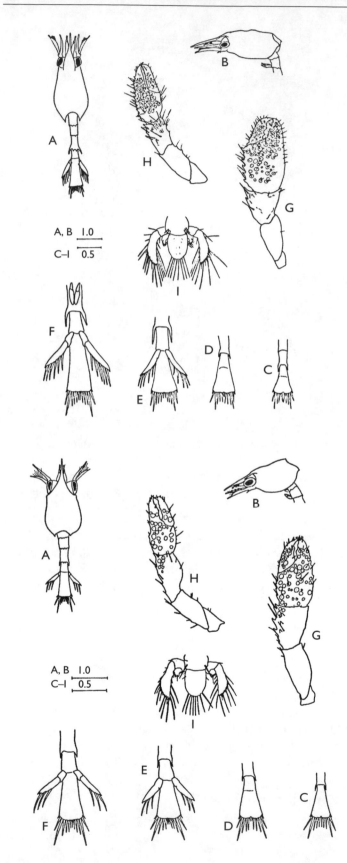

Fig 25. *Orthopagurus schmitti*. (A, B) Zoea IV, generalized. (C–F) Telson (ventral view), zoea I–IV, respectively. (G–I) Right and left cheliped, telson of megalopa (dorsal view). Scales in millimeters. (From Lough, 1975, Fig. 7)

Fig 26. *Pagurus beringanus*. (A, B) Zoea IV, generalized. (C–F) Telson (ventral view), zoea I–IV, respectively. (G–I) Right and left cheliped, telson of megalopa (dorsal view). Scales in millimeters. (From Lough, 1975, Fig. 8)

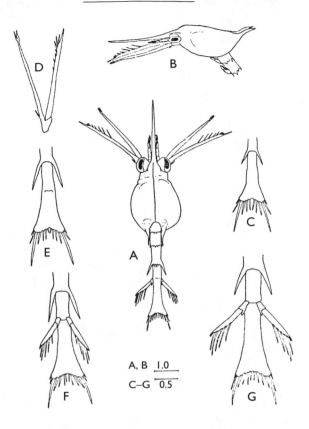

Fig 27. *Pagurus tanneri*. (A, B) Zoea IV, generalized. (C–F) Telson (ventral view), zoea I–IV, respectively. (G–I) Right and left cheliped, telson of megalopa (dorsal view). Scales in millimeters. (From Lough, 1975, Fig. 9)

2a. Telson posterior margin with 5 + 5 spines, fourth spine longest; no additional secondary setae on uropod inner margin; total length, 5.60 mm .... (Fig. 22) *Pagurus granosimanus*

2b. Telson posterior margin with usual 7 + 7 spines, fifth spine longest; 2 small secondary setae on inner margin of uropods; total size, 4.08 mm .... Fig. 29) *Pagurus hirsutiusculus*

3a. Telson posterior margin square; antennal exopodite broad; carapace posterior processes short; stout appearance 4

3b. Telson posterior margin concave; antennal exopodite elongate; carapace posterior processes elongate; slender appearance 6

4a. Uropod inner margin with 6 setae; total size, 6.80 mm ....
.... (Fig. 25) *Orthopagurus schmitti*

4b. Uropod inner margin with 3 setae; smaller size 5

5a. Lateral spines on fifth abdominal segment short; antennae broader; total size, 3.92 .... (Fig. 26) *Pagurus beringanus*

5b. Lateral spines on fifth abdominal segment elongate, extending to sixth segment; antennae more slender; total size, 5.32 mm .....
.... (Fig. 30) *Pagurus* sp. J

6a. Antennal endopodite with distinct double-pronged spine at apical end; antennal exopodite with distinct short spines; uropod inner margin with 5 setae; carapace with prominent mid-ridge; total size, 9.36 mm .... (Fig. 27) *Pagurus tanneri*

6b. Antennal exopodite and endopodite not as above 7

Fig 28. *Pagurus* sp. I. (A, B) Zoea IV, generalized. (C–F) Telson (ventral view), zoea I–IV, respectively. Scales in millimeters. (From Lough, 1975, Fig. 10).

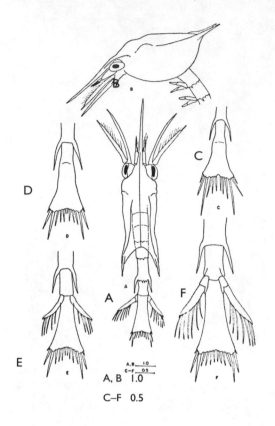

Fig. 29. *Pagurus hirsutiusculus.* (A, B) Zoea IV, generalized. (C–F) Telson (ventral view), zoea I–IV, respectively. (G–I) Right and left cheliped, telson of megalopa (dorsal view). Scales in millimeters. (From Lough, 1975, Fig. 11)

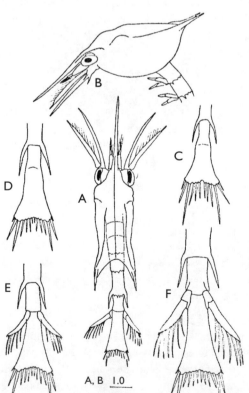

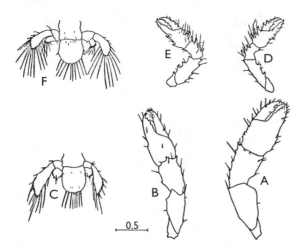

Fig. 30. (A–C) Right and left cheliped, telson of *Pagurus* sp. J. megalopa (dorsal view). (D–F) Right and left cheliped, telson of *Pagurus turgidus* megalopa (dorsal view). Scales in millimeters. (From Lough, 1975, Fig 12)

7a. Uropod inner margin with 7 long setae; carapace posterior processes elongate, with widely .... (Fig. 28) *Pagurus* sp. I

7b. Uropod inner margin with 4–5 setae 8

8a. Uropod inner margin with 4 setae, fourth seta weakly developed; carapace posterior processes comparatively short with small central notch; total size, 8.88 mm .... (Fig. 23) *Pagurus ochotensis*

8b. Uropod inner margin with 4–5 setae, fifth seta, if present, weakly developed; carapace posterior processes more elongate; telson posterior margin more convex; total size, 8.96 mm .... (Fig. 24) *Pagurus* sp. C

Key to pagurid megalopae

1a. Telson posterior margin with 6 setae; right uropod smaller then left; left cheliped smaller than right, chela smooth, moderately broad and hairy; carapace 1.64 mm long by 1.40 mm wide; total length, 3.40 mm .... (Fig. 22) *Pagurus granosimanus*

1b. Telson posterior margin with 10 setae; left uropod slightly smaller than right; both chelipeds ca equal size, narrow, heavily toothed and hairy; antennal scale scimitar-shaped; carapace 1.16 mm long by 1.00 mm wide; total length, 2.36 mm .... (Fig. 30) *Paguristes turgidus*

1c. Telson posterior margin with 8 setae 2

2a. Telson and uropods comparatively small and reduced; comparatively small chelipeds with few hairs, no teeth, and smooth surface; carapace 1.32 mm long by 1.00 mm wide; total length, 2.36 mm .... (Fig. 29) *Pagurus hirsutiusculus*

2b. Not as above 3

3a. Uropods ca equal in size; telson posterior margin rounded; chela broad with blunt tips, serrated margins 4

3b. Uropods unequal, left slightly larger; telson more rectangular; chela long and narrow with pointed tips, margins not serrated 5

4a. Carapace, 1.84 mm long by 1.80 mm wide, total length, 4.36 mm; antennal scale with 6 spines; chela and carpus margins heavily toothed .... (Fig. 25) *Orthopagurus schmitti*

4b. Carapace, 1.52 mm long by 1.36 mm wide, total length, 3.12 mm; antennal scale with 4 spines: chela and carpus margins not as heavily toothed .... (Fig. 26) *Pagurus beringanus*

5a. Antennae short, not reaching beyond chelipeds; antennal scale with 7 spines; carpus of right cheliped inner margin with several small teeth and short spines; telson with rounded posterior margin; carapace, 1.60 mm long by 1.20 mm wide; total size, 3.28 mm .... (Fig. 30) *Pagurus* sp. J

5b. Antennae long, reaching beyond chelipeds 6

6a. Carapace, 2.40 mm long by 2.28 mm wide, total size, 5.20 mm; telson more rectangular; uropods more elongate with narrower tips; chelipeds longer and narrower, inner margin of right carpus with ca 6 distinct teeth and several long spines; antennal scale with 9 spines .... (Fig. 24) *Pagurus* sp. C

6b. Carapace, 1.88 mm long by 1.64 mm wide, total size, 4.40 mm; telson posterior margin more rounded; uropods stouter, tips broader; chelipeds shorter and stouter, inner margin of right carpus with 4-5 moderate-sized spines; antennal scale with spines .... (Fig. 23) *Pagurus ochotensis*

Family Porcellanidae

Key to porcellanid zoeae (illustrations from Gonor, 1970)

1a. Zoea without pleopods on abdomen, or just visible beneath cuticle; all setae on telson paired .... Zoea I, 2

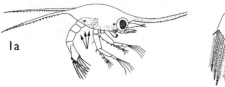

1b. Zoea with pleopods (a) on abdomen, 1 unpaired medial seta (b) on telson .... Zoea II, 5

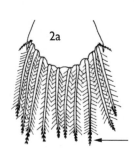

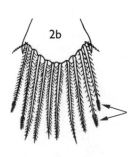

2a. All 5 pairs of major telson setae with terminal brushes of heavy spines .... *Petrolisthes*, 3

2b. Only outer 2 pairs of major telson setae with terminal brushes of heavy spines .... *Pachychele*, 4

3a. I distal seta on the inner margin of segment I of endopod on
   maxilliped I .... *Petrolisthes cinctipes*

3b. 3 distal setae on the inner margin of segment I of endopod on
   maxilliped I .... *Petrolisthes eriomerus*

3a

3b

4a. 9 setae on endopod (a) of maxilla 2; 4 setae on distal lobe of
   coxal endite (b) of maxilla .... *Pachycheles rudis*

4b. 8 setae on endopod (a) of maxilla 2; 6 setae on distal lobe of
   coxal endite (b) of maxilla 2 .... *Pachycheles pubescens*

5a. All 5 pairs of major telson setae with terminal brushes of heavy
   spines .... *Petrolisthes*, 6

5b. Only outer 2 pairs of major telson setae with terminal brushes
   of heavy spines .... *Pachychele*, 7

4a

6a. I distal seta on outer margin (a), 2 distal setae on inner margin
   (b) of segment I of endopod on maxilliped 2 ....
   .... *Petrolisthes cinctipes*

6b. No setae on outer margin (a), 2 distal setae on inner margin
   (b) of segment I of endopod on maxilliped 2 ....
   .... *Petrolisthes eriomerus*

4b

7a. Distal seta present on outer margin of segment I of endopod
   on maxilliped I and 2 .... *Pachycheles pubescens*

7b. Not as above .... *Pachycheles rudis*

5a

6a

6b

7a

7b

5b

Pachycheles pubescens, **Pubescent Porcelain Crab.** *Pachycheles pubescens* typically grows to a carapace width of 22 mm as an adult. Its range extends from the Queen Charlotte Islands to Cabo Thurloe, Baja California. Adults are found from the low intertidal to 55 m under rocks on the open coast or in inshore waters in areas with strong currents (Jensen, 1995). Females bearing eggs were collected off the central Oregon coast in May and June. *Pachycheles pubescens* molts through two zoeal stages and one megalopa. Information on the timing of hatching and length of the larval period were not found. Larvae described in Gonor and Gonor (1973a).

Pachycheles rudis, **Thick-clawed Porcelain Crab.** *Pachycheles rudis* typically grows to a carapace width of 19 mm as an adult. Its range extends from Kodiak, Alaska, to Bahía Magdalena, Baja California. Adults are found from the low intertidal to 29 m under rocks or nestled in holes (Jensen, 1995). Females bearing eggs were collected off the central Oregon coast in May and June. *Pachycheles rudis* molts through two zoeal stages and one megalopa. Information on the timing of hatching and length of the larval period were not found. Larvae described in Gonor and Gonor (1973a).

Petrolisthes cinctipes, **Flat Porcelain Crab.** *Petrolisthes cinctipes* typically grows to a carapace width of 24 mm as an adult. Its range extends from Porcher Island, British Columbia, to Santa Barbara, California. Adults are found in the upper and middle intertidal under rocks on or near the open coast (Jensen, 1995). Females bearing eggs were collected off the central Oregon coast in May and June. *Petrolisthes cinctipes* molts through two zoeal stages and one megalopa. Information on the timing of hatching and length of the larval period were not found. Larvae described in Gonor and Gonor (1973a).

Petrolisthes eriomerus, **Flattop Crab.** *Petrolisthes eriomerus* typically grows to a carapace width of 19 mm as an adult. Its range extends from Chicagof Island, Alaska, to La Jolla, California. Adults are found from the low intertidal to 86 m under rocks on the open coast or in sheltered waters (Jensen, 1995). Females bearing eggs were collected off the central Oregon coast in May and June. *Petrolisthes eriomerus* molts through two zoeal stages and one megalopa. Brood laying occurs February through mid-April, hatching May through mid-August. Second broods are produced mid-May through August, hatching August through early October. Larvae described in Gonor and Gonor (1973a).

Family Cancridae

Zoeae of the genus *Cancer* are morphologically very similar. Below is a general description of the five zoeal stages and megalopa applicable to all local *Cancer* species (Fig. 31). Species details follow.

Zoea I (Fig. 31A): Carapace with thin pointed rostrum ca as long as dorsal spine and pair of smaller lateral spines. First and second maxilliped pairs with four swimming setae on exopods. Abdomen composed of five segments and forked telson (Fig. 31B). Second abdominal segment with small lateral knobs. Each abdominal segment with small posterior lateral extensions that lengthen in successive stages. Each furca of telson with three inner setae and 1–2 exospines.

Zoea II (Fig. 31C): First and second maxilliped pairs with six swimming setae on exopods. Each furca of telson with three or four inner setae.

Zoea III (Fig. 31D): First and second maxilliped pairs with eight swimming setae on exopods. Each furca of the telson now with four or five inner setae.

Zoea IV (Fig. 31E): First and second maxilliped pairs with 10 swimming setae on exopods. Pleopod buds on abdominal segments 2–5 and uropod buds on segment 6. Each furca of telson now with five inner setae (Fig. 31F).

Zoea V (Fig. 31G): First and second maxilliped pairs with 11 and 12 swimming setae on exopods, respectively. Pleopod buds on abdominal segments increased in size.

Megalopa (illustrated in Key B, the megalopa key): Carapace with prominent rostrum and dorsal spine. Maxillipeds and

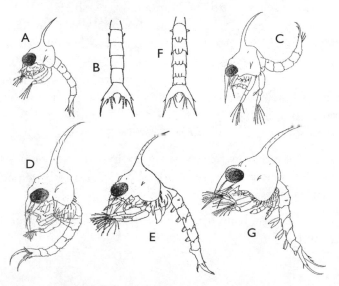

Fig. 31. *Cancer* spp. (From Ally, 1975)

pereopods now resemble adult form. First pereopod pair chelate, pereopod pairs 2–5 walking legs. Three long setae on dactyl (last segment) of fifth pereopod pair.

Cancer antennarius, **Pacific Rock Crab.** *Cancer antennarius* females usually grow to a carapace width of 148 mm; males reach a width of 178 mm. The range extends from Queen Charlotte Sound, British Columbia, to Cabo San Lucas, Mexico. Adults are found in the low intertidal under large rocks and subtidally to 91 m on gravel bottoms and in kelp beds (Jensen, 1995). Females bearing eggs were collected from South Humboldt Bay, California, in April. Information on the timing and length of larval period was not found. Sizes below are from laboratory-reared larvae. Larvae described in Roesijadi (1976).

Zoea I: Length from tip of telson to tip of rostrum, 1.8 mm. Telson with two pairs of lateral exospines.

Zoea II: Length from tip of telson to tip of rostrum, 2.0 mm.

Zoea III: Length from tip of telson to tip of rostrum, 2.3 mm. Each furca of telson now with four inner setae.

Zoea IV: Length from tip of telson to tip of rostrum, 3.1 mm. Each furca of telson now with five inner setae.

Zoea V: Length from tip of telson to tip of rostrum, 4.4 mm.

Megalopa: Carapace 2.3–3.3 mm long, 1.4–2.4 mm wide. Single stout spine on ischiopodite of cheliped. *Cancer antennarius* megalopae similar in appearance to those of *C. gracilis;* former slightly stouter.

Cancer gracilis, **Graceful Crab.** *Cancer gracilis* females typically grow to a carapace width of 87 mm; males reach 115 mm. The range extends from Prince William Sound, Alaska, to Bahía Playa María, Mexico. Adults are found primarily subtidally to 143 m on sand or mud (Jensen, 1995). Females carrying eggs were collected from San Pedro Bay, California, in September. The eggs hatched in late September and reached the first juvenile instar 42–47 days later. Sizes below are from laboratory-reared larvae. Larvae described in Ally (1975).

Zoea I: Length from tip of rostrum to tip of dorsal spine, 1.1 mm. Telson with one pair of dorsal exospines and one pair of lateral exospines.

Zoea II: Length from tip of rostrum to tip of dorsal spine, 1.5 mm.

Zoea III: Length from tip of rostrum to tip of dorsal spine, 1.9 mm. Each furca of telson now with four inner setae.

Zoea IV: Length from tip of rostrum to tip of dorsal spine, 2.5 mm. Each furca of telson now with five inner setae.

Zoea V: Length from tip of rostrum to tip of dorsal spine, 3.3 mm.

Megalopa: Carapace 2.3–3.3 mm long, 1.4–2.4 mm wide. *Cancer gracilis* megalopae similar in appearance to those of *C. antennarius*; former with more slender appearance.

Cancer magister, **Dungeness Crab.** *Cancer magister* typically grows to a carapace width of 190–230 mm (rarely 330 mm). Its range extends from the Pribilof Islands to Santa Barbara, California. Adults are most commonly found subtidally on sandy bottoms and in eelgrass beds (Jensen, 1995). Strathmann (1987) reports egg deposition October through December, but mostly November and December, in Wasington and Oregon. Hatching occurs January through early March, but mostly January and February, off northern California and Oregon. The larval period is 80–160 days. Sizes below are from laboratory-reared larvae. Larvae described in Poole (1966); lengths for zoea II and zoea IV not provided.

Zoea I: Length from tip of rostrum to tip of telson, 2.5 mm. Telson with one pair of dorsal exospines and one pair of lateral exospines.

Zoea II: Each furca of telson now with four inner setae.

Zoea III: Length from tip of rostrum to tip of telson, 4 mm. Each furca of telson now with five inner setae.

Zoea V: Length from tip of rostrum to tip of telson, 9 mm.

Megalopa: *Cancer magister* is the largest cancrid. Carapace length from tip of rostrum to posterior margin of carapace, 5.3–6.6 mm; width at widest point, 3.5–4.6 mm.

Cancer oregonensis, **Pygmy Rock Crab.** *Cancer oregonensis* typically grows to a carapace width of 53 mm. Its range extends from the Pribilof Islands to Palos Verdes, California. Adults are found in the low intertidal and subtidal to 436 m (Jensen, 1995), often under rocks and in holes, crevices, and kelp holdfasts (Strathmann, 1987). In southern Puget Sound, Washington, ovigerous females are found November through March. Information on the length of the larval period was not found. Sizes below are from larvae taken in plankton samples. Larvae illustrated in Lough (1975).

Zoea I: Length from tip of rostrum to tip of dorsal spine, 1.64 mm. Telson with two pairs of lateral exospines.

Zoea II: Length from tip of rostrum to tip of dorsal spine, 2.6 mm.

Zoea III: Length from tip of rostrum to tip of dorsal spine, 3.36 mm. Each furca of telson now with four inner setae.

Zoea IV: Length from tip of rostrum to tip of dorsal spine, 4.48 mm. Each furca of telson now with five inner setae.

Zoea V: Length from tip of rostrum to tip of dorsal spine, 5.28 mm.

Megalopa: Carapace length from tip of rostrum to posterior margin of carapace ca 3.4 mm; width at widest point, ca 2.2 mm. *Cancer oregonensis* is similar in appearance and size to *C. productus*.

Cancer productus, **Red Rock Crab.** *Cancer productus* females grow to a carapace width of 158 mm; males reach 200 mm. The range extends from Kodiak, Alaska, to Isla San Martín, Baja California. Adults are found from the middle intertidal to 79 m, most commonly in protected boulder beaches or gravelly bottoms (Jensen, 1995). Females in the laboratory produced egg masses between December and April. Information on the length of the larval period was not found. Sizes below are from laboratory-reared larvae. Larvae described in Trask (1970).

Zoea I : Length from tip of rostrum to tip of telson, 2.5 mm. Telson with one pair of lateral exospines.

Zoea II : Length from tip of rostrum to tip of telson, 3.0 mm.

Zoea III : Length from tip of rostrum to tip of telson, 3.5 mm. Each furca of telson now with five inner setae.

Zoea IV : Length from tip of rostrum to tip of telson, 4.0 mm.

Zoea V: Length from tip of rostrum to tip of telson, 5.5 mm.

Megalopa: Carapace length from tip of rostrum to posterior margin of carapace, 3.4–3.6 mm; width at widest point, 2.0–2.1 mm. *Cancer productus* is similar in appearance and size to *C. oregonensis*.

Family Grapsidae
Hemigrapsus zoeae are very similar in appearance. General descriptions of the five zoeal stages and the megalopa are given below and followed by species descriptions.

Zoea I: (Fig. 32A) Carapace with rostrum and dorsal spine of ca equal length, and smaller pair of lateral spines. Rostrum ca one-third longer than antennae. First and second maxilliped pairs with four swimming setae on exopods. Abdomen composed of five segments and telson. Second segment with pair of lateral knobs (in *H. nudus*, third segment with pair of smaller lateral knobs). Each furca of telson with three inner setae. No exospines present on telson.

Zoea II (Fig 32B): Rostrum now two-thirds longer than antennae. First and second maxilliped pairs with six swimming setae on exopods. Abdomen now composed of six segments.

Zoea III (Fig. 32C): First and second maxilliped pairs with eight swimming setae on exopods. Each furca of telson now with four inner setae.

Zoea IV (Fig. 32D): First and second maxilliped pairs with 10 swimming setae on exopods. Pleopod buds on abdominal segments 2–5, and uropod buds on segment 6. Each furca of telson now with five inner setae.

Zoea V (Fig. 32E): First and second maxilliped pairs with 12 swimming setae on exopods. Pleopod buds on abdominal segments 2–5 and uropod buds on segment 6 now larger.

Megalopa (illustrated in Key B, the megalopa key): Carapace smooth, with small rostrum bent downward. Maxillipeds and pereopods now resemble adult form. First pereopod pair chelate, pereopod pairs 2–5 walking legs. Three long setae present on dactyl (last segment) of fifth pereopod.

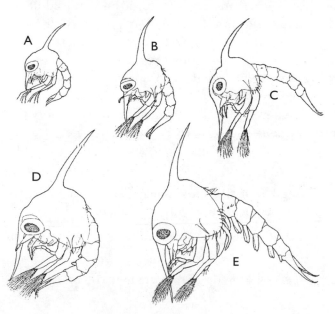

Fig. 32. *Hemigrapsus* spp.
(From Hart, 1935)

Hemigrapsus nudus, **Purple Shore Crab.** *Hemigrapsus nudus* females grow to a carapace width of 34 mm; males typically reach 56 mm. The range extends from Yakobi Island, Alaska, to Bahía de Tortuga, Mexico. Adults are found among rocks in the high and middle intertidal on exposed coasts (Jensen, 1995). Around Departure Bay, British Columbia, females carrying eggs were found by Hart (1935) in April and May. Females in berry have been reported as late as June and July at Friday Harbor, Washington. Information on the length of the larval period was not found. Sizes below are from laboratory-reared larvae. Larvae described in Hart (1935).

Zoea I: Length from tip of rostrum to tip of dorsal spine, 1.2 mm. Second segment with pair of lateral knobs, and third segment with pair of smaller lateral knobs.

Zoea II: Length from tip of rostrum to tip of dorsal spine, 1.6 mm.

Zoea III: Length from tip of rostrum to tip of dorsal spine, 2.2 mm.

Zoea IV: Length from tip of rostrum to tip of dorsal spine, 2.8 mm.

Zoea V: Length from tip of rostrum to tip of dorsal spine, 3.5 mm.

Megalopa: Carapace length front to back, 1.8 mm; width at widest point, 1.5 mm. Posterior margin of telson with short setae (other than uropod setae).

Hemigrapsus oregonensis, **Yellow Shore Crab.** *Hemigrapsus oregonensi* can reach a carapace width of 50 mm as an adult. Its range extends from Resurrection Bay, Alaska, to Baja California. Adults are found in mudbank burrows in estuaries and throughout the intertidal among rocks on mud or gravel bottoms (Jensen, 1995). Females carrying eggs were collected near Vancouver in March. At Departure Bay, British Columbia, eggs hatch from mid-May until August. In the laboratory the first instar juvenile was reached in four to five weeks. Sizes below are from laboratory-reared larvae. Larvae described in (Hart 1935).

Zoea I: Length from tip of rostrum to tip of dorsal spine, 1.1 mm. Second segment only with pair of lateral knobs.

Zoea II: Length from tip of rostrum to tip of dorsal spine, 1.6 mm.

Zoea III: Length from tip of rostrum to tip of dorsal spine, 2.0 mm.

Zoea IV: Length from tip of rostrum to tip of dorsal spine, 2.5 mm.

Zoea V: Length from tip of rostrum to tip of dorsal spine, 2.5 mm.

Megalopa: Carapace length front to back, 1.7 mm; width at widest point, 1.2 mm. Carapace narrower than in *H. nudus*. Posterior margin of telson without setae (other than uropod setae).

Pachygrapsus crassipes, **Striped or Lined Shore Crab.**
Pachygrapsus crassipes reaches a carapace width of 48 mm as an adult. Its range extends from Ecola State Park, Oregon, to the Gulf of California. Adults are found in the upper and middle intertidal of rocky shores and also in estuaries (Jensen, 1995). Females carrying eggs were collected from Seal Beach, California, from May through November. *Pachygrapsus crassipes* molts through five zoeal stages (there may be a sixth and seventh zoeal stages, Claudio DiBacco pers. comm.) and one megalopa. In the laboratory the larval period from hatching through the fifth zoea was 95 days. Sizes below are from laboratory-reared larvae. Larvae described in Schlotterbeck (1976).

Zoea I: Length from tip of rostrum to tip of dorsal spine, 1.0 mm (Fig. 33A). Carapace with rostral and dorsal spines only, small lateral spines not present until zoea II. Rostral spine equal in length to antennae. First and second maxilliped pairs with four swimming setae on exopods. Abdomen composed of six unequal segments and forked telson (Fig. 33B). Abdominal segments 1–5 with small posterior lateral

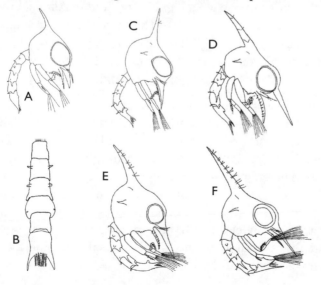

Fig. 33. *Pachygrapsus crassipes*. (From Schlotterbeck, 1976)

extensions which lengthen in successive stages. Second and third abdominal segments with lateral knobs. Each furca of telson has three inner setae. No exospines present on telson, but knob projecting from ventral surface of telson.

Zoea II: Length from tip of rostrum to tip of dorsal spine, 1.2 mm (Fig. 33C). Carapace now with small lateral spines. Antennae now two-thirds length of rostrum. First and second maxilliped pairs with six swimming setae.

Zoea III: Length from tip of rostrum to tip of dorsal spine, 1.5 mm (Fig. 33D). First and second maxilliped pairs with eight swimming setae.

Zoea IV: Length from tip of rostrum to tip of dorsal spine, 1.75 mm (Fig. 33E). First and second maxilliped pairs with 10 swimming setae.

Zoea V: Length from tip of rostrum to tip of dorsal spine, 2.5 mm (Fig. 33F). First and second maxilliped pairs with 12 swimming setae.

Megalopa: Carapace length front to back, 4.1 mm; width at widest point, 3 mm (size from field samples). Second-largest megalopa found off the coast of Oregon.

Family Majidae

Chionoecetes bairdi, **Tanner Crab.** *Chionoecetes bairdi* females typically grow to a carapace width of 81 mm; males can reach 140 mm. The range extends from the Bering Sea to Winchester Bay, Oregon. Adults are found at 6–474 m on sand or mud bottoms (Jensen, 1995). *Chionoecetes bairdi* molts through two zoeal stages and one megalopa. In the Gulf of Alaska, broods are produced April and May and hatch ca 11 months later. The larval period is ca 90 days (Strathmann, 1987). Sizes below are from larvae taken in plankton samples. Larvae described in Haynes (1973, 1981a) and Jewett and Haight (1977).

Zoea I: Length from tip of rostrum to tip of dorsal spine, 4.0–4.5 mm (Fig. 34A). Carapace with relatively short lateral spines. Rostrum long, thin, and spinulate. First and second maxilliped pairs with four swimming setae on exopods. Abdomen composed of five segments and forked telson (Fig. 34B). Abdominal segments 2 and 3 with small lateral spines; abdominal segments 3–5 with long posteriolateral spines. Each furca of telson with three inner setae, lateral exospine, smaller dorsal exospine, and a dorsal exospine located between previous two.

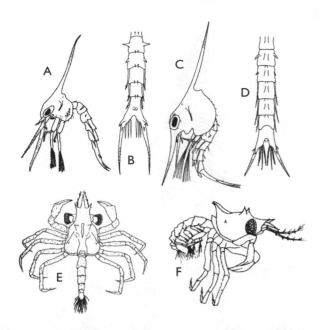

Fig. 34. *Chionoecetes
bairdi.* (From Haynes,
1973, 1981a; Jewett and
Haight, 1977)

Zoea II: Length from tip of rostrum to tip of dorsal spine,
6.0–6.4 mm (Fig. 34C). First and second maxilliped pairs with
six swimming setae on exopods. Abdomen now composed
of six segments, and placement of spines on segments
unchanged. Spine lengths increased (Fig. 34D). Each half of
telson with additional fine seta added medially. Pleopod
buds now present on the ventral side of abdominal segments
2–5.

Megalopa: Carapace length 3.1–3.5 mm; width at widest
point, 1.8–2.1 mm. Carapace with pair of anteriolateral
spines, pair of mid-dorsal spines, small pair of lateral spines,
and set of posterior spines (Fig. 34E, F). Maxillipeds and
pereopods now resemble adult form. First pereopod pair
chelate, pereopod pairs 2–5 walking legs.

Oregonia gracilis, **Graceful Decorator Crab.** *Oregonia gracilis*
reaches 39 mm carapace width as an adult. Its range extends
from the Bering Sea to Monterey, California. Adults are found
from the intertidal to 436 m on bottoms of mixed composition
(Jensen, 1995). Hart (1960) collected females bearing eggs from
March through September in British Columbia. *Oregonia gracilis*
molts through two zoeal stages and one megalopa. In the
laboratory, the larval period from hatching to but not including
the megalopa was four weeks. Sizes below are from laboratory-
reared larvae. Larvae described in Hart (1960).

Zoea I: Length from tip of rostrum to tip of dorsal spine, 3.5
mm (Fig. 35A, B). Carapace with relatively short lateral
spines. Rostrum long, thin, and spinulate. First and second

Fig. 35. *Oregonia gracilis.*
(From Hart, 1960)

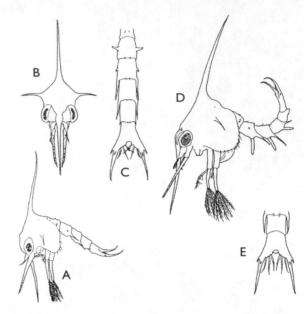

maxilliped pairs with four swimming setae on exopods. Abdomen composed of five segments and forked telson (Fig. 35C). Abdominal segments 2 and 3 with small lateral spines, abdominal segments 3 and 4 with long posteriolateral spines, and segment 5 with shorter posteriolateral spines. Each furca of telson with three inner setae, lateral exospine, and smaller dorsal exospine.

Zoea II: Length from tip of rostrum to tip of dorsal spine, 5 mm (Fig. 35D). First and second maxilliped pairs with six swimming setae on exopods. Abdomen now composed of six segments, and placement of spines on segments unchanged. Spines longer than in previous stage. Each half of telson with additional fine setae medially (Fig. 35E). Pleopod buds now present on the ventral side of abdominal segments 2–5.

Megalopa (illustrated in Key B, the megalopa key): Carapace length from tip of rostrum to posterior margin of carapace, 3.3 mm; width at widest point, 1.3 mm. Carapace with pair of anteriolateral spines, pair of mid-dorsal spines, small pair of lateral spines, and elongated posterior spine. Maxillipeds and pereopods now resemble adult form. First pereopod pair chelate, pereopod pairs 2–5 walking legs.

Family Pinnotheridae

Fabia subquadrata, **Mussel Crab**. *Fabia subquadrata* lives commensally within molluscs. Females are typically 22 mm across, males only 7.3 mm. The range extends from Akutan Pass, Alaska, to Ensenada, Baja California. Adults are found

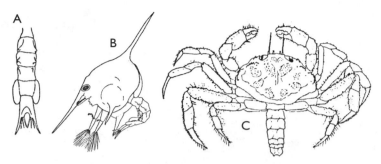

Fig. 36. *Fabia subquadrata.*
(From Lough, 1975)

most commonly in the mussel *Modiolus modiolus* in the intertidal and to 220 m (Jensen, 1995). In Puget Sound, Washington, females carrying eggs were rare in August and increased in numbers to a peak from November to January (Morris et al., 1980). *Fabia subquadrata* molts through five zoeal stages and one megalopa. In the laboratory, the larval period from hatching through the megalopa was 52 days. Sizes below are from laboratory-reared larvae. Larvae described in Lough (1975).

Zoea I: Length between tips of rostrum and dorsal carapace spine, 1.4–1.5 mm. Carapace globular, with pointed rostrum, long dorsal spine, and two posteriolateral spines. First and second maxilliped pairs with four swimming setae on exopods. Abdomen composed of five segments and telson. Fifth abdominal segment laterally expanded (Fig. 36A). Telson forked with three inner setae on each furca.

Zoea II: Length between tips of rostrum and dorsal carapace spine, 2.2–2.4 mm. First and second maxilliped pairs with six swimming setae on exopods. Abdomen now composed of six segments.

Zoea III: Length between tips of rostrum and dorsal carapace spine, 3.4–3.6 mm. First and second maxilliped pairs with eight swimming setae on exopods.

Zoea IV: Length between tips of rostrum and dorsal carapace spine, 5.2–5.6 mm (Fig. 36B). First and second maxilliped pairs with 9–10 swimming setae on exopods. Small pleopod buds present on abdominal segments 2–5.

Zoea V: Length between tips of rostrum and dorsal carapace spine, 6.6–7.1 mm. First and second maxilliped pairs with 11–12 swimming setae on exopods. Pleopod buds larger.

Megalopa: Carapace oval, wider than long, with "teeth" on lateral edges. Carapace, 2.5 by 3.8 mm (Fig. 36C).

Fig. 37. *Pinnotheres taylori.*
(From Hart, 1935)

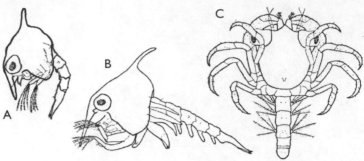

Fig. 37. *Pinnotheres taylori.*
(From Hart, 1935)

Pinnotheres taylori. *Pinnotheres taylori* lives commensally in the tunic of transparent ascidians. Females are typically 7.5 mm across, males only 4.8 mm. The range extends from Quatsino Sound, Vancouver Island, British Columbia, to Puget Sound, Washington. Adults can be found subtidally at 11–64 m (Hart, 1982). A female *P. taylori* bearing eggs was collected from a tunicate in March. Eggs hatched in early May reached the first juvenile instar in four weeks. Sizes below are from laboratory-reared larvae. Larvae described in Hart (1935).

Zoea I: Length from tip of rostrum to tip of dorsal spine, 1 mm (Fig. 37A). Rostrum and dorsal spine thin, relatively short, and blunt at tips. First and second maxilliped pairs with four swimming setae on exopods. Abdomen composed of five segments and telson. Second and third segments with small lateral knobs. Each furca of telson with three inner setae.

Zoea II: Length from tip of rostrum to tip of dorsal spine, 1.3 mm (Fig. 37B). Exopods of first and second maxilliped pairs with six swimming setae. Pleopod buds on abdomen long and thin.

Megalopa (Fig. 37C): Carapace 0.7 mm long, 0.5 mm wide. Rostrum small, almost blunt. Carapace with single small, blunt mid-dorsal, spine.

Family Portunidae

Carcinus maenas, **Shore Crab, Green Crab.** *Carcinus maenas* typically grows to a carapace width of 79 mm. It is native to Europe but has been introduced widely. It was first introduced to the west coast of North America in 1989 in San Francisco Bay and has since been reported as far north as Willapa Bay, Washington. Adults are found in estuaries in intertidal and subtidal waters to 6 m (Jensen, 1995). Females carrying eggs were collected in April and May in Great Britain at Plymouth, Devon, and at Brighton, Sussex, respectively. *Carcinus maenas* molts through four zoeal stages and one megalopa. At 12 °C the larval period was ca 58 days (Williams, 1967). Sizes below are from laboratory-reared larvae. Larvae described in Rice and Ingle (1975).

Zoea I: Length from tip of rostrum to posterior margin of telson, ca 1.4 mm (Fig. 38A). Carapace with large rostral and dorsal spines. Rostral spine more than twice length of antennae. First and second maxilliped pairs with four swimming setae on exopods. Abdomen composed of five segments and forked telson (Fig. 38B). Abdominal segment 2 with dorsolateral spines. Telson posterior margin with three pairs of setae. Each furca of telson with one large dorsal exospine, one small dorsal exospine, and very small thin lateral exospine.

Zoea II: Length from tip of rostrum to posterior margin of telson, 1.6–1.9 mm (Fig. 38C). First and second maxilliped pairs with six swimming setae on exopods. Small dorsal exospines and small lateral exospines now reduced or absent (Fig. 38D).

Zoea III: Length from tip of rostrum to posterior margin of telson, 2.1–2.2 mm (Fig. 38E). First and second maxilliped pairs with eight swimming setae on exopods. Abdomen now composed of six segments. Abdominal segments 2–5 bearing well-developed pleopod buds.

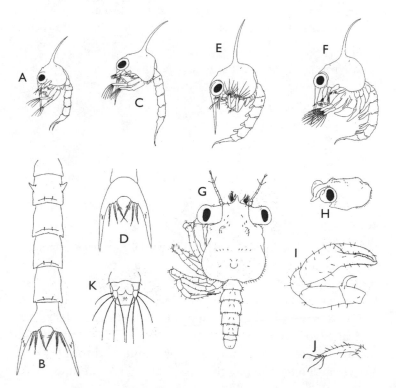

Fig. 38. *Carcinus maenas.* (From Rice and Ingle, 1975)

Zoea IV: Length from tip of rostrum to posterior margin of telson, 2.2–2.5 mm (Fig. 38F). First and second maxilliped pairs with 10 swimming setae on exopods. Pleopods now long and slender.

Megalopa: Length from tip of rostrum to posterior margin of carapace, 1.3–1.4 mm (Fig. 38G). Rostrum small and bent slightly downward (Fig. 38H). Maxillipeds and pereopods now resemble adult forms. Chelipeds with prominent ischiobasal hook (Fig. 38I). Dactyl (last segment) of fifth pereopod with three long setae (Fig. 38J). Telson narrows posteriorly and has uropods with setae (Fig. 38K).

Family Xanthidae

Key to xanthid zoeae (drawings from Hart, 1935; Connolly, 1925)

1a. Pleopod buds absent from abdomen 2
1b. Pleopod buds present on abdomen 3

2a. First and second maxillipeds with 4 swimming setae on
    exopods .... Zoea I, 4
2b. First and second maxillipeds with 6 and 7 swimming setae on
    exopods, respectively .... Zoea II, 4

3a. First and second maxillipeds with 8 and 9 swimming setae on
    exopods, respectively .... Zoea III, 4
3b. First and second maxillipeds with 9 and 10–11 swimming setae
    on exopods, respectively .... Zoea IV, 4

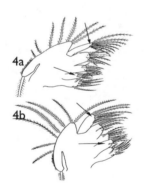

4a. Rostal spines ca same size, dorsal spine may be slightly longer
    than rostal spines .... *Lophopanopeus bellus*, 5
4b. Rostrum longer than dorsal spines .... *Rhithropanopeus harrisii*

5a. In zoeae I and II, 2 tiny, inconspicuous lateral exospines on
    telson (in addition to one on dorsal exospine) ....
    .... *Lophopanopeus bellus diegensis**
5b. In all four zoea stages, only 1 dorsal exospine ....
    .... *Lophopanopeus bellus bellus**

*Note that the subspecies of *Lophopanopeus* can be separated only during the first two zoeal stages.

Lophopanopeus bellus bellus, **Black-clawed Crab.** *Lophopanopeus bellus bellus* females typically grow to a carapace width of 24 mm; males can reach 40 mm. The range extends from Resurrection Bay, Alaska, to Point Sur, California. Adults can be found in the low intertidal and subtidal to 80 m under rocks, typically half-buried in sand or gravel (Jensen, 1995). In Puget Sound, females become ovigerous January through early March, and hatching begins in May and peaks in June. Two-thirds of females produce a second brood that hatches in the fall (Strathmann 1987). Larvae described in Hart (1935) and Knudsen (1959).

Rhithropanopeus harrisii, **Brackish-water Crab.** *Rhithropanopeus harrisii* females typically grow to a carapace width of 11 mm; males can reach 19 mm. Originally an Atlantic species, *R. harrisii* was reported in San Francisco Bay in 1940 and has since spread north to Coos Bay, Oregon. Adults live in the brackish water of estuaries and can tolerate trips into freshwater. They are found under rocks on sand (Jensen, 1995). Larvae described in Connolly (1925).

References

Ally, J. R. R. (1975). A description of the laboratory-reared larvae of *Cancer gracilis* Dana, 1852 (Decapoda, Brachyura). Crustaceana 28:231–46.

Berkeley, A. A. (1930). The post-embryonic development of the common Pandalids of British Columbia. Contri. Can. Biol. Fish. 6:79–163.

Bliss, D. E., Ed. (1982). The Biology of Crustacea. Academic Press, New York.

Boyd, C. M. (1960). The larval stages of *Pleuroncodes planipes* Stimpson (Crustacea, Decapoda, Galatheidae). Biol. Bull. 118:17–30.

Brusca, R. C. and G. J. Brusca (1990). Invertebrates. Sinauer Associates, Sunderland, Mass.

Coffin, H. G. (1960). The ovulation, embryology and developmental stages of the hermit crab *Pagurus samuelis* (Stimpson). Walla Walla College Publications of the Department of Biological Sciences and the Biological Station 25:1–30.

Connolly, C. J. (1925). The larval stages and megalops of *Rhithropanopeus harrisii* (Gould). Contrib. Canad. Biol., Toronto 2:327–33.

Cook, H. L. and M. A. Murphy (1971). Early developmental stages of the brown shrimp, *Peaneus aztecus* Ives, reared in the laboratory. Fish. Bull. 69:223–39.

Fitch, B. M. and E. W. Lindgren (1979). Larval development of *Pagurus hirsutiusculus* (Dana). reared in the laboratory. Biol. Bull. 156:76–92.

Gonor, S. L. (1970). The larval histories of four Porcellanid Anomurans (Crustacea, Decapoda) from Oregon. Master's Thesis, Oregon State University, Corvallis. 106 pp.

Gonor, S. L. and J. J. Gonor (1973a). Descriptions of the larvae of four North Pacific Porcellanidae (Crustacea: Anomura). Fish. Bull. (U.S.) 71:189–223.

———— (1973b). Feeding, cleaning, and swimming behavior in larval stages if porcellanid crabs (Crustacea, Anomura). Fish. Bull. (U.S.) 71:225–34.

Gore, R. H. (1979). Larval development of *Galathea rostrata* under laboratory conditions, with a discussion of larval development in the Galatheidae (Crustacea, Anomura). Fish. Bull. (U.S.) 76:781–808.

Gurney, R. (1942). Larvae of Decapod Crustacea. Ray Society, London.

Hart, J. F. L. (1935). The larval development of British Columbia Brachyura, I. Xanthidae, Pinnotheridae (in part) and Grapsidae. Can. J. Res. 12:411–32.

———— (1937). Larval and adult stages of British Columbia Anomura. Can. J. Res. 15:179–220.

———— (1960). The larval development of British Columbia Brachyura, II. Majidae, subfamily Oregoniinae. Can. J. Zool. 38:539–45.

———— (1965). Life history and larval development of *Cryptolithodes typicus* Brandt (Decapoda, Anomura) from British Columbia. Crustaceana 8:255–77.

———— (1971). Key to planktonic larvae of families of decapod Crustacea of British Columbia. Syesis 4:227–34.

———— (1982). Crabs and Their Relatives of British Columbia. British Columbia Provincial Museum, Victoria.

Haynes, E. (1973). Description of the prezoea and stage I zoea of *Chionoecetes bairdi* and *C. opilio* (Oxyrhyncha, Oregoniinae). Fish. Bull. (U.S.) 71:769–75.

———— (1978a). Description of larvae of a hippolytid shrimp, *Lebbeus groenlandicus,* reared *in situ* in Kachemak Bay, Alaska. Fish. Bull. (U.S.) 76:457–65.

———— (1978b). Description of larvae of the humpy shrimp, *Pandalus goniurus,* reared *in situ* in Kachemak Bay, Alaska. Fish. Bull. (U.S.) 76:235–48.

———— (1980). Larval morphology of *Pandalus tridens* and a summary of the principal morphological characteristics of North Pacific pandalid shrimp larvae. Fish. Bull. 77:625–41.

———— (1981a). Description of the stage II zoea of snow crab, *Chionoecetes bairdi* (Oxyrhyncha, Majidae), from plankton of lower Cook Inlet, Alaska. Fish. Bull. (U.S.) 79:177–82.

———— (1981b). Early zoeal stages of *Lebbeus polaris, Eulalus suckleyi, E. fabricci, Spirontocaris arcueta, S. ochotensis,* and *Heptacarpus camtschaticus* (Crustacea, Decapoda, Caridea, Hippolytidae) and morphological characterization of zoeae of *Spirontocaris* and related genera. Fish. Bull. (U.S.) 79:421–40.

———— (1985). Morphological development, identification, and biology of larvae of Pandalidae, Hippolytidae, and Crangonidae (Crustacea, Decapoda) of the northern North Pacific Ocean. Fish. Bull. 83:253–88.

Haynes, E. B. (1984). Early zoeal stages of *Placetron wosnessenskii* and *Rhinolithodes wosnessenski* (Decapoda, Anomura, Lithodidae) and review of lithodid larvae of the northern North Pacific Ocean. Fish. Bull. 82:315–24.

Israel, H. R. (1936). A contribution toward the life histories of two California shrimp, *Crango franciscorum* (Stimpson) and *Crango nigricauda* (Stimpson). Fish Bull. (Calif.) 46:1–28.

Iwata, F. and K. Konishi (1981). Larval development in the laboratory of *Cancer amphioetus* Rathbun, in comparison with those of seven other species of Cancer (Decapoa, Brachyura). Publ. Seto Mar. Biol. Lab. 26:369–91.

Jensen, G. C. (1995). Pacific Coast Crabs and Shrimp. Sea Challengers, Monterey, California.

Jewett, S. C. and R. E. Haight (1977). Description of megalopa of snow crab *Chionoecetes bairdi* (Majidae, subfamily Oregoniinae). Fish. Bull. 75:459–63.

Johnson, M. W. and W. M. Lewis (1942). Pelagic larval stages of the sand crabs *Emerita analoga* (Stimpson), *Blepharipoda occidentalis* Randall, and *Leipdopa myops* Stimpson. Biol. Bull. 83:67–87.

Knudsen, J. W. (1959). Life cycle studies of the Brachyura of Western North America. II. The life cycle *Lophopanopeus bellus diegensis* Rathbun. Bull. So. Calif. Acad. Sci. 58 (part 2):57–64.

Kurata, H. (1964). Larvae of decapod Crustacea of Hokkaido. 3. Pandalidae. Hokkadio Reg. Fish. Res. Lab Bull. 28:23–24 (English translation 1996, Fish. Res. Bd. Can. Trans. No. 1693).

Lough, R. G. (1975). Dynamics of crab larvae (Anomura, Brachyura) off the central Oregon coast, 1969–1971. PhD. Dissertation, Oregon State University, Corvallis. 299 pp.

MacDonald, J. D., R. B. Pike, and D. I. Williamson (1975). Larvae of the British species of *Diogenes, Pagurus, Anapagurus,* and *Lithodes* (Crustacea, Decapoda). Proc. Zool. Soc. Lond. 128:209–57.

MacGinitie, G. E. (1938). Movements and mating habits of the sand crab, *Emerita analoga*. Amer. Mid. Nat. 19:471–81.

MacMillan, F. E. (1971). The larvae of *Pagurus samuelis* (Decapoda: Anomura) reared in the laboratory. Bull. So. Calif. Acad. Sci. 70:58–68.

Martin, J. W. (1984). Notes and bibliography on the larvae of xanthid crabs, with a key to the known xanthid zoeas of the western Atlantic and Gulf of Mexico. Bull. Mar. Sci. 34:220–39.

McCrow, L. T. (1972). The ghost shrimp, *Callianassa californiensis*. Oregon State University, Corvallis. 55pp.

Modin, J. C. and K. W. Cox (1967). Post-embryonic development of laboratory-reared ocean shrimp, *Pandalus jordani* Rathbun. Crustaceana 13:197–219.

Morris, R. H., D. P. Abbott, and E. C. Haderlie (1980). Intertidal Invertebrates of California. Stanford University Press, Stanford.

Needler, A. B. (1938). The larval development of *Pandalus stenolepis*. J. Fish. Res. Bd. Can. 4:88–95.

Nyblade, C. (1974). Coexistence in sympatric hermit crabs. Ph.D. dissertation. University of Washington, Seattle. 241 pp.

Pike, R. B. and D. I. Williamson (1960). The larvae of *Spirontocaris* and related genera (Decapoda, Hippolytidae). Crustaceana 2:13–208.

Poole, R. L. (1966). A description of laboratory reared zoeae of *Cancer magister* Dana, and maegalopa taken under natural conditions (Decapoda, Brachyura). Crustaceana 11:83–97.

Price, V. A. and K. K. Chew (1972). Laboratory rearing of spot shrimp larvae (*Pandalus platyceros*) and descriptions of stages. J. Fish. Res. Bd. Canada 29:413–22.

Rice, A. L. and R. W. Ingle (1975). The larval development of *Carcinus maenas* (L.) and *C. mediterraneus* Czerniavsky (Crustacea, Brachyura, Portunidae) reared in the laboratory. Bull. Br. Mus. (Nat. Hist.) Zool 28:103 19.

Roesijadi, G. (1976). Descriptions of the prezoeae of *Cancer magister* Dana and *Cancer productus* Randall and the larval stages of *Cancer antennarius* Stimpson (Decapoda, Brachyura). Crustaceana 31:275–95.

Schlotterbeck, R. E. (1976). The larval development of the lined shore crab, *Pachygrapsus crassipes* Randall, 1840 (Decapoda Brachyura, Grapsidae) reared in the labratory. Crustaceana 30:184–200.

Strathmann, M. F. (1987). Reproduction and Development of Marine Invertebrates of the Northern Pacific Coast. University of Washington Press, Seattle.

Trask, T. (1970). A description of laboratory-reared larvae of *Cancer productus* Randall (Decapoda, Brachyura) and a comparison to larvae of *Cancer magister* Dana. Crustaceana 18:133–46.

Williams, B. G. (1967). Laboratory rearing of the larval stages of *Carcinus maenas* (L.) (Crustacea: Dacapoda). J. Nat. Hist. 2:121-26.

Williamson, D. I. (1957). Cruatacea, Decapoda: larvae, V. Caridea, family Hippolytidae. Fiches Identif. Zooplancton 68:1–5.

———— (1982). Larval morphology and diversity. In: The Biology of Crustacea, Vol. 2, Embryology, Morphology, and Genetics (L. G. Abele, ed; D. E. Bliss, ed.-in-chief), pp. 43–110. Academic Press, New York.

Appendix: Sources of illustrations used in keys

decapod zoeal key A illustrations

1a. (from Hart, 1971)
1b. (adapted from Johnson and Lewis, 1942; Gonor, 1970; Lough, 1975;d Williamson, 1982)
1c. (from Gurney, 1942)
3a. (from Hart, 1971)
4a. (from Hart, 1971)
4b. (from Schlotterbeck, 1976)
7a. (from Hart, 1971)
9a. (from Johnson and Lewis, 1942)
10a. (from Gonor, 1970), Gonor and Gonor, 1973b)
11a. (from Nyblade, 1974)
11b. (adapted from Haynes, 1985)
12a. (adapted from Boyd, 1960)
13a. (from Lough, 1975)
13b. (from Lough, 1975)
14a. (from McCrow, 1972)
15a. (from Hart, 1937)
17a. (from Gurney, 1942)
17b. (adapted from Gurney, 1942)
19a. (from Hart, 1971)
19b. (from Hart, 1971)
19c. (from Hart, 1971)

megalopae key B

1a. (adapted from Berkeley, 1930)
1b. (adapted from Lough, 1975)
1c. (from Gonor, 1970; Hart, 1937)
2a. (from Lough, 1975)
4a. (from Lough, 1975)
4b. (from Lough, 1975)
4c. (from Lough, 1975)
4d. (from Jewett and Haight, 1977)
5a. (from Lough, 1975)
6a. (adapted from Lough, 1975)
6b. (from Lough, 1975)
7a. (from Lough, 1975)
8a. (from Lough, 1975)
11a. Illustrations adapted from Hart, 1935)
11b. (adapted from Hart, 1935)
11c. (from Rice and Ingle, 1975)
12a. (from Connolly, 1925)
12b. (adapted from Lough, 1975)
13a. (from Gonor, 1970)
13b. (from Hart, 1937)
15a. (from Gonor, 1970)
15b. (from Gonor, 1970)
16a. (from Gonor, 1970)
16b. (from Gonor, 1970)
17a. (from Lough, 1975)
17b. (from Lough, 1975)
18a. (from Lough, 1975)
18b. (from Lough, 1975)
19a. (from Lough, 1975)
19b. (from Lough, 1975)

xanthidae key

4a. (from Hart, 1935)
4b. (from Connolly, 1925)

15

Phoronida

Kevin B. Johnson

Reproduction and Development

Phoronids reproduce primarily sexually and may be either hermaphroditic or dioecious. Asexual reproduction is well documented in only one species of *Phoronis* but reported present in all representatives of the genus. Internal fertilization is probably the dominant fertilization strategy in phoronids; fertilized primary oocytes are typically released into the water column, where larval development is completed.

Almost all phoronids produce a planktonic actinotroch larva. In some species (e.g., *Phoronis vancouverensis*) early embryos, attached to the female adult's tentacles, are brooded to an early actinotroch stage and then released. In others (e.g., *Phoronopsis viridis*, *Phoronis pallida*, and *Phoronis architecta*) all development takes place in the water column. Whether release into the plankton occurs shortly after fertilization or after a period of brooding, planktonic actinotroch larvae tend to be planktotrophic and reside in the plankton for extended periods.

The actinotroch larva bears a ring of tentacles arising from the cylindrical larval body just posterior to the hood, the most anterior major body region. It is propelled through the water by cilia lining the anterior tentacles and by a posterior telotrochal ciliary band. Additional details on the biology and development of the Phoronida are summarized in Brusca and Brusca (1990).

Identification and Description of Local Taxa

The actinotroch larvae illustrated herein are advanced stages approaching competence for metamorphosis. Early-stage larvae are morphologically simple and difficult to differentiate (see p. 258). As actinotroch larvae develop, their body size increases, the number of tentacles increases, they gain pigmentation, and blood corpuscle mass increases. A key based on the fully developed stages can be misleading if one attempts to apply it to earlier, less developed stages. It is recommended that the investigator collect plankton over a period of days or weeks to observe developmental changes in a cohort of actinotroch larvae. By maintaining actinotroch larvae in the laboratory, metamorphosis can be observed and an advanced developmental state confirmed; see Strathmann (1987) for rearing techniques. Developmentally advanced actinotroch larvae can then be reliably compared to illustrations for identification. Phoronid actinotroch larvae are among the most beautiful and intriguing planktonic larval forms; with this guide one should be able to identify the advanced actinotroch larvae found in the Pacific Northwest.

This following list of phoronid species known or likely in the Pacific Northwest (USA) was compiled from a combination of published distributions (Austin, 1985; Kozloff; 1993), knowledge of the latitudinal distributions of West Coast phoronids (R. Zimmer, pers. comm.), and firsthand local observations. Including the previously undescribed *Actino-trocha D*, which has no known adult counterpart, there are five species of actinotroch larvae likely present in the Pacific Northwest plankton (Table 1). Descriptions are provided below for all five of these species.

Fully developed actinotroch larvae are most easily identified on the basis of larval size and number of tentacles. These attributes, however, change during development making it necessary to use all available information to identify an actinotroch larva of unknown developmental stage accurately. As mentioned previously, it is recommended that actinotroch larvae be raised through metamorphosis, ensuring that the most advanced larval stage possible is used for identification.

Phoronis vancouverensis. This species is known from along the North American west coast from British Columbia to southern California. It is commonly thought to be synonymous with the Japanese species *Phoronis ijimai* (e.g., Austin, 1985). This is likely incorrect, however, since there is no actinotroch larva in the region of Japan fitting the description of *P. vancouverensis* (R. Zimmer, pers. comm.).

Table 1. Species in the phylum Phoronida from the Pacific Northwest

Actinotrocha D
Phoronis architecta
Phoronis pallida
Phoronis vancouverensis
Phoronopsis viridis

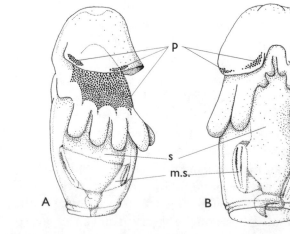

Fig. 1. Actinotroch larva of *Phoronis vancouverensis*, lateral (A) and dorsal (B) views. Pigment specks (p.) visible at the margin of the preoral hood and in the ventral collar region. s, stomach; m.s., metasomal sac. Length ca 600 μm. (Adapted from Zimmer, 1964)

The actinotroch (Fig. 1) is relatively small compared to many other species (~600 µm) and opaque. There are blackish-brown pigment specks on the margin of the preoral hood and the ventral side of the body between the tentacles and the hood margin. Tentacles number 12–14 at metamorphic competence.

Phoronis pallida. This phoronid, first described from Europe, is well known in central and southern California (Austin, 1985). It has also been observed in the waters of British Columbia (R. Zimmer, pers. comm.) and is known to be abundant in Coos Bay, Oregon.

Like *Phoronis vancouverensis*, *P. pallida* is smaller than many other actinotroch larvae (~600 µm) and opaque (Fig. 2). It is easily distinguished from *P. vancouverensi* by a girdle of yellowish-brown pigmentation just posterior to the tentacles, and by the circumesophageal blood corpuscle mass, appearing bow tie–shaped from a ventral, or frontal, view. Also, refractile globules highlight the wall of the stomach, which extends from the bow tie–shaped blood corpuscle mass to slightly posterior of the point at which the tentacles connect to the body. Larval tentacles number 10 at metamorphic competence. In addition to Zimmer (1994), information on the larval development of *P. pallida* is available in Silén (1954).

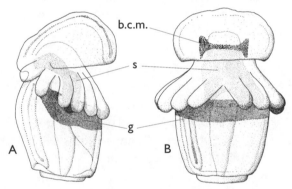

Fig. 2. Actinotroch larva of *Phoronis pallida*, lateral (A) and dorsal (B) views. Bow tie–shaped blood corpuscle mass (b.c.m.), refractile globules lining stomach wall (s.), and girdle (g.) of yellowish-brown pigmentation are visible. Length ca 600 µm. (Aadapted from Zimmer, 1964)

Phoronis architecta. This species was first described from the east coast of North America but is common along the coast of southern California. Less known north of California, the actinotroch larva of this species has been observed in plankton samples taken in Puget Sound, and *P. architecta* adults have been identified from near Tacoma, Washington (R. Zimmer, pers. comm.). *Phoronis architecta* has been lumped with *Phoronis muelleri* but is better treated as a separate species. *Phoronis muelleri*, found only rarely in California (Zimmer, pers. comm.), is not known from the Pacific Northwest coast.

The average length of the actinotroch larvae of *P. architecta* is 1.03 mm. The larva bears a pointed apical sense organ that adorns the preoral hood and can be quite pronounced as

Fig. 3. Actinotroch larva of *Phoronis architecta*, lateral view. s.o., apical sense organ; p.l., pigmented lobe of the stomach; b.c.m., blood corpuscle masses; s., stomach; a.t., adult tentacles; l.t., larval tentacles; m.s., metasomal sac. Length ca 1 mm. (Drawn from photographs supplied by R. Zimmer)

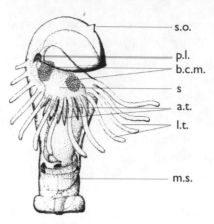

metamorphosis approaches (Fig. 3). A pair of pigmented evaginations, termed stomach lobes, arise from the dorsolateral surface of the stomach's extreme anterior. The stomach may span most of the length of the trunk cavity, with a relatively short intestine connecting it to the posterior anus. Two pairs of blood corpuscle masses are present in mature larvae on the ventrolateral and dorsolateral surfaces of the stomach, anterior to the tentacles. At metamorphosis, this actinotroch usually has 26 larval tentacles (13 pairs). Adult tentacles, numbering eight pairs and usually arising after all 26 larval tentacles are present, appear as laterally paired fans of separate tentacles underneath and posterior to the larval tentacles (similar to an actinotroch larva described in Ikeda, 1901). Adult tentacles tend to be more pointed at the distal end compared to the rounded larval tentacles. For more information on the early development of *P. architecta*, see Brooks and Cowles (1905). The larva of *P. architecta* is similar to the actinotroch larva of *P. muelleri* (= *Actinotrocha branchiata*), which is well described from the coast of northern Europe (Selys-Longchamps, 1902, 1903, 1907; Silén, 1954; Siewing, 1974; Hay-Schmidt, 1989).

Brooks and Cowles (1905) examined development in *P. architecta*. Their Fig. 34 illustrates an actinotroch larva they label "Species A," which they identify as the larva of *P. architecta*. This animal is not the larva of *P. architecta*, however, for it lacks the apical sense organ, the pigmented stomach lobe, and adult tentacles separate from the larval tentacles, and it has the wrong number of tentacles and blood corpuscle masses. Only this one illustration of *Actinotrocha* Species A is offered and compared to an unidentified *Actinotrocha* Species B, which ironically is probably what we now know to be *P. architecta*.

Phoronopsis viridis. This species is known from British Columbia to southern California (Austin, 1985). Although it is generally considered synonymous with *Phoronopsis harmeri*, there is a

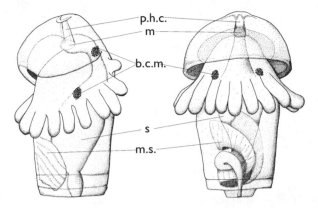

Fig. 4. Actinotroch larva
of *Phoronopsis viridis*,
lateral (A) and dorsal (B)
views. p.h.c., preoral
hood coelom; m., mouth;
b.c.m., blood corpuscle
masses; s., stomach, m.s.,
metasomal sac. Length ca
1.1. (Adapted from
Zimmer, 1964)

distinct phenotypic basis for the separation of these two species. Adult phoronids described as *P. viridis* are green and generally occur south of an otherwise similar white phoronid occurring near British Columbia and originally described as *P. harmeri*. Though this issue is unresolved, these are currently considered the same species.

In *P. viridis*, the preoral hood coelom (the anterior space between the coelomic lining and the epidermis) is a rather obvious boxlike space between the brain and the esophagus (Fig. 4). This actinotroch larva may be as large as 1 mm or slightly larger, and the body wall is relatively transparent. There are two pairs of blood corpuscle masses. One pair is in the posterior "flange" or "corner" of the preoral hood and is most easily visible from the dorsal view. The other mass is on either side of the "collar," slightly anterior to the larval tentacles. When ready to metamorphose, this larva has ca 20 tentacles (10 pairs). Several studies deal with the development of *P. viridis*, including a comparison of its development with a *Phoronis* species from Monterey Bay (Rattenbury, 1951), observations of early development (Fairfax, 1977), and detailed descriptions of "*Actinotrocha A*" (Zimmer, 1964).

Actinotrocha D. An unidentified actinotroch larva, referred to here as *Actinotrocha D*, is commonly found in plankton samples from Coos Bay, Oregon. Development has been observed in all known adult phoronid species in the Pacific Northwest. Since *Actinotrocha D* resembles no previously described larva, it is likely the larva of an undescribed species.

Actinotrocha D is easily recognized by its enormous size, commonly 2–3 mm and as large as 4 mm (Fig. 5). There are three pair of blood corpuscle masses. Two sets are paired laterally and located at the base of the tentacles. One pair lies within the oral hood (Fig. 6). Larvae approaching meta-morphosis may have 30 or more tentacles. The stomach of this larva is distended and fluid-filled, taking up most of the body

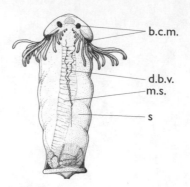

Fig. 5. *Actinotrocha D*, dorsal view. The metasomal sac (m.s.) is situated on the ventral side of the animal, running between the large fluid-filled stomach (s.) and the body wall. b.c.m., blood corpuscle masses; d.b.v., dorsal blood vessel. (Drawn from video and live observations, Coos Bay, Oregon)

Fig. 6. *Actinotrocha D*, anterior dorsolateral view.

volume. A pleated dorsal blood vessel extends posteriorly from the base of the tentacles to the telotroch. Because the stomach is transparent and nearly the same diameter as the body itself, it is not the most obvious structure in the body posterior of the tentacles. Rather, the metasomal sac, a flattened tube extending the length of the body and coiling upon itself several times at the posterior end, is located on the ventral side of the animal and may be mistaken for the stomach.

Early Pelagic Stages in Phoronida

Postembryonic phoronid stages preceding the actinotroch stage have fewer distinguishing features than an advanced actinotroch larva. Consequently, identifying early developmental stages to genus or species is difficult. Many earlier stages are distinctly phoronid, however, and generalized diagrams of early developmental stages (Fig. 7) can aid identification of larvae as phoronids.

Fig. 7. Generalized phoronid developmental stages following gastrula but preceding the advanced actinotroch larva. Phases G and H represent an early actinotroch with small tentacles. All stages are presented in lateral view except phase H, which is shown in ventral view. Size varies with species and stage of development but generally ranges from <100 to >1,000 mm. (Adapted from Rattenbury, 1951, 1954; Silén, 1954; Forneris, 1957; Zimmer, 1964)

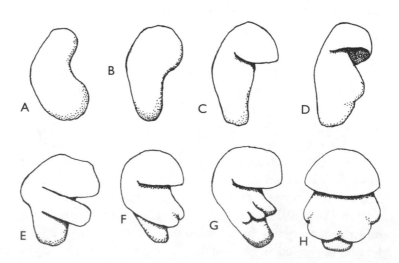

Acknowledgments

The author expresses sincere appreciation to Russel Zimmer of the University of Southern California for his invaluable suggestions for actinotroch identification and comments on the manuscript.

References

Austin, W. C. (1985). An Annotated Checklist of Marine Invertebrates in the Cold Temperate Northeast Pacific, Khoyatan Marine Laboratory, Cowachin Bay, B.C.

Brooks, W. K. and R. P. Cowles (1905). *Phoronis architecta:* its life history, anatomy, and breeding habits. Mem. Nat. Acad. Sci. 10:72–113.

Brusca, R. C. and G. J. Brusca (1990). Invertebrates. Sinauer Associates, Inc., Sunderland.

Fairfax, R. E. (1977). The embryology and reproductive biology of *Phoronopsis viridis.* Master's Thesis, Humboldt State University.

Forneris, L. (1957). Phoronidea: Family Phoronidae, Actinotrocha Larvae. Cons. Int. Explor. Mer Zooplankton sheet 69:1–4.

Hay-Schmidt, A. (1989). The nervous system of the actinotroch larva of *Phoronis muelleri* (Phoronida). Zoomorph. 108:333–51.

Ikeda, I. (1901). Observations on the development, structure and metamorphosis of Actinotrocha. J. Coll. Sci. Tokyo Imp. Univ. 13:507–92.

Kozloff, E. N. (1993). Seashore Life of the Northern Pacific Coast. University of Washington Press, Seattle.

Rattenbury, J. C. (1951). Studies of embryonic and larval development in California Phoronidea. Doctoral Dissertation, University of California.

——— (1954). The embryology of *Phoronopsis viridis.* J. Morph. 95:289–333.

Selys-Longchamps, M. (1902). Recherches sur le development des Phoronis. Arch. Biol. 18:495–603.

——— (1903). Über Phoronis und Actinotrocha bei Helgoland. Wissensch. Meeresunters., N.F., Bd. 6, Abt. Helgoland.

——— (1907). Development postembryonaire et affinites des *Phoronis.* Acad. d Sci. Belgique. Cl. d Sci. Ser. 2:1–157.

Siewing, V. R. (1974). Morphological investigations about the archicoelomate-problem 2. The body segmentation in *Phoronis muelleri* de Selys-Longchamps (Phoronidea). Zool. Jb. Anat. Bd. 92:275–318. In German.

Silén, L. (1954). Developmental biology of Phoronidea of the Gullmar Fjord area (West Coast of Sweden). Acta Zoologica 35:215–57.

Strathmann, M. F. (1987). Reproduction and Development of Marine Invertebrates of the Northern Pacific Coast. University of Washington Press, Seattle.

Zimmer, R. L. (1964). Reproductive biology and development of Phoronida. Doctoral Dissertation, University of Washington.

16

Bryozoa

Katherine Rafferty

Two classes of Bryozoa are found in the marine environment, the Gymnolaemata and the Stenolaemata, with the majority of the marine species in the Gymnolaemata. Species descriptions, mostly from scientific collecting expeditions during the late 1880s and early 1900s, remain a primary source of information on bryozoan larvae. Many bryozoan species are found in the Pacific Northwest, but the larvae of only a handful have been described. Because of the limited number of larval descriptions, this chapter does not include a key to the larvae. Instead we offer a general description of the larvae of marine bryozoans, illustrations of a diversity of bryozoan larvae from around the world, and a brief account of local larvae for which we have published descriptions.

All stenolaemates and most gymnolaemates retain their embryos and release lecithotrophic larvae of short pelagic duration (Woollacott and Zimmer, 1978). A few gymnolaemate species release zygotes that develop into planktotrophic cyphonautes larvae (Fig. 1A); most, however, release lecithotrophic coronate larvae (Fig 1B). All species in the class Stenolaemata produce lecithotrophic coronate larvae (Fig. 1C). The pelagic duration of the lecithotrophic larvae in both classes is so brief that they seldom appear in zooplankton samples. The cyphonautes larvae, with their much longer pelagic duration, are common components of plankton samples collected in coastal waters.

Fig. 1. Bryozoan larval forms: (A) Planktotrophic cyphonautes larva (Gymnolaemata). (B) Lecithotrophic coronate larva (Gymnolaemata). (C) Lecithotrophic coronate larva (Stenolaemata). (A from Rupert and Barnes, 1994, Fig. 19-16; B from Hyman 1959, Fig. 131F; C from Nielsen 1970, Fig. 2A)

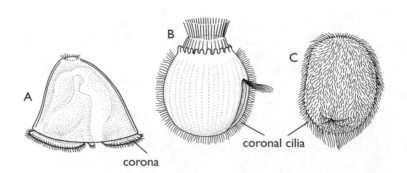

corona

coronal cilia

Class Stenolaemata

All stenolaemates release coronate larvae. The larvae are characteristically small, on the order of 100 µm wide by 150 µm long, ovoid, lecithotrophic, completely covered with cilia, and possess a pyriform gland (or pyriform complex) as well as an apical disc. Some gymnolaemates also release coronate larvae (see Figs. 1B, 5); they can be differentiated from stenolaemate coronate larvae by the presence of an apical tuft of longer cilia. Table 1 presents a list of the stenolaemate species present in the Pacific Northwest and indicates species for which a larval description is available; the larvae of only two (*Tubilipora pulchra* and *Disporella hispida*) of ca 33 local species have been described (Barrois, 1877; Nielsen, 1970). Drawings of representative coronate larvae are presented in Fig. 2.

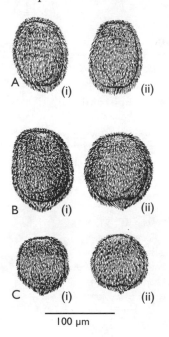

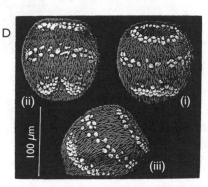

Fig. 2. Coronate larvae from the class Stenolaemata. (A) *Cresiella producta*. (B) **Crisia** *eburnea*. (C) **Disporella hispida**. (D) **Tubulipora** *phalangea*. Genera and species in bold are found locally. (i) larvae in profile; (ii) larvae from flattened side; (iii) larvae creeping on the substrate. (from Nielsen, 1970)

100 µm

Class Gymnolaemata

Although there are obvious gross morphological differences between coronate and cyphonautes larva of the Gymnolaemata, there are consistent structural characteristics common to both types of larva in this class (Fig. 3). The larvae all have a ciliated girdle (the corona), an anterior tuft of long cilia and the external (or metasomal) sac (Zimmer and Woollacott, 1977).

The type of larva released (feeding or non-feeding) is dependent on the development mode of the species. Because of their long pelagic duration, they are usually the most abundant bryozoan larvae in plankton tows. Local genera displaying this pattern of development include *Membranipora*,

Table 1. Species in the class Stenolaemata from the Pacific Northwest

Family Oncousoeciidae
Stomatopora granulata
Proboscina incrassata

Family Diastoporidae
Diaperoecia californica
Diaperoecia intermedia
Diaperoecia johnstoni
Diaperoecia obelium
Plagioecia patina

Family Tubuliporidae
Tubulipora flabellaris
Tubulipora pacifica
Tubulipora tuba
Tubilipora pulchra*

Family Fondiporidae
Fulifascigera fasciculata

Family Crisiidae
Bicrisia edwardsiana
Crisidia cornuta
Crisia elongata
Crisia occidentalis
Crisia operculata
Crisia serrulata
Crisia pugeti
Filicrisia franciscana
Filicrisia geniculata

Family Cytididae
Discocytis canadensis

Family Heteroporidae
Heteropora alaskensis
Heteropora magna
Heteropora pacifica

Family Lichenoporidae
Disporella hispida*
Disporella fimbriata
Disporella separata
Lichenopora verrucaria
Lichenopora novae-zelandiae
Oncousoecia ovoidea
Ciaperoecia californica
Ciaperoecia major

*Published larval description available.

Fig. 3. Hypothetical gymnolaemate larva, showing the components common to both coronate and cyphonautes larval forms. (Zimmer and Woollacott, 1977, Fig. 1)

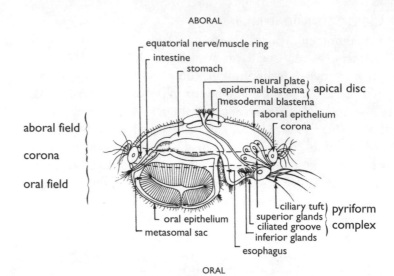

Table 2. Species in the class Gymnolaemata from the Pacific Northwest

Family Flustrellidridae
Flustrellidra corniculata

Family Alcyonidiidae
Alcyonidium mammilatum
Alcyonidium parasiticum
Alcyonidium pedunculatum
Alcyonidium polyoum

Family Clavoporidae
Calvopora occidentalis

Family Triticellidae
Triticella pedicellata

Family Arachnidiidae
Nolella stipata

Family Immergentiidae
Immergetia sp.

Family Vesiculariidae
Bowerbankia gracilis

Family Buskiidae
Buskia nitens

Family Penetrantiidae
Penetrantia sp.

Family Aeteidae
Aetea sp.

Family Membraniporidae
Conopeum reticulum*
Membranipora membranacea*

Family Electridae
Electra crustulenta*

Family Hincksinidae
Cauloramphus brunea
Cauloramphus echinus

Cauloramphus spiniferum
Ellisina levata
Hincksina alba
Hincksina pallida

Family Alderinidae
Callopora circumclathrata
Callopora corniculifera
Callopora horrida
Callopora lineata
Callopora armata
Alderina brevispina
Copidozoum protectum
Copidozoum tenuirostre
Doryporella alcicornis
Tegella aquilirostris
Tegella armifera
Tegella robertsonae

Family Chapperiellidae
Chapperiella condylata
Chapperiella patula

Family Microporidae
Micropora coriacea
Microporina borealis

Family Cellariidae
Cellaria diffusa
Cellaria mandibulata

Family Scrupocellariidae
Caberea boryi
Caberea ellisi
Scrupocellaria californica
Scrupocellaria varians
Tricellaria gracilis
Tricellaria occidentalis*

Tricellaria ternata
Tricellaria praescuta

Family Epistomiidae
Synnotum aegyptiacum

Family Bicellariellidae
Bugula californica
Bugula pugeti
Bugula cucllifera
Bugula pacifica*
Bugula flabellata*
Caulibugula californica
Caulibugula ciliata
Caulibugula occidentalis
Dendrobeania curvirostrata
Dendrobeania laxa
Dendrobeania lichenoides
Dendrobeania longispinosa
Dendrobeania murrayana

Family Cribrilinidae
Cribrilina radiata
Cribrilina annulata
Cribrilina corbicula
Colletosia radiata
Lyrula hippocrepis
Puellina setosa
Reginella furcata
Reginella nitida

Family Hippothoidae
Hippothoa divaricata
Hippothoa hyalina
Trypostega claviculata

Family Umbonulidae
Umbonula arctica

Table 2 continued. Species in the class Gymnolaemata from the Pacific Northwest

Family Stomachetosellidae
Stomachetosella cruenta
Stomachetosella limbata
Stomachetosella sinuosa

Schizoporellidae
Dakaria dawsoni
Dakaria ordinata
Dakaria pristina
Hippodiplosia insculpta
Hippodiplosia reticulato-punctata
Schizomavella auriculata
Schizoporella cornuta
Schizoporella unicornis
Schizoporella linearis

Family Hippoporinidae
Gemelliporella inflata
Hippomonavella longirostrata
Hippoporella nitescens
Lacerna fistulata
Stephanosella vitrea

Family Microporellidae
*Fenestrulina malusii**

Microporella californica
Microporella ciliata
Microporella setiformia
Microporella umbonata
Microporella vibraculifera

Family Eurystomellidae
Eurystomella bilabiata

Family Smittinidae
Codonellina cribriformis
Mucronella ventricosa
Parasmittina collifera
Porella columbiana
Porella concinna
Porella porifera
Rhamphostomella cellata
Rhamphostomella costata
Rhamphostomella curvirostrata
Smittina cordata
Smittina landsborovi

Family Reteporidae
Lepraliella bispina
Phidolopora labiata
Rhynchozoon tumulosum

Family Cheiloporinidae
Cheilopora praelonga
Cryptosula pallasiana

Family Phylactellidae
Lagenipora punctulata
Lagenipora socialis

Family Celleporidae
Holoporella brunnea
Costazia costazia
Costazia robertsoniae
Costazia ventricosa

Family Myriozoidae
Myriozoum coarctatum
Myriozoum subgracile
Myriozoum tenue

*Published larval description available.

Alcyonium, Conopeum, and *Electra* (Atkins, 1955a, b; Ryland, 1965; Reed, 19917). The production of a few large eggs that develop into lecithotrophic larvae with short pelagic durations is a second developmental pattern. It is seen locally in the genera *Bowerbankia, Schizoporella,* and *Hippodiplosia* (Reed, 1978, 1980; Nielson, 1981). A third pattern of development is characterized by extraembryonic nutrition of larvae reared in a brood chamber. These larvae are also lecithotrophic and reside only briefly in the plankton. Species in the genus *Bugula* display this developmental pattern (Reed, 1987). Few of the larvae of species in the Gymnolaemata have been described; of ca 128 local species in this class, only eight have been described (Table 2).

Cyphonautes Larvae

The most obvious diagnostic features of cyphonautes larvae (Fig. 4) are their triangular shell and extreme lateral compression (Ryland, 1964). Food particles are removed from a continuous current of water that the ciliary action drives through the mantle cavity or vestibule (Atkins, 1955b).

Fig. 4. Cyphonautes
larvae. (A)
**Membranipora
membranacea**. (B)
Electra pilosa. (C)
Electra crustulenta. (D)
Conopeum reticulum.
(E) **Flustrellidra** hispida;
1, apical organ; 3,
adhesive sac; 5, shell; 12,
vibratile plume; 14,
pyriform organ; 15, ciliary
girdle; 17, muscle strands.
(F) **Alcyonidium** mytili; 5,
apical organ; 6, groove; 7,
ciliary girdle; 8, vibratile
plume; 9, ciliated cleft; 10,
adhesive sac. Genera and
species in bold are found
locally. A, B present
lateral, apical, and
anterior views (clockwise
from left). C, D present
lateral and apical views.
(A–D from Ryland, 1965;
E, F from Hyman, 1959)

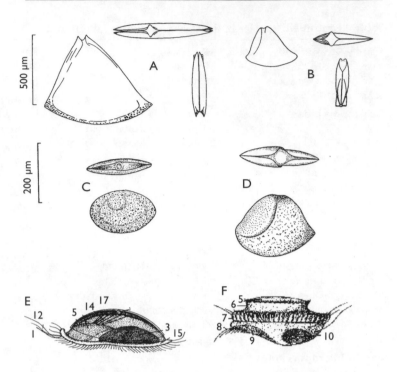

Membranipora membranacea. This species produces a large cyphonautes with an ornamented shell (Fig. 4A). The shell outline is roughly triangular. Faint striations (growth lines) may show near the apex and down the posterior margin. The anterior and posterior margins of the valves are of equal length, or sometimes the anterior margin is slightly longer. The posterior flange is narrow. The notch for the apical organ is deep and flared but relatively small. The shell and internal organs are transparent. The larvae reach a basal width of 440 μm. They are most abundant in coastal waters that have nearby rocky shores and are found in the plankton throughout the year, most abundantly in May through September (Atkins, 1955b; Ryland, 1964, 1965; Yoshioka, 1982).

Conopeum reticulum. The small cyphonautes measures ca 250-290 μm in length and 180–200 μm high. The shell is roughly bell-shaped (Fig. 4D). The valves are flat, truncated, and strongly flared at the apical organ. The anterior and posterior margins are asymmetrical and proportionally taller than in most other small cyphonautes. There is a broad flange bordered by a prominent ridge running from the apex toward the posterior margin. The shell is gray and lightly encrusted with small dark particles. The internal organs are not visible (Ryland, 1965).

Electra crustulenta. The larvae measure 160–200 µm long by 120–170 µm tall The shell outline is roughly oval (Fig 4C). The shell is encrusted with fine particles that give it a brownish or gray, granular appearance. The internal organs are barely visible, often only the stomach showing clearly (Ryland, 1965).

Coronate Larvae

The morphology of coronate larvae is more diverse than that of the cyphonautes. The structures common to all coronate larvae are (see Fig. 3) the corona, the apical disc, the blastema (or pallial sinus) associated with the apical disk, the pyriform complex, and the metasomal (or internal) sac (Zimmer and Woollacott, 1977; Reed, 1980). Zimmer and Woollacott (1977) group coronate larvae into five categories based on the size

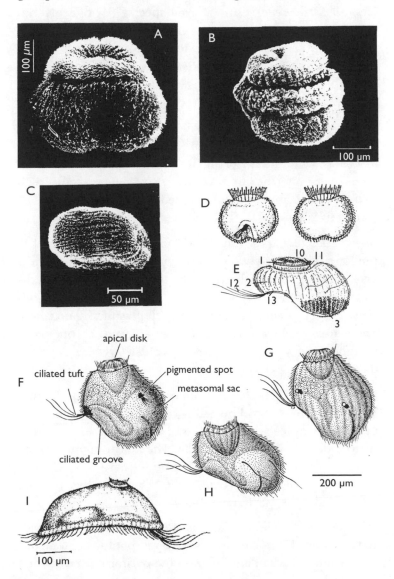

Fig. 5. Coronate larvae in the class Gymnolaemata. (A) **Hippodiplosia insculpta**. (B) *Watersipora arcuata* (type E larva). (C) **Bowerbankia gracilis** (type AEO/PS larvae). (D) **Bugula flabellata** (type AEO/ps larva). (E) **Cellepora** sp. (Type AE larva); I, apical organ; 2, coronal cells; 3, adhesive sac; 10, sensory bristles; I I, groove; 12, vibratile plume; 13, ciliated cleft. (F) **Scrupocellaria** *reptans* (Type AEO/ps larva). (G) **Bugula** *plumosa* (Type AEO/ps larva). (H) *Bicellariella ciliata*. (I) *Scruparia* sp. (Type O larva). Genera and species in bold are found locally. (A from Nielsen, 1981; B from Reed, 1991; C from Reed and Cloney, 1982; D from Grave, 1930; E from Hyman, 1959; F–H from Woollacott and Zimmer, 1978; I from Barrois, 1877)

and position of the corona, and the size of the pallial sinus. The categories are as follows (definitions from Reed, 1980); cxamples of each category are presented in Fig. 5.

Type O. Larva with narrow corona located orally, e.g., *Scuparia* sp. and *Alcyonidium duplex* (Zimmer and Woollacott, 1977).

Type E. Larva with narrow equatorial corona, e.g., *Alcyonidium polyoum* (Zimmer and Woollacott, 1977).

Type AE. Larva with expanded corona that is aboral and equatorial in position, e.g., *Cellepora pumicosa* (Zimmer and Woollacott, 1977).

Type AEO/ps. Larva with extensive corona that is aboral, equatorial, and oral in position and with small pallial sinuses. This type is characteristic of all cellularioid cheilostome larvae, e.g., all *Scrupocellaria* and *Bugula* species (Woollacott and Zimmer, 1978).

Type AEO/ps. Larva with extensive corona that is aboral, equatorial, oral in position and with well-developed pallial sinuses, e.g., *Bowerbankia gracilis* (Reed and Cloney, 1982).

Hippodiplosia insculpta. The larva is roughly spherical (slightly compressed along the apical axis) and 330–350 μm diameter (Fig. 5A). It is light reddish orange with darker coloration associated with the apical organ and below the corona. There are no eyespots. The larva is densely covered with cilia, with two pairs of compound cilia located laterally just below the corona (Nielsen, 1981).

Bugula pacifica. Larvae are 110–120 μm diameter and 150–165 μm long. They have two pairs of photoreceptors. One pair is anterolateral, midway between the aboral epithelium and the equator. The other pair is posterolateral at the equator. *Bugula simplex* and *B. pacifica* have identical cytoplasmic organization. The sensory cells are flush with the adjacent coronal cells (Hughes and Woollacott, 1980). Typical *Bugula* larvae are depicted in Fig. 5D, G.

Bugula flabellata. The larva is almost spherical (170–190 μm) and transparent to yellowish in color (Fig. 5D). There is a prominent apical organ at the anterior end of the larva surrounded by a "circlet" of cilia. The apical organ is often delimited by a circular groove the outer margin of which is frequently scalloped (Hyman, 1959). There are two pairs of pigmented eye spots (Grave, 1930).

Tricellaria occidentalis. The larva is ca 135 μm diameter and 140–160 μm long. There are three orange-red photoreceptors, one on the anteromedian line between the pyriform complex and the vesiculated epithelium and the other two on the larval equator about 135 degrees on either side of the anteromedian line (Hughes and Woollacott, 1980).

Fenestrulina malusii. The larvae are small and yellowish white. They swim rapidly and settle within an hour of liberation (Nielsen, 1981).

Scruparia sp. This larva is known from a single illustration (Barrois, 1877). It is similiar to *Flustrellidra hispida* (shelled lecithotrophic larva) but lacks a shell (Fig. 5I). No other *Scruparia* has been examined (Zimmer and Woollacott, 1977).

Bowerbankia gracilis. The larvae are coronate type AEO/PS described by Reed and Cloney (1982). They lack pigment spots and are light yellow in color. They are barrel-shaped, elongated along their aboral-oral axis (Fig. 5C). The corona, which covers most of the larval surface, consists of 32 long, narrow cells that extend from the aboral pole to the oral pole on the posterior and lateral sides of the larva. The coronal cells on the anterior side are shorter and extend from the aboral pole to the pyriform organ. The pyriform organ consists of a superior glandular field with a glandular pit, a papilla that bears a ciliary tuft, an inferior glandular field, and a ciliated cleft. A glandular region of the oral epithelium is infolded to form a small internal sac. The infolded glandular epithelium consists of a single layer of columnar cells virtually filled with large bipartite secretary granules. The internal sac opens as a narrow slit in the median plane of the larva (Reed and Cloney, 1982).

Alcyonidium gelatinosum. The corona on the free-swimming larvae of *Alcyonidium gelatinousm* forms a narrow equatorial band of ciliated cells that separates the aboral and oral hemispheres (Fig. 4F). The apical disc is large, covering most of the aboral hemisphere (Reed 1991).

References
Atkins, D. S. (1955a). The cyphonautes larvae of the Plymouth area and the metamorphosis of *Membranipora membranacea*. J. Mar. Biol. Ass. (UK) 34:441–49.
——— (1955b). The ciliary feeding mechanism of the cyphonautes larva [Polyzoa ectoprocta]. J. Mar. Biol. Ass. (UK) 34:451–66.
Barrois, J. (1877). Recherches sur l'embryologie des Bryozoairs (= memoire sur l'embryologie des Bryozoairs). Trav. Stn. Zool. Wimereux. 1:1–30.
Grave, B. H. (1930). The natural history of *Bugula flabellata* at Woods Hole, Massachusetts, including the behavior and attachment of the larva. J. Morph. 49:355–79.
Hughes, R. L., Jr. and R. M. Woollacott (1980). Photoreceptors of bryozoan larvae (Cheilostomata, Cellularioidea). Zoologica Script. 9:129–38.

Hyman, L. H. (1959). The Invertebrates: Smaller Coelomate Groups. McGraw-Hill, London and New York.

Nielsen, C. (1970). On metamorphosis and ancestrula formation in cyclostomatous bryozoans. Ophelia 7:212–56.

——— (1981). On morphology and reproduction of *Hippodiplosia insculpta* and *Fenestrulina maulusii* (Bryozoa, Cheilostomata). Ophelia 20:91–125.

Reed, C. G. (1978). Larval morphology and settlement of the bryozoan, *Bowerbankia gracilis* (Vesicularioidea, Ctenostomata): structure and eversion of the internal sac. In: Settlement and metamorphosis of marine invertebrate larvae, F. S. Chia and M. E. Rice (eds.). Elsevier/North-Holland Biomedical Press, New York. pp. 41–48.

——— (1980). The reproductive biology, larval morphology, and metamorphosis of the marine bryozoan, *Bowerbankia gracilis* (Vesiculaarioidea, Ctenostomata). Ph.D. Dissertation, University of Washington, Seattle. 291pp.

——— (1991). Bryozoa. In: Reproduction of Marine Invertebrates Vol. VI, A.C. Giese, J. S. Pearse, and V. B. Pearse (eds.), pp. 86–246.

Reed, C. G. and R. A. Cloney (1982). The larval morphology of the marine Bryozoan *Bowerbankia gracilis* (Ctenostomata: Vesicularioidea). Zoomorph.100:23–54.

Ruppert, E. E. and R. D. Barnes. (1994). Invertebrate Zoology. Saunders College Publishing, New York.

Ryland, J. S. (1964). The identity of some cyphonautes larvae (polyzoa). J. Mar. Biol. Ass. (UK) 44: 645–54.

——— (1965). Polyzoa (Bryozoa) order cheilostomata cyphonautes larvae. Cons. Int. Explor. Mer. Sheet 106:1–6.

——— (1974). Behavior, settlement and metamorphosis of bryozoan larvae: A review. Thalassia Jugo. 10:239–62.

Woollacott, R. M. and R. L. Zimmer (1972). Fine structure of a potential photoreceptor organ in the larva of *Bugula neritina* (bryozoa). Z. Zeliforsch. 123:458–69.

——— (1977). Structure and classification of Gymnolaemate larvae. In: Biology of Bryozoans, R. M. Woollacott and R. L. Zimmer (eds.), pp. 57–89. Academic Press, New York.

——— (1978). Metamorphosis of cellularioid bryozoans. In: Settlement and Metamorphosis of Marine Invertebrate larvae, G.-S. Chia and M. E. Rice (eds), pp. 49–63. Elsevier/North-Holland Biomedical Press, New York.

Yoshioka, P. M. (1982). Role of planktonic and benthic factors in the population dynamics of the bryozoan *Membranipora membranacea*. Ecology. 63:457–68.

17

Brachiopoda

Alan L. Shanks

The brachiopods are a small phylum of sessile filter feeders with bivalved shells. Superficially they look like clams, but they can easily be distinguished from clams by noting that the brachiopod is attached to the substratum by a peduncle that passes through one of the valves. Brachiopods are composed of three distinct regions: the mantle, which secrets the bivalved shells and encloses the internal mantle cavity; the lophophore, a ciliated tentaculate feeding organ found within the mantle cavity, and the pedicle, which attaches the brachiopod to the substratum. The phylum is composed of two classes, the Articulata, without shell or cirri, and the Inarticulata, with both shell and cirri. Locally there are four species of articulate and only one species of inarticulate brachiopod (Table 1). More detailed descriptions of the Brachiopoda can be found in Hyman (1959) and Brusca and Brusca (1990).

Table 1. Species in the phylum Brachiopoda from the Pacific Northwest (from Kozloff, 1996)

Class Articulata
Hemithiris psittacea
Terrebratulina unguicula
Terebratalia transersa
Laqueus californianus

Class Inarticulata
Crania californica

Class Articulata

The Articulata have indirect development. Their larvae are lecithotrophic, demersal, and generally have a short free-swimming stage. Brachiopod larvae are divided into three body regions: the apical (or anterior) lobe, the mantle lobe, and the pedicle lobe (Fig. 1A). When articulate larvae are competent to settle, the apical lobe is uniformally ciliated with an apical tuff. An annular band of long locomotory cilia surrounds the posterior margin of the apical lobe (Fig. 1A, B). The mantle lobe extends like a skirt over the pedicle lobe. Early in development, the mantle lobe is evenly ciliated, but this ciliation is lost during developlment. At competence, ciliation on the mantle lobe has been reduced to a ventromedial longitudinal band of cilia and a pair of ciliated pits near the apical lobe. The posterior margin of the mantle lobe bears four bundles of long chaetae (Fig. 1A). When disturbed, these chaetae are splayed out. At settlement, the larva attaches to the substratum via the pedicle, and the mantle lobe folds up and over the apical lobe, enclosing it (Fig. 1C). The mantle lobe looses its ciliation and secretes the bivalve shell. After settlement, the tentacles (cirri) develop on the lophophore.

In the four local species, larvae are retained in the mantle cavity and released as demersal non-feeding larvae. Meta-

Fig. 1. Articulate brachiopod larvae. (A) Larva of *Waltonia*. (B) Larva of *Terebratalia*; there is one species in this genus locally. (C) Schematic of metamorphosis in articulate brachiopods; the mantle lobes flex up to cover the apical lobe, and the pedicle attaches to the substratum. (A,C from Brusca and Brusca, 1990, Fig. 21; B from Nielsen, Fig. 19)

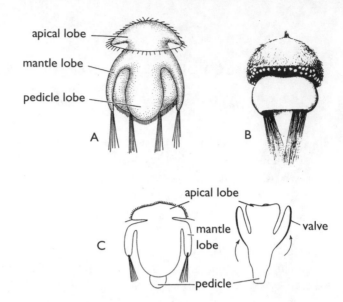

morphosis occurs within 100 hours (*Hemithiris psittacea, Terrebratulina unguicula,* and *Laqueus californianus*) to 200 hours (*Terebratalia transersa*).

Class Inarticulata

The Inarticulata have direct development. Their larvae are planktotrophic, pelagic, and the free-swimming stage can be quite long. Brachiopods are rare in plankton samples and, when they are present, it is usually larvae from the Inarticulata. The lophophores, mantle, and shell develop early in the inarticulate brachiopods and the pedicle develops late or after settlement. The lophophores consists of a variable number of ciliated tentacles, or cirri. The cirri are added in pairs at the anterior edge of the lophophore (Fig. 2B, C). After the acquisition of two pairs of cirri, the lophophore becomes a trocholophe. The trocholophe can be everted for feeding or locomotion or retracted between the valves of the shell (Fig. 2B). There is considerable variation in the number of cirri present at the time of settlement. For example, in *Lingula*, if a suitable settlement substratum is encountered early in the pelagic phase, settlement can occur with as few as eight cirri, but settlement can be delayed until a suitable substratum is encountered during which time the larva adds cirri (Chuang, 1990). At settlement, *Lingula* larvae can have as many as 20 cirri. See Reed (1987) and Chuang (1990) for a more detailed presentation of the characteristics of larval brachiopods.

Crania californica is the single local inarticulate species. Its eggs are freely shed. The larvae are bilobed, demersal, and non-feeding.

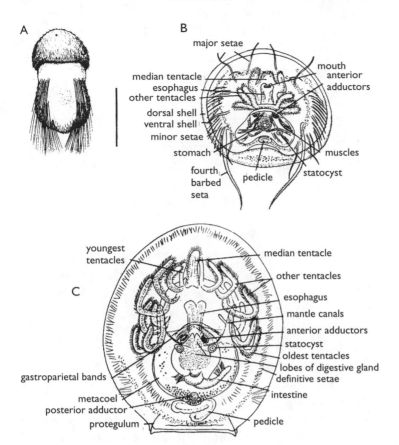

Fig. 2. Inarticulate brachiopod larvae. (A) Larva of *Crania;* there is one species in this genus locally. Scale = 100 mm. (B) Late-stage larva of *Pelagodiscus atlanticus.* (C) Late-stage larva of *Lingula* (A from Nielsen, 1990, Fig. 19; B, C from Hyman, 1959, Fig. 207)

References

Brusca, R. C. and G. J. Brusca (1990). Invertebrates. Sinauer Associates, Inc., Sunderland.

Chuang, S. H. (1990). Brachiopoda. In: Reproductive Biology of Invertebrates. Vol. IV, Part B. Fertilization, Development, and Parental Care, K. G. Adiyodi and R. G. Adiyodi (eds.), pp. 212–54. John Wiley & Sons, New York.

Hyman, L. H. (1959). The Invertebrates: Smaller Coelomate Groups. McGraw-Hill, London and New York.

Kozloff, E. N. (1996). Marine Invertebrates of the Pacific Northwest.Univeristy of Washington Press, Seattle. 539 pp.

Nielsen, C. (1990). The development of the brachiopod *Crania* (*Neocrania*) *anomala* (O. F. Müller) and its phylogenetic significance. Acta Zool. Stockh. 72:7–28.

Reed, C. G. (1987). Phylum Brachiopoda. In: Reproduction and Development of Marine Invertebrates of the Northern Pacific Coast, M. F. Strathmann (ed.), pp. 486–93. University of Washington Press, Seattle.

18

Echinodermata

Bruce A. Miller

The phylum Echinodermata is a morphologically, ecologically, and taxonomically diverse group. Within the nearshore waters of the Pacific Northwest, representatives from all five major classes are found—the Asteroidea (sea stars), Echinoidea (sea urchins, sand dollars), Holothuroidea (sea cucumbers), Ophiuroidea (brittle stars, basket stars), and Crinoidea (feather stars). Habitats of most groups range from intertidal to beyond the continental shelf; this discussion is limited to species found no deeper than the shelf break, generally less than 200 m depth and within 100 km of the coast.

Reproduction and Development

With some exceptions, sexes are separate in the Echinodermata and fertilization occurs externally. Intraovarian brooders such as *Leptosynapta* must fertilize internally. For most species reproduction occurs by free spawning; that is, males and females release gametes more or less simultaneously, and fertilization occurs in the water column. Some species employ a brooding strategy and do not have pelagic larvae. Species that brood are included in the list of species found in the coastal waters of the Pacific Northwest (Table 1) but are not included in the larval keys presented here.

The larvae of echinoderms are morphologically and functionally diverse and have been the subject of numerous investigations on larval evolution (e.g., Emlet et al., 1987; Strathmann et al., 1992; Hart, 1995; McEdward and Jamies, 1996) and functional morphology (e.g., Strathmann, 1971, 1974, 1975; McEdward, 1984, 1986a,b; Hart and Strathmann, 1994). Larvae are generally divided into two forms defined by the source of nutrition during the larval stage. Planktotrophic larvae derive their energetic requirements from capture of particles, primarily algal cells, and in at least some forms by absorption of dissolved organic molecules. These larvae are characterized by a bilaterally symmetrical arrangement of ciliated arms that form an elaborate feeding structure to capture, sort, and direct food particles toward the mouth (see Figs. 1, 3, 4, 7). These ciliated structures also provide propulsion and steering functions. Ingested food particles are processed within a simple gut, and fecal pellets are expelled through a

Table 1. Species in the phylum Echinodermata from the Pacific Northwest

	Larval Development Mode	Spawning Period
Class Echinoidea		
Order Clypeasteroida		
Dendraster excentricus (Eschscholtz, 1831)	pelagic, planktotrophic	Feb-Sep
Order Echinoida		
Allocentrotus fragilis (Jackson, 1912)	pelagic, planktotrophic	Late-Spr.
Strongylocentrotus franciscanus (A. Agassiz, 1863)	pelagic, planktotrophic	Mar-Jul
Strongylocentrotus purpuratus (Stimpson, 1857)	pelagic, planktotrophic	Dec-Jun
Strongylocentrotus droebachiensis (O.R. Muller, 1776)		
Order Spatangoida		
Brisaster latifrons (A. Agassiz, 1898)	pelagic, lecithotrophic	Spr.-Sum.
Class Asteroidea		
Order Forcipulatida		
Evasterias troschelii (Stimpson, 1862)	pelagic, planktotrophic	Apr-Jun
Leptasterias hexactis (Stimpson, 1862)	brooder, lecithotrophic	Nov-Apr
Orthasterias koehleri (de Loriol, 1897)	pelagic, planktotrophic	May-Aug
Pisaster brevispinus (Stimpson, 1857)	pelagic, planktotrophic	Apr-Jun
Pisaster giganteus (Stimpson, 1857)	pelagic, planktotrophic	Apr-Jun
Pisaster ochraceus (Brandt, 1835)	pelagic, planktotrophic	Apr-Jun
Pycnopodia helianthoides (Brandt, 1835)	pelagic, planktotrophic	Mar-Jul
Stylasterias forreri (de Loriol, 1887)	pelagic, planktotrophic	Apr-Jun
Order Platyasterida		
Luidia foliolata Grube, 1886	pelagic, planktotrophic	Apr-Jun
Order Spinulosida		
Solaster dawsoni Verrill, 1880	pelagic, lecithotrophic	Mar-Jun
Solaster stimpsoni Verrill, 1880	pelagic, lecithotrophic	Mar-Jun
Pteraster tesselatus Ives, 1888	pelagic, lecithotrophic	Jun-Aug
Crossaster borealis (Fisher, 1906)		
Henricia spp.	brooders/pelagic, lecithotrophic	Win.-Spr.
Order Valvatida		
Ceramaster articus (Verrill, 1909)		
Ceramaster patagonicus (Sladen, 1889)		
Hippasteria spinosa Verrill, 1909	pelagic, planktotrophic	
Mediaster aequalis Stimpson, 1857	pelagic, lecithotrophic	Apr-Jun
Asterina miniata (Brandt, 1835)	pelagic, planktotrophic	May-Jul
Dermasterias imbricata (Grube, 1857)	pelagic, planktotrophic	Apr-Aug
Class Holothuroidea		
Order Aspidochirotida		
Parastichopus californicus (Stimpson, 1857)	pelagic, planktotrophic	Apr-Aug
Parastichopus leukothele Lambert, 1986	pelagic, planktotrophic	?
Parastichopus parvimensis H. L. Clark, 1913	pelagic, planktotrophic	?
Order Dendrochirotida		
Psolidium bulatum Ohshima, 1915		
Psolus squamatus (Koren, 1844)		
Psolus chitonoides H. L. Clark, 1901	pelagic, lecithotrophic	Mar-May

table continues

Table 1 continued. Species in the phylum Echinodermata from the Pacific Northwest

	Larval Development Mode	Spawning Period
Eupentacta pseudoquinquesemita Deichmann, 1938		
Eupentacta quinquesemita (Selenka, 1867)	pelagic, lecithotrophic	Mar-May
Pentamera lissoplaca (H.L. Clark, 1924)		
Pentamera populifera (Stimpson, 1864)		
Pentamera pseudocalciegera Deichmann, 1938		
Pentamera trachyplaca (H.L. Clark, 1924)		
Havelockia benti (Deichmann, 1937)		
Cucumaria miniata (Brandt, 1835)	pelagic, lecithotrophic	Mar-May
Cucumaria piperata (Stimpson, 1864)	pelagic, lecithotrophic	Mar-May
Cucumaria pseudocurata Deichmann, 1938	brooder, lecithotrophic	Dec-Jan
Cucumaria fallax Ludwig, 1894		
Cucumaria lubrica H.L. Clark, 1901		

Order Molpadiida

Molpadia intermedia (Ludwig, 1894)		
Paracaudina chilensis (J. Müller, 1850)	pelagic, lecithotrophic	Mar-Apr

Order Apodida

Leptosynapta clarki Heding, 1928
Leptosynapta roxtona Heding, 1928
Leptosynapta transgressor Heding, 1928

Class Ophiuroidea
Order Ophiurida

Ophiopholis aculeata (Linnaeus, 1767)	pelagic, planktotrophic	Spr.-Fall
Amphipholis squamata (Delle Chiaje, 1829)	ovoviviparous brooder	Sum.-Fall
Amphipholis pugetana (Lyman, 1860)	pelagic, planktotrophic	?
Amphioplus macraspis (H.L. Clark, 1911)		
Amphioplus strongyloplax (H.L. Clark, 1911)		
Amphiodia occidentalis (Lyman, 1860)	benthic, lecithotrophic	Spr.-Sum.
Amphiodia urtica (Lyman, 1860)	pelagic, planktotrophic	?
Gorgonocephalus eucnemis (Müller & Troschel, 1842)	benthic, lecithotrophic	Winter
Ophiopteris papillosa (Lyman, 1875)	pelagic, planktotrophic	?
Ophiura leptoctenia H.L. Clark, 1911		
Ophiura lütkeni (Lyman, 1960)		
Ophiura sarsi Lütkeni, 1855		

Class Crinoidea

Florometra serratissima (A. H. Clark, 1907)	pelagic, lecithotrophic	all year

ventral anus. Because development of these larvae is dependent on an external food source, few lipid reserves are provided in the egg, and species with this type of larval form generally produce relatively large numbers of small eggs (80–170 µm in echinoids, 120–200 µm in asteroids). Temperature and food supply influence larval development rate, which determines the length of time required to reach the stage at which larvae are competent to settle and metamorphose to the benthic juvenile stage. In the northeast Pacific, this larval period may last from three weeks to several months. Planktotrophy is common within the echinoids, asteroids, and ophiuroids but rare within the holothuroids and absent in the crinoids.

Lecithrotrophic, or non-feeding, larvae derive their nutritional requirements from yolk and lipid reserves supplied in the egg. Because these reserves must be sufficient to supply all the energetic requirements until metamorphosis, eggs from lecithotrophic species are larger (400–625 µm in holothuroids, 750–1,500 µm in asteroids) and fewer in number than eggs of species with feeding larvae. At least in the echinoids and ophiuroids, lecithotrophic larvae have evolved from species with feeding larval forms and, in some cases, have retained internal vestiges of larval feeding arms (Hendler, 1978, 1982; Amemiya and Emlet, 1992; Emlet, 1995). Because feeding from external sources does not occur, these larvae do not develop elaborate feeding and propulsion structures. The larval body either is evenly covered with cilia or has well-defined rows of cilia, providing swimming propulsion (see Figs. 5, 6, and Emlet, 1994a). This larval form is found throughout each class of echinoderms but is most prevalent within the holothuroids and is found exclusively within the crinoids. Larvae that are brooded by the adult are also non-feeding. Brooding is most common in the ophiuroids, crinoids, and some groups of asteroids.

A third type of larval form, the facultative planktotroph, has characteristics of both feeding and non-feeding larvae. This larval form may develop by feeding in the plankton, but if deprived of food it is capable of developing through metamorphosis on energy and material provided in the egg (Emlet, 1986; Hart, 1996). This type of larva is represented in local waters by one echinoid species, the heart urchin, *Brisaster latifrons* (Strathmann, 1979; Hart, 1996).

Although the type of larval form (feeding or non-feeding) is known for most nearshore species of echinoderms, published descriptions of larval morphology for some groups are incomplete. Larvae described in the literature include most of the shallow-water echinoids and holothuroids found off the Pacific Northwest. The majority of ophiuroid larvae have not been described. The larvae of some asteroid species have been described, particularly the non-feeding and brooding forms, but little has been published on species with planktotrophic forms.

The biology of echinoderms is reviewed in Hyman (1955), and reproduction and development is reviewed in Okazaki (1975) and Giese et al. (1991). Most of the following discussion on development and larval morphology of local species is taken from Strathmann (1987), which provides the most comprehensive reference on reproduction and development of echinoderms found in the northeast Pacific.

Collecting and Observing Echinoderm Larvae

Echinoderm larvae, particularly species with an extended spawning period (e.g., *Dendraster excentricus*), can be found in the nearshore plankton from late winter through fall. Echinoplutei are most abundant at depths of 5–10 m. Nearshore, both asteroid and ophiuroid larvae can be found at most depths but are frequently most abundant slightly deeper (10–20 m) in the water column, closer to the pycnocline. Early stages of most holothuroid larvae are positively buoyant and, hence, are most abundant near the surface. Holothuroid larvae at advanced stages of development may be found at all depths (Miller, 1995).

A plankton net and cod-end bucket with 202 µm mesh will capture larvae as small as four-arm and most prism-stage pluteus larvae. Handling live larvae in the laboratory should be done with 100 µm mesh or finer, to minimize damage to the arm epithelia and surface cilia. Undamaged bipinnariae, brachiolariae, and pluteus larvae may be kept in the laboratory for several days without feeding. For observation of later stages, larvae must be fed and the water in culture jars stirred and changed periodically. Refer to Strathmann (1987) and Leahy (1986) for laboratory culture of echinoderm larvae. With careful culture techniques (and, perhaps, a bit of luck) larvae (feeding and non-feeding forms) may complete development through metamorphosis in the laboratory. Competent stages of most echinoid and holothuriod larvae, and some asteroid larvae, can be induced to metamorphose in the laboratory if provided with a substrate covered with a bacterial film. For many species, natural rock covered with coralline algae is sufficient to induce metamorphosis (Miller, 1995).

Material that is to be preserved should be immediately transferred to 1.5% buffered formaldehyde (5% formalin). An excess of sodium borate (widely available as Borax) added to the stock formaldehyde solution is an adequate buffer. The calcareous parts of echinoderm larvae dissolve quickly in formalin solutions that are not buffered. Larvae preserved in this manner will retain their color for several weeks and remain in good condition for at least several months. Larvae that are to be kept for longer periods should be transferred to buffered ethanol after fixing in formalin.

The calcareous arm spicules and ossicles of echinoderm larvae can be difficult to see under normal transmitted light. To highlight the birefringent skeletal rods and ossicles, larvae are best observed with substage illumination and cross-polarized light. Polarizing filters are placed above and below the sample and oriented such that the planes of light transmission are 90° to each other. Polarizing filter accessories are available from some microscope companies; however, a more economical approach is to mount a standard rotating polarizing filter designed for camera lenses onto the bottom of the microscope objective. A mount can be readily made by cutting a section of PVC pipe that will slip over the outside of the objective lens and gluing the filter onto the bottom of the PVC section. Make sure the glue does not prevent the filter from rotating. The mount can be tapped for a set screw to hold it in place on the microscope objective. There must also be polarizing film below the material to be viewed. Plastic polarizing film is available at most photography suppliers and can be ordered from Edmund Scientific Company, Barrington, New Jersey. Cut a piece of this film to fit under the microscope stage. If heat from substage illumination is a problem, place the filter on top of the stage and cover it with a sheet of glass. In use, the filter mounted on the objective is rotated until the background is dark and calcareous structures become refractive, that is, until they appear illuminated against a dark background. Under cross-polarized light, other characteristics such as color are difficult to discern, so it may be necessary to alternate between polarized and incident light by rotating the filter on the microscope objective. A secondary fiber optic light for incident light is helpful. This method is useful for sorting and identification of any larvae or material with calcareous parts, such as mollusc larvae and some sponge spicules.

Morphology

Class Echinoidea

Local echinoids include three species of sea urchins (*Strongylocentrotus franciscanus*, *S. purpuratus*, *Allocentrotus fragilis*), a sand dollar (*Dendraster excentricus*), and a heart urchin (*Brisaster latifrons*) (Table 1). All are free-spawning and have pelagic larvae called echinoplutei (Fig. 1). Larvae from at least one of these species can be found from late winter through fall. Natural spawning of each species occurs over a period of one to several months, and some populations may shed gametes several times during a season (Chatlynne, 1969; Gonor, 1973; Cameron and Rumrill, 1982). With the exception of *B. latifrons*, local species have small eggs, 80–170 μm diameter.

Cleavage is radial and holoblastic. At 12°C, a ciliated blastula hatches at ca 24 hours after fertilization. The hatched swimming blastula has a thick posterior vegetal pole where gastrulation by invagination begins. Gastrulation proceeds until the archenteron fills almost three-fourths the length of the blastula. The archenteron then turns toward one side, which becomes the flat ventral side of the prism stage of the larva. At the base

Fig. 1. Echinoplutei.
(A) *Dendraster excentricus*, four-armed stage. (B) *Dendraster excentricus*, eight-armed stage. (C) *Brisaster latifrons*, six-armed stage. (D) Fenestrated arm rod of *D. excentricus*. (E) *Strongylocentrotus* sp., late prism–early pluteus stage. (F) *Strongylocentrotus purpuratus*, eight-armed stage. (G) *Strongylocentrotus franciscanus*, eight-armed stage, only posterior region and posterior pedicellaria shown. (A–D, F–G, ventral view; E, right lateral view. Abbreviations: anus (a), anterolateral arm (ala), anterolateral rod (alr), body rod (br), ciliated band (cb), epaulette (ep), intestine (i), juvenile rudiment (jr), oesophagus (oe), pedicellariae (ped), posterodorsal arm (pda), postoral arm (poa), postoral rod (por), preoral arm (pra), posterior process (pp). (Adapted from Strathmann, 1971, 1979, 1987)

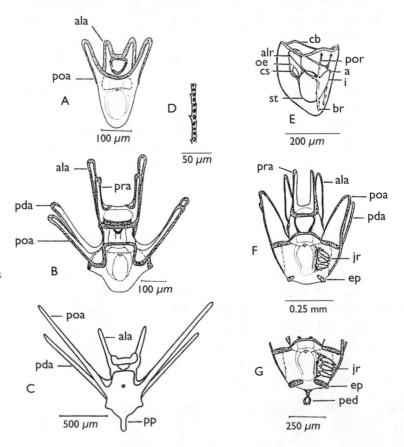

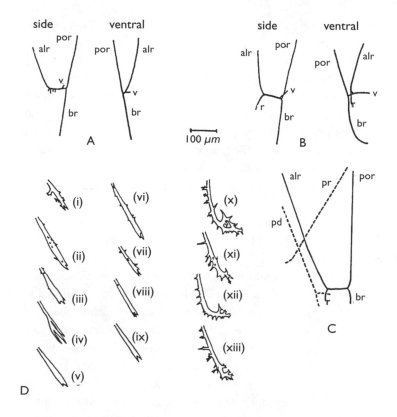

side   ventral   side   ventral

A

100 μm

B

D   (i) (vi) (x)
     (ii) (vii) (xi)
     (iii) (viii) (xii)
     (iv) (ix) (xiii)
     (v)

C

Fig. 2. Calcareous skeletal rods of strongylocentrotids in the four-armed stage, side view on left, ventral view on right. (A) Arrangement in *Strongylocentrotus purpuratus* and *Allocentrotus fragilis*, four-armed stage. (B) Arrangement in *Strongylocentrotus franciscanus*, four-armed stage. (C) Arrangement in *Strongylocentrotus franciscanus*, eight-armed stage. (D) Distal ends of body rods in four-armed and early six-armed stages; (i)–(v), *S. purpuratus*; (vi)–(ix), *A. fragilis*; (x)–(xiii), *S. fransicanus*. Abbreviations: anterolateral rod (alr), body rod (br), posterodorsal rod (pd), postoral rod (por), preoral rod (pr), recurrent rod (r), ventral transverse rod (v). (Adapted from Strathmann, 1979)

of the archenteron, the primary mesenchyme cells aggregate into two masses within which the triradiate primary spicules of the larval skeleton are secreted. These calcareous spicules, which eventually form the supportive structures for the larval arms, are key diagnostic traits for distinguishing between echinoid species (Fig. 2).

By four to seven days (depending on species and temperature), an early four-armed pluteus larva has formed. Larval growth involves both the addition of arms and increase in their size. The anterolateral and postoral pairs of arms appear first, with a third rod extending posteriorly from each primary spicule to form a body rod (Fig. 1E). Small medially projecting rods, the ventral transverse rods, also develop from the primary spicules. Next, the larva develops a pair of posterodorsal arms and a dorsal skeletal arch with arms that extend into the oral lobe and become the preoral arms of the eight-armed echinopluteus (e.g., Fig. 1B, 1F). The body rods thicken and elaborate posteriorly; and in *Dendraster excentricus* they form a basket-like structure. The larva of *Brisaster latifrons* also develops a single posterior process supported by a calcareous rod (Fig. 1C).

The larval arms and the preoral lobe support a band of ciliated cells that form a continuous loop that extends over the ciliated oral region and mouth. This ciliated band is used for feeding and swimming. Larvae of the three species of sea urchin also develop two additional bands of cilia located posterially, epaulettes, which are important for locomotion.

After development of eight arms, a juvenile rudiment begins to form on the left side of the larva, between the postoral and posterodorsal arms. The juvenile rudiment is the developing cell mass that forms all the structures of the juvenile sea urchin. The rudiment forms within an invagination of the body wall, the vestibule. Five primary tube feet (podia) of the developing juvenile can extend from the vestibule and are used to attach to the substratum when the larva is ready to metamophose. The primary spines of the juvenile may also be seen protruding from the vestibule.

In larvae competent to metamorphose, the rudiment may be viewed as an "inside-out" juvenile sea urchin, attached to the larval body. The primary tube feet extend from what will become the oral surface, and the primary spines originate from what will become the aboral surface of the juvenile. Upon settlement and metamorphosis, most of the larval body is resorbed, including the arm tissues, and the rudiment tissues evert in such a manner to cover the aboral surface with the calcareous plates associated with each primary spine. For a detailed description of echinoid metamorphosis, see Chia and Burke (1978).

Key to echinoid larvae (modified from Strathmann, 1979)

1a. Postoral and posterodorsal arm rods fenestrated (Fig. 1D) 2
1b. Arm rods not fenestrated 3

2a. Larvae opaque, with orange-red pigment spots (when living, and for several weeks when preserved in buffered formalin); posterior process present by late 4-armed stage ....
.... *Brisaster latifrons*
2b. Most of larval surface transparent, typically with pale green pigment spots; no posterior process; posterior distal ends of body and arm rods curve inward to form basket-like structure in 8-armed stage; no posterior epaulettes in 4-armed stage (Fig. 1B) .... *Dendraster excentricus*

3a. With 4 or 6 arms 4
3b. With 8 arms 6

4a. Ventral transverse rod prominent and meeting at midline in 4- and early 6-armed stage (Fig. 2B); recurrent rods prominent; distal end body rods very thorny and curved toward midline (before late 6-armed stage, Fig. 2D); small pigment spots deep red .... *Strongylocentrotus franciscanus*

4b. Ventral transverse rods short and not meeting at midline;
recurrent rods absent or not prominent; body rods not curved
toward midline; no posterior pedicellariae; small pigment spots
may be other than deep red 5

5a. Pigment spots (deep red) over most of larval surface ....
.... *Strongylocentrotus purpuratus*

5b. Pigment spots few and inconspicuous; pink or orange pigment
spots only in arms, or near ciliated band .... *Allocentrotus fragilis*

6a. Recurrent rod still present in early 8-armed stage; I pedicellaria
forms at posterior and 2 on right side during 8-armed stage;
posterior transverse rod present and associated with
developing posterior pedicellaria; with epaulettes; pigment spots
deep red .... *Strongylocentrotus franciscanus*

6b. Recurrent rods absent or not prominent; no pedicellaria form
on larva prior to metamorphosis; posterior spicules (if present)
not extended as transverse rod; with epaulletes; pigment spots
may not be deep red 7

7a. Pigment spots deep red .... *Strongylocentrotus purpuratus*
7b. Pigment spots orange or pink .... *Allocentrotus fragilis*

Class Asteroidea

Sea stars found in nearshore waters of the Pacific Northwest
are represented by four orders: Platyasterida, Valvatida,
Spinulosida, and Forcipulatida (Table 1). Two species brood

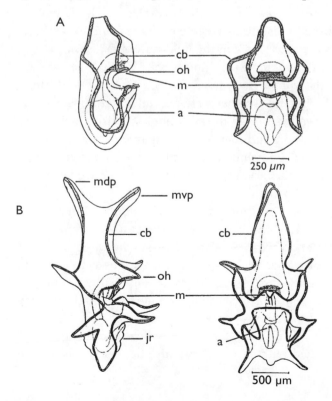

Fig. 3. Asteroid bipinnaria,
lateral (left) and ventral
(right) views. (A) *Asterina
miniata* (general form of
forcipulate bipinnaria).
(B) *Luidia foliata.*
Abbreviations: anus (a),
ciliated band (cb),
juvenile rudiment (jr),
mouth (m), median
dorsal process (mdp),
median ventral process
(mvp), oral hood (oh).
(Adapted from
Strathmann, 1979)

Fig. 4. Brachiolaria of *Pisaster ochraceus* (general form of forcipulate brachiolaria), (A) lateral and (B) ventral views. Abbreviations: anus (a), adhesive disc (add), anterodorsal arm (ada), anterolateral arm (ala), brachiolar arm (ba), ciliated band (cb), mouth (m), median dorsal process (mdp), posterodorsal arm (pda), posterolateral arm (pla), postoral arm (poa), preoral arm (pra). (Adapted from Strathmann, 1971).

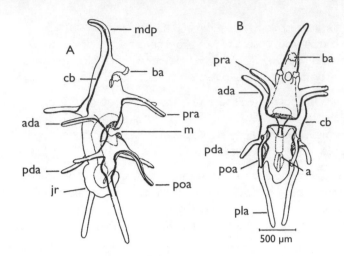

(*Leptasterias hexactis* and *Henricia* sp.), and the remainder have pelagic larvae. Both feeding and non-feeding larval forms are found. Feeding larvae are pelagic for one to several months; non-feeding larvae develop on the order of two to four weeks. Oguro (1989) provides a discussion of development and larval types in asteroids.

The early stage of feeding larvae is the bipinnaria (Fig. 3), and a later stage with attachment arms is the brachiolaria (Fig. 4). The bipinnaria swims and feeds by means of a ciliated band that forms a continuous loop along the sides of the body, the larval arms, and around the anus. An anterior ventral band of cilia becomes a separate loop, the preoral loop. This latter trait may be used to distinguish asteroid larvae from the similarly shaped auricularia larva of the holothuroid *Parastichopus* sp., which has only a single continuous band of cilia (see Fig. 6A). At 12°C, the common ochre seastar, *Pisaster ochraceus*, forms a feeding bipinnaria five days after fertilization (Fraser et al., 1981). In advanced larvae, the posterior left hydrocoel develops five lobes that are rudiments of the hydrocoels of the rays of the juvenile. The juvenile rudiment develops posteriorly on the left side of the larval body with the oral surface facing the left side of the body.

In the orders Forcipulatida and Valvatida, and in some of the Spinulosida, the brachiolaria develops three anterior brachiolar arms with glandular tips surrounding a central adhesive disc (Fig. 4). These structures are used to attach to the substratum when the larva is metamorphically competent. Upon settlement, the developing juvenile has a single primary tube foot and ocellus (eye spot) on the end of each ray. In the forcipulates, six primary calcareous ossicles (one central, plus one associated with each ray) can also be seen on the aboral surface.

Pelagic non-feeding larvae are found in the orders Spinulosida (*Pteraster tesselatus*, *Henricia* spp., *Solaster stimpsoni*, *S. dawsoni*) and Valvatida (*Mediaster aequalis*). The large, yolky eggs (0.75–1.5 mm diam.) and early larvae are usually positively buoyant and can be found near the surface. The early ciliated larva is opaque and in some species colored bright red-orange. A local species of the genus *Henricia*, *M. aequalis*, *S. stimpsoni*, and *S. dawsoni* develop into modified brachiolarias with anterior brachiolar arms and an adhesive disc that differ from those found in the forcipulates (Fig. 5). *Pteraster tesselatus* does not develop brachiolar arms.

The forcipulate *Lepasterias hexactis* broods its larvae. Large yolky eggs are shed under the female body, where they are fertilized by sperm shed into the surrounding water. Larvae of *L. hexactis* are brooded under the female. They develop tube feet by 40 days and crawl away as small juveniles at two months (Chia, 1966, 1968).

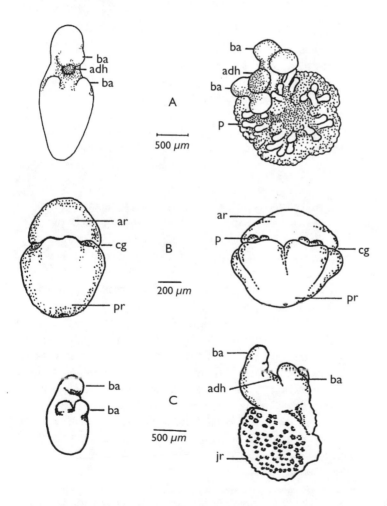

Fig. 5: Larval forms of lecithrotrophic (non-feeding) asteroids. (A) Brachiolaria of a species of *Henricia*, frontal view (left) and during metamorphosis (right), left side of larva and ventral (oral) side of juvenile. (B) *Pteraster tesselatus*, at six days (left), lateral view, and at eight days (right), lateral view. (C) *Mediaster aequalis*, brachiolaria at 10 days (left), frontal view, and advanced brachiolaria at ~25–30 days (right), right side of larva and dorsal (aboral) side of juvenile. Abbreviations: adhesive disc (add), anterior region (ar), brachiolar arm (ba), circumferential groove (cg), juvenile rudiment (jr), podium (p), posterior region (pr). A adapted from Strathmann, 1987; B adapted from McEdwards, 1992; C adapted from Birkeland et al., 1971)

Key to asteroid larvae

1a. Larval form with paired feeding arms or lobes, with 2 continuous bands of cilia along arms (e.g., Figs. 3, 4); mouth and anus present; transparent, some surface pigmentation may be present; no internal calcareous skeleton or ossicles 5

1b. No feeding structures or mouth; opaque 2

2a. Color usually orange, may vary from yellow to dark red 3

2b. Color light tan or pale olive-green; with modified brachiolar arms; with groups of podia on juvenile rudiment in advanced larvae .... *Solaster dawsoni* or *Solaster stimpsoni*

3a. With modified brachiolar arms (e.g., Figs. 5A, 5C) 4

3b. No brachiolar arms; color usually orange (may vary from light yellow to dark red); 1-day old larvae ovoid, evenly ciliated; older larvae with circumferential groove, forming anterior and posterior regions; 5 bulges around circumference on posterior region; 5 clusters of podia (1 terminal podium, 2 pairs adjacent podia) within circumferential groove, each centered under bulge on posterior region (Fig. 5B) .... *Pteraster tesselatus*

4a. 3 brachiolar arms (one median anterolateral, 2 ventrolateral, Fig. 5C); advanced larvae with disk of developing juvenile (1.2–1.3 mm diam.) at posterior end; no podia form prior to settlement; color orange .... *Mediaster aequalis*

4b. 4 brachiolar arms (2 anterodorsal, 2 ventrolateral) and single stalked adhesive disc between paired brachiolar arms (Fig. 5A); advanced larvae with juvenile disk on left side and 5 groups of podia on oral (left) surface of juvenile disk; color orange .... *Henricia* sp.

5a. Bipinnaria larger than same stage in other groups (>2.6 mm long); no brachiolaria stage; elongate median dorsal and ventral process (region anterior to oral hood one-half total body length) (Fig. 3) .... *Luidia foliatata*

5b. Bipinnaria not as above, or with feeding brachiolaria stage; long, transparent feeding and locomotory arms in brachiolaria stage, with band of cilia along preoral arms and surrounding mouth, and separate band along all other arms (e.g., Fig. 4); some pigmentation may be present (pale green, gold, orange red), usually on distal ends of arms; advanced larvae with juvenile disk on left side at posterior end (pigmented same color as arms) and brachiolar arms and adhesive disk at anterior end; calcareous ossicles and early juvenile spines may be present on juvenile disk in advanced larvae .... valvatid or forcipulate brachiolariae (*Asterina miniata*, *Dermasterias imbricata, Evasterias troschelii Orthasterias koehleri, Pisaster brevispinus, P. giganteus, P. ochraceus, Pycnopodia helianthoides, Stylasterias forreri*)[1]

[1]Distal ends of arms and juvenile disk are pigmented pale gold in *P. ochraceus* and orange-red in *P. helianthoide;* other species may have similar pigmentation. Coloration in *A. miniata* may be variable (S. Rumrill, pers. comm.).

Class Holothuroidea

The nearshore sea cucumbers are represented by nine species within three orders (Table 1). Three of these species, all within the genus *Parastichopus*, have a pelagic feeding stage (auricularia larva) that develops into a non-feeding doliolaria larva before metamorphosis. Two of these species, *P. leukothele* and *P. parvimensis*, are generally found in deeper water on the continental shelf, but because the feeding larva of these species may be pelagic for one to two months their larvae may occur nearshore. The dendrochirotid *Cucumaria pseudocurata* is a brooder, and thus larvae of this species are not found in the plankton. All other species produce pelagic, non-feeding, doliolaria larvae that metamorphose to benthic juveniles in six to thirteen days. Holothuroid reproduction and development are reviewed in Hyman (1955) and Giese et al. (1991). Descriptions of larval forms of northeast Pacific holothuroids have been made by Mortenson (1921), Johnson and Johnson (1950), Young and Chia (1982), McEuen and Chia (1985), Cameron (1985), and McEuen (1986).

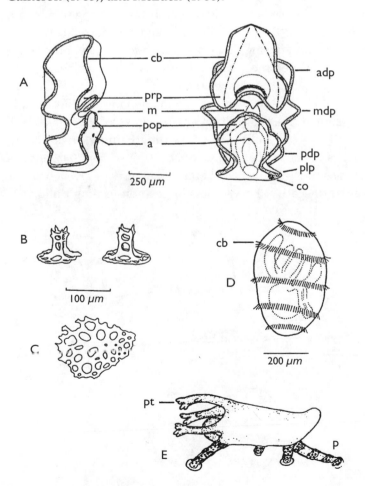

Fig. 6. Larval forms of holothuroids. (A) Auricularia stage of *Parastichopus californicus*, lateral view on left, ventral view on right. (B) Fenestrated table ossicles from pentacula stage larva of *Parastichopus* sp. (C) General form of flat ossicles from pentacula stage larvae. (D) Doliolaria stage of *Parastichopus californicus*. (E) General form of early pentacula stage larva. Abbreviations: anus (a), anterodorsal process (adp), ciliated band (cb), calcareous ossicle (co), mouth (m), median dorsal process (mdp), podium (p), posterodorsal process (pdp), posterolateral process (plp), postoral process (pop), preoral process (prp), primary feeding tentacles (pt). (A adapted from Strathmann, 1971; B adapted from Lambert, 1986; D adapted from Strathmann, 1987; E adapted from Johnson, 1931)

Pelagic planktotrophic development within the northeast Pacific holothuroids is restricted to the genus *Parastichopus*. The early larva, the auricularia, develops a looped band of cilia for both feeding and swimming (Fig. 6A). Further growth produces arms and lobes over which the single, continuous band is looped. This single band of cilia distinguishes the auricularia larva from the similar early bipinnaria stage larva of asteroids, which develops two separate ciliated bands. A second trait useful for distinguishing auricularia is the presence of a calcareous ossicle in the left posterolateral lobe. Further, the anterior colomic cavities of the auricularia do not enlarge and extend forward on both sides of the esophagus as they do in asteroids. Larvae are transparent with occasional tinting along the ciliary band. In preparation for metamorphosis, the larva of *Parastichopus* develops into a simplified, barrel shape, the doliolaria larva (Fig. 6D). The ciliary band rearranges into five transverse rings of cilia. Five primary tentacles develop and push through the oral indentation. Protrusion of the tentacles marks the onset of the pentacula stage (Fig. 6E). The tentacles are used to attach to the substrate during settlement, when the benthic juvenile stage begins. Calcareous ossicles begin to form during the pentacula stage and cover the larval surface at metamorphosis. These ossicles can be used to distinguish between some species.

In other groups of holothuroids, pelagic lecithrotrophic development begins with large yolky eggs (267–627 μm diam.), usually brightly pigmented, extruded singly or bound in strings or pellets that disperse soon after release. Larger eggs are positively buoyant, whereas smaller eggs are neutral or slightly negative. Larvae develop as non-feeding doliolaria,

Fig. 7. Doliolaria stage larval forms of holothuroids. (A) *Eupentacta quinquesemita*, ventral view. (B) *Psolus chitonoides*. (C) *Cucumaria* sp., advanced stage, ventral view. (D) *Paracaudina chilensis*, ventral view on left, left side view on right. Abbreviations: ciliated band (cb), oral indentation (o), podial pit (ppt), primary tentacle (pt), vestibule (v). (A, B adapted from Strathmann, 1987; D adapted from Inaba, 1930)

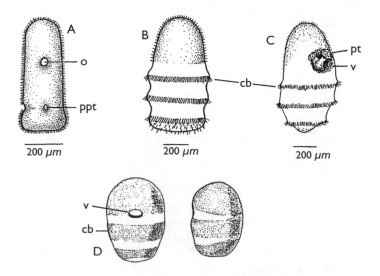

usually retaining the color of the egg (Fig. 7). Early doliolaria are evenly ciliated, but older larvae develop two or three transverse rings of cilia. These rings first form as epidermal ridges along the posterior two-thirds of the body, and cilia between the ridges are eventually lost. The preoral lobe, which is the pigmented yolk reservoir anterior to the oral indentation, retains its uniform ciliation. As part of metamophosis, larvae develop and extend five primary podia through the oral indentation, marking onset of the pentacula stage. After settlement in the dendrochirotids, two podia also develop from the podial pits posterior to the oral indentation.

Key to holothuroid larvae

1a. Transparent larva (auricularia) with single, continuous band of cilia along arms and lobes; calcareous ossicle present in left posterolateral lobe (Fig. 6A) .... *Parastichopus* spp.
1b. Opaque, pigmented, barrel-shaped larva (doliolaria) (Fig. 6D) ....
.... 2

2a. 5 transverse ciliary bands present (Fig. 6D); yellow-tan pigmentation; late pentacula stage covered with fenestrated "table" ossicles (Fig. 6B), which protrude from body surface ....
.... *Parastichopus* spp.
2b. Evenly ciliated or 2–3 ciliary bands present 3

3a. No ciliary bands (evenly ciliated); pigmentation of preoral lobe light green, posterior body semitranslucent white; pentacula semiopaque white or light tan; late pentacula with small ossicles perforated with fine pores (Fig. 7A) .... *Eupentacta quinquesemita*
3b. 2–3 ciliary bands 4

4a. Doliolaria and pentacula red-orange, with yellow-orange posterior lobe in older larvae; 3 ciliated bands by 7 days of age (Fig. 7B) .... *Psolus chitonoides*
4b. Pigmentation not red-orange 5

5a. 2 ciliary bands present (1 in early doliolaria); pigmentation brown-red in doliolaria and pentacula (Fig. 7D) ....
.... *Paracaudina chilensis*
5b. 3 ciliary bands present, pigmentation of preoral lobe olive green, light green posterior; late pentacula with large ossicles perforated with large pores (Fig. 7C) .... *Cucumaria* sp.

Class Ophiuroidea

Ophiuroids display a wide variety of reproductive strategies: (1) pelagic development with planktotrophic larvae; (2) pelagic, demersal, or benthic development with lecithotrophic larvae; (3) external brooding; (4) ovoviviparous or viviparous bursal brooding; and (5) asexual development. The seven species of ophiuroids found in nearshore waters off Oregon display at least three of these strategies (Table 1). Four of these species have pelagic, planktotrophic larvae. One species (*Amphipholis squamata*) is an ovoviviparous brooder. The lecithotrophic larvae of the basket star, *Gorgonocephalus eucnemis*, have not been fully described, but development is probably by external brooding or as a benthic or demersal larva on or in soft corals. Pelagic development has been inferred from egg size for larvae of *Amphiodia occidentalis* (Rumrill, 1982), but further work has shown this species to have lecithotrophic larvae that probably develop by external brooding or as benthic or demersal larvae (R. Emlet, pers. comm.). Development is similar among species with small eggs and planktotrophic larvae. Patterns of development are modified in species with lecithotrophic larvae and in species with internal development.

Species with planktotrophic development have small eggs (70–200 μm diam.) that are free-spawned, with external fertilization. Larvae develop into an ophiopluteus with a pelagic period of 20–90 days. Cleavage is radial, holoblastic, and equal. Larvae hatch as a ciliated coeloblastula. Primary mesenchyme cells migrate inward at the vegetal pole, and gastrulation takes place as an invagination of the flattened vegetal plate. The larval mouth forms where the invaginating archenteron fuses with a stomodeal depression in the body wall. The archenteron also gives rise to the larval gut, with the anus at the site of the blastopore. Groups of primary mesenchyme cells on either side of the larval archenteron secrete a pair of triradiate or tetraradiate calcareous spicules. The skeletal rods that support the larval body and arms develop from these spicules.

The ophiopluteus larva is distinguished primarily by the long widely spreading posterolateral arms, which develop first. Anterolateral arms develop next, followed by postoral arms that form as a pair of extensions that branch from the junction between the posterolateral and anterolateral arms. The postoral arms are the last to develop, completing the development of the eight-armed ophiopluteus (Fig. 8A). Continuous bands of cilia loop around the arms and larval body of planktotrophic ophioplutei. The cilia are used for swimming and to collect and transport food particles to the mouth.

The juvenile rudiment develops in a mid-ventral position in all planktotrophic ophioplutei, but metamorphosis occurs by two different patterns. In type 1 metamorphosis, the anterolateral, posterdorsal, and postoral arms are gradually resorbed. The rudiment continues to develop while attached

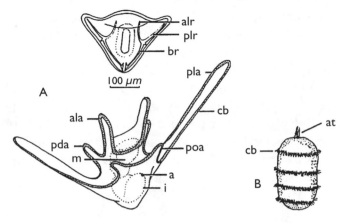

Fig. 8. Larval forms of ophiuroids and crinoids. (A) Ophioplutei of *Ophiopholis aculeata*, two-armed stage (above), ventral view, and eight-armed stage (below), oblique view. (B) Doliolaria of the crinoid *Florometra serratissima*. Abbreviations: anus (a), anterolateral arm (ala), anterolateral rod (alr), apical tuft (ap), body rod (br), ciliated band (cb), intestine (i), mouth (m), posterodorsal arm (pda), posterolateral arm (pla), postoral arm (poa). (A adapted from Strathmann, 1987; B adapted from Emlet, 1999b)

to the long posterolateral arms, which are finally shed just prior to or following settlement. The larvae of *Ophiopholis aculeata* display this type of metamorphosis. In type 2 metamorphosis, all four pairs of arms are resorbed into the rudiment. In both types of metamorphosis, a terminal podium and pairs of podia develop along each radial extension of the hydrocoel. Six branched spicules form on the right side of the larva which will become the central and radial plates on the aboral surface of the juvenile disc. Five spicules also form on the left side of the larva and are rudiments of the terminal plates adjacent to the juvenile arms.

Ophiopholis aculeata is the only species with pelagic planktotrophic larvae found in Pacific Northwest waters that has been thoroughly described (Olsen, 1942, Fig. 7A). Although ophioplutei develop characteristic skeletal traits and pigmentation on the arms and ciliated band which may permit differentiation between species, descriptions of the larvae of local species have not been made. Because larvae may be pelagic for several months, it is also possible that nearshore collections of plankton may include larvae from species found in the deeper waters of the continental slope. Little is known of development modes of these deep-water species.

Class Crinoidea

Only one species of crinoid, the feather star, *Florometra serratissima*, is found in nearshore northeastern Pacific waters. Adults may be found in water ranging from 11 to 1,200 m depth, but this species has been observed to be abundant on rocky reef habitat at depths 50–150 m off Oregon (Oregon Department of Fish and Wildlife, unpub. data). Although early development of *F. serratissima* has been described, nothing is known of the embryology of the dozens of species of crinoids found in deeper waters.

Sexes are separate and fertilization is external. In British Columbia, *F. serratissima* is reported to spawn year-round (Mladenov, 1986). Eggs are pale pink and 207 μm in diameter. Cleavage is radial, holoblastic, and asynchronous. Gastrulation occurs by invagination. Larvae hatch at about one and a half days as actively swimming, evenly ciliated gastrulae with an apical tuft of cilia at the anterior end (Mladenov and Chia, 1983). By four days, larvae have developed four circumferential bands of cilia, a vestibular invagination on the mid-ventral surface, and an adhesive pit on the anteroventral surface (Fig. 8B). Larvae at this stage are termed doliolaria. Settlement occurs at ca five to seven days.

References

Amenia, S. and R. B. Emlet (1992). The development and larval form of an Echinothuroid Echinoid, *Asthenosoma ijimai*. Biol. Bull. 182:15–30.

Birkeland, C., F.-S. Chia, and R. R. Strathmann (1971). Development, selection, delay of metamorphosis and growth in the seastar, *Mediaster aequalis* Stimpson. Biol. Bull. 141:99-108.

Cameron, R. A. and S. S. Rumrill (1982). Larval abundance and recruitment of the sand dollar *Dendraster excentricus* in Monterey Bay, California, USA. Mar. Biol. 71:197–202.

Chatlynne, L. G. (1969). A histochemical study of oogenesis in the sea urchin *Strongylocentrotus purpuratus*. Biol. Bull. 136:167–84.

Chia, F.-S. (1966). Brooding of a six-rayed starfish, *Leptasterias hexactis*. Biol. Bull. 130:304–5.

———— (1968). The embryology of a brooding starfish, *Leptasterias hexactis* (Stimpson). Acta. Zool. Bd. XLIX:1–44.

Chia, F.-S. and R. D. Burke (1978). Echinoderm metamorphosis: fate of larval structures. In: Settlement and Metamorphosis of Marine Invertebrate Larvae, F.-S. Chia and M. E. Rice (eds.), pp. 219–34. Elsevier-North Holland Biomedical Press, New York.

Emlet, R. B. (1986). Facultative planktotrophy in the tropical echinoid *Clypeaster rosaceus* (Linnaeus) and a comparison with obligate planktotrophy in *Clypeaster subdepressus* (Gray) (Clypeasteroida: Echinoidea). J. Exp. Mar. Bio. Ecol. 95:182–202.

———— (1994a). Functional consequences of simple cilia in the mitraria of oweniids: an anomalous larva of an anomalous polychaete and comparisons with other larvae. In: Reproduction and Development of Marine Invertebrates, W. H. Wilson, S. A. Stricker, and G. L. Shinn (eds.). John Hopkins Univ. Press, Maryland.

———— (1994b). Body form and patterns of ciliation in nonfeeding larvae of Echinoderms: functional solutions to swimming in the plankton? Amer. Zool. 34:570-85.

———— (1995). Larval spicules, cilia, and symmetry as remnants of indirect development in the direct developing sea urchin *Heliocidaris erythrogramma*. Devel. Biol.167:405–15.

Emlet, R. B., L. R. McEdward, and R. S. Strathmann (1987). Echinoderm larval ecology viewed from the egg. In: Echinoderm Studies, M. Jangoux and J. M. Lawrence (eds.). Balkema, Rotterdam, Vol. 2: 55–136.

Fraser, A., J. Gomez, E. B. Hartwick, and M. J. Smith (1981). Observations on the reproduction and development of *Pisaster ochraceus* (Brandt). Can. J. Zool. 59:1700–7.

Giese, A. C., J. S. Pearse, and V. B. Pearse (1991). Reproduction of Marine Invertebrates. Vol. VI, Echinoderms and Lophophorates. pp. 247–759. The Boxwood Press, Pacific Grove, California.

Gonor, J. J. (1973). Reproductive cycles in Oregon populations of the echinoid, *Strongylocentrotus purpuratus* (Stimpson). J. Exp. Mar. Biol. Ecol. 12:45–78.

Hart, M. W. (1995). What are the costs of small egg size for a marine invertebrate with feeding planktonic larvae? Am. Nat. 146(3):415–26.

——— (1996). Evolutionary loss of larval feeding in a facultatively feeding larva, *Brisaster latifrons*. Evolution 50(1):174–87.

Hart, M. W. and R. R. Strathmann (1994). Functional consequences of phenotypic plasticity in echinoid larvae. Biol. Bull. 186:291–99.

Hendler, G. (1982). An echinoderm vitellaria with a bilateral larval skeleton: evidence for the evolution of ophiuroid vitellariae from the ophioplutei. Biol. Bull. 163:431–37 .

Hyman, L. H. (1955). The Invertebrates, Vol IV: Echinodermata. McGraw-Hill Book Co., New York.

Inaba, D. (1930). Notes on the development of a holothurian, *Caudian chilensis* (J. Muller). Sci. Reports Tohoku Imp. Univ. (Ser. 4) 5(2):215–48.

Johnson, L. T. C. (1931). Rearing and identification of certain Holothurian larvae. MS Thesis. University of Washington, Seattle. 29 pp.

Johnson, M. W. and L. T. Johnson (1950). Early life history and larval development of some Puget Sound echinoderms. Studies Honoring Trevor Kincaid, Seattle. Scripps Inst. Oceanogr. Contr. 439:73–84.

Leahy, P. S. (1986). Laboratory culture of *Strongylocentrotus purpuratus* adults, embryos, and larvae. In: Schroeder TE (ed.) Methods in Cell Biology, Vol. 27. Academic Press, New York. pp. 1–13.

McEdward, L. R. (1984). Morphometric and metabolic analysis of the growth and form of and echinopluteus. J. Exp. Mar. Biol. Ecol. 82:259–87.

——— (1986a). Comparative morphometrics of echinoderm larvae. I. some relationships between egg size and initial larval form in echinoids. J. Exp. Mar. Biol. Ecol. 96:251–65.

——— (1986b). Comparative morphometrics of echinoderm larvae. II. larval size, shape, growth, and the scaling of feeding and metabolism in echinoplutei. J. Exp. Mar. Biol. Ecol. 96:267–86.

——— (1992). Morphology and development of a unique type of pelagic larva in the starfish *Pteraster tesselatus* (Echinodermata: Asteroidea). Biol. Bull. 182:177-87.

McEdward, L. R. and D. A. Jamies (1996). Relationships among development, ecology, and morphology in the evolution of Echinoderm larvae and life cycles. Biol. J. Linn. Soc. 60:381–400.

McEuen, F. S. (1986). The reproductive biology and development of twelve species of holothuroids from the San Juan Islands, Washington. Ph.D. Dissertation, University of Alberta, Edmonton, Alberta. 286 pp.

McEuen, F. S. and F.-S. Chia (1985). Larval development of the molpadiid holothuroid *Molpadia intermedia* (Ludwig 1897) (Echinodermata). Can. J. Zool. 63:2553–59.

Miller, B. A. (1995). Larval abundance and early juvenile recruitment of echinoids, asteroids, and holothuroids on the Oregon coast. M.S. Thesis, University of Oregon. 110 pp.

Mladenov, P. V. (1986). Reproductive biology of the feather star *Florometra serratissima*:gonadal structure, breeding pattern, and periodicity of ovulation. Can. J. Zool. 64:1642–51.

Mladenov, P. V. and F.-S. Chia (1983). Development, settling behavior, metamorphosis and pentacrinoid feeding and growth of the feather star *Florometra serratissima*. Mar. Biol. 73:309–23.

Mortenson, T. (1921). Studies of the development and larval forms of echinoderms. Gad. Copenhagen. 261 pp.

Oguro, C. (1989). Evolution of the development and larval types in asteroids. Zool. Sci. 6:199–210.

Okazaki, K. (1975). Normal development to metatmorphosis. In: Czihak (1975), pp. 177–232.

Olsen, H. (1942). Development of a brittle star *Ophiopholis aculeata*, with a short report on the outer hyaline layer. Bergens. Mus. Aarbok. Natur. 6:1–107.

Rumrill, S. S. (1982). Contrasting reproductive patterns among ophiuroids (Echinodermata) from southern Monterey Bay, USA. M.S. Thesis, University of California, Santa Cruz. 260 pp.

Strathmann, M. F. (1987). Reproduction and Development of Marine Invertebrates of the Northern Pacific Coast. pp. 511–606. University of Washington Press, Seattle.

Strathmann, R. R. (1971). The feeding behavior of plantotrophic echinoderm larvae: mechanisms, regulation, and rates of suspension-feeding. J. Exp. Mar. Biol. Ecol. 6:109–60.

——— (1974). Introduction to function and adaptation in echinoderm larvae. Thallasia Jugo. 10(1/2):321–39.

——— (1975). Larval feeding in echinoderms. Amer. Zool. 15:717–30.

——— (1979). Echinoid larvae from the northeast Pacific (with a key and comment on an unusual type of planktotrophic development). Can. Jour. Zool. 57(3):610–16.

Strathmann, R. R., L. Fenaux, and M. F. Strathmann (1992). Heterochronic developmental plasticity in larval sea urchins and its implications for evolution of nonfeeding larvae. Evolution 46(4):972–86.

Young, C. M. and F.-S. Chia (1982). Factors controlling spatial distribution of the sea cucumber *Psolus chitonoides*: settling and post-settling behavior. Mar. Biol. 69:195–205.

19

Hemichordata, Class Enteropneusta: The Acorn Worms

Alan L. Shanks

The Enteropneusta are marine, benthic, bilateral, enterocoelous vermiform invertebrates. They are common components of the infauna of all soft-bottom benthic habitats and can be locally abundant (Hadfield, 1975). Their burrowing and feeding mix the sediment and alter benthic biochemistry. Adults accumulate toxic halogenated organics, which are used in predator defense, fouling control, burrow conditioning, and bacteriostasis (reviewed in King et al., 1995).

The class Enteropneusta includes about 70 species distributed into four families, the Protoglossidae, Harri-maniidae, Spengelidae, and Ptychoderidae (Benito and Pardos, 1997). Kozloff (1996) points out that acorn worms, as they are known, can be found at a half a dozen or more locations around the Pacific Northwest. At some of these sites the worms are abundant. At least several enteropneust species are present in the Pacific Northwest, though only two species have been described (Table 1).

Enteropneusts are broadcast spawners with external fertilization. In the spawning events that have been observed, females initiate spawning, releasing their eggs in long strings of mucus. Males follow, releasing sperm also bound in mucus. The mucus surrounding sperm fairly rapidly dissolves, releasing the sperm. Observed spawning events have been epidemic. In some species, eggs in a "cocoon" of mucus are retained in the burrow, where fertilization occurs (Hadfield, 1975).

Direct development has been observed in several *Saccoglossus* species. Species with large eggs in the genera *Harrimania* and *Protobalanus* may also be direct developers. In some *Saccoglossus* species, larvae hatch out of the egg after one to two days. The embryos are rounded, with the anterior end bearing an apical tuft and a broad telotrochal ciliar band that provides propulsion (Fig. 1A). These larvae swim briefly before settling to the bottom. During this period the anterior end of the larvae elongates into a proboscis (Fig. 1B). Other *Saccoglussus* species settle immediately to the bottom after hatching (Hadfield, 1975).

Table 1. Species in the class Enteropneusta from the Pacific Northwest

Family Ptychoderidae
Glossobalanus berkeleyi (Wiley, 1931)

Family Harrimaniidae
Saccoglossus bromophenolosus (King, 1994)
Saccoglossus spp. (Kozloff, 1996)

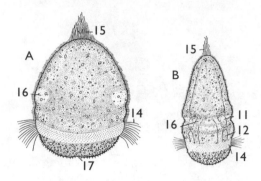

Indirect development via a tornaria larva occurs in the families Ptychoderidae and Spengelidae. At hatching, the larvae of species in these families are generally in the late gastrula stage of development and are uniformly covered with cilia. With development, ciliation becomes concentrated into a single band that loops above the mouth and connects at the apical plate (Fig. 2A). A ventral loop of cilia runs posteriorly on each side of the body and meets across the ventral surface between the mouth and the anus. A telotroch with long cilia develops, encircling the anus (Fig. 2A). These cilia provide most of the propulsion of the larva. The larva at this stage of development is know as a tornaria and, depending on the species, is achieved after several days to several weeks. Tornariae are generally small, a millimeter or less in length, but some species have tornariae up to a centimeter in length. They are highly transparent, allowing clear inspection of internal structures (Hadfield, 1975).

With continued development, the ciliar band becomes increasingly more complex (Fig. 2B), looping extensively about the body of the larva. In *Ptychodera* species, small freely projecting lappets develop on the ciliar band (see Hyman, 1959, Fig 43E; Strathmann and Bonar, 1976, Fig. 1). These larvae are referred to as tentaculate tornariae. In the next stage of development, the ciliary band begins to regress, a circular constriction forms about the middle of the larva, and the body elongates anterior to the constriction (Fig. 2C, D). Metamorphosis begins in the plankton just prior to settlement. During this process, the ciliary band vanishes, the circular constriction develops into a collar, and the animal elongates posterior and anterior to the collar (Fig. 2E). Anterior to the collar, the body develops into a proboscis. A long trunk develops posterior to the collar. At this stage, the larvae remain mobile due to the ciliated telotroch. Culture in the laboratory suggests that the development of tornariae can be quite long, up to several months.

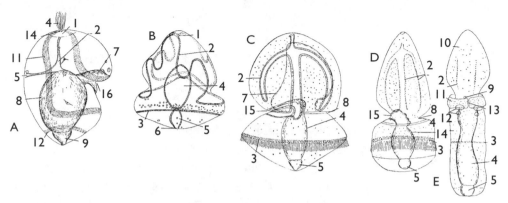

(A) 1, apical thickening; 2, protocoel; 4, apical tuft; 5, hydropore; 7, esophagus; 8, stomach; 9, intestine; 11, ciliary band; 12, telotroch; 14, eye; 15, muscle fibers. (B–E) 1, eye; 2, ciliary band; 3, telotroch; 4, stomach; 5, intestine; 6, anus; 7, protocoel; 8, proboscis-collar groove; 9, hydropore; 10, proboscis; 11, collar; 12, first and second gill pores; 13 groove of invagination of collar cord; 14, truck; 15, mouth.

Fig. 2. Indirect development in the enteropneusts, as exemplified by the tornaria larva of *Balanoglossus clavigerus* (family Ptychoderidae). (A) Beginning development of the ciliary bands. (B) Fully developed tornaria. (C) Ciliary bands regressing, midbody circular constriction forming, and elongation of the body anterior to the constriction. (D) Ciliary band disappearing as collar develops. (E) Larva just prior to settlement; ciliary band nearly gone, collar developed, and body elongated. (From Hyman, 1959, Figs. 42, 43)

References

Hadfield, M. G. (1975). Hemichordata. In: Reproduction of Marine Invertebrates Vol. II, A. C. Giese and J. S. Pearse (eds.). Academic Press, New York. pp. 1–42.

Hyman, L. H. (1959). The Invertebrates: Smaller Coelomate Groups. McGraw-Hill, London and New York.

King, G. M., C. Giray, and I. Kornfield (1994). A new hemichordata, *Saccoglossus bromophenolosus* (Hemichordata: enteropneusta: Harrimaniidae) form North America. Proc. Biol. Soc. Wash. 107:383–90.

King, G. M., C. Giray, and I. Kornfield (1995). Biogeographical, biochemical, and genetic differentiation among North American saccoglossids (Hemichordata; Enteropneusta; Harrimaniidae). Mar. Biol. 123:369-77.

Kozloff, E. N. (1996). Marine Invertebrates of the Pacific Northwest. University of Washington Press, Seattle.

Strathmann, R. and D. Bonar (1976). Ciliary feeding of tornaria larvae of *Ptychodera flava* (Hemichordata: Enteropneusta). Mar. Biol. 34:317–24.

Willey, A. (1931). *Glossobalanu berkeleyi*, a new enteropneust from the west coast. Trans. Ro, Soc Canada, Sect. 5, Ser. 3, 24:19–28.

20

Urochordata:Ascidiacea

Steven Sadro

Widespread interest in the study of tunicates began after Kowalevsky's publications (1886–1871) describing the chordate nature of the ascidian tadpole larva. Since then, larvae from ten families worldwide of the class Ascidiacea have been described (Cloney, 1982). There are ca 60 species of tunicates from ten families found in the waters of the Pacific Northwest (see Tables 1 and 2).

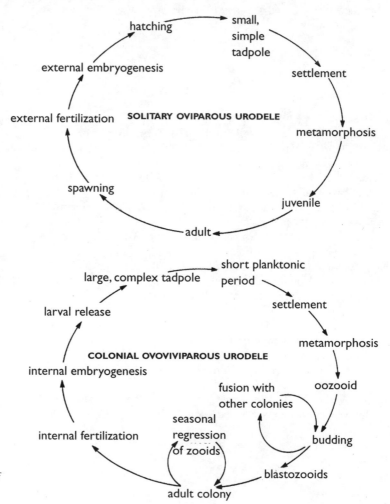

Fig. 1. Generalized life cycles of typical solitary ascidians (top) and compound ascidians (bottom). The diagrams incorporate characteristics of many species, not all aspects of which are present in the life cycle of any given species. (From Svane and Young, 1989)

Reproduction and Development

Ascidian species that reproduce sexually are considered simple ascidians and are solitary. Species that reproduce both sexually and asexually (e.g., through budding) are considered compound ascidians and have a colonial growth form (Berrill, 1975; Strathmann, 1987). Most ascidians are simultaneous hermaphrodites, and examples of both self-fertile species and species that are not self-fertilizing exist (Berrill, 1975). Ascidians may be oviparous (typically solitary forms), ovoviviparous (typically compound forms), or viviparous (Fig. 1). Most solitary oviparous ascidians produce large numbers of relatively small eggs that undergo development into tadpole larvae (Berrill, 1950). Compound ascidians generally produce only a few large eggs that develop into relatively complex tadpole larvae. All ascidian larvae are lecithotrophic, though some are direct-developing and bypass the tadpole stage completely to hatch as small juveniles (e.g., *Molgula pacifica* and *Pelonia corrugata*).

Ascidian tadpole morphology has received considerable study and review (see Millar, 1971; Berrill, 1975; Cloney, 1978, 1982; Katz, 1983; Svane and Young, 1989; Burighel and Cloney, 1997). There is large variation in ascidian size, color, and complexity of internal structures. Compound ascidians have the largest larvae (to 4.5 mm long) and most complex internal anatomy (Fig. 2). Solitary ascidians typically have smaller larvae (to 1.0 mm long) and possess less complex internal anatomy. The surface of all ascidian larvae is covered by a transparent tunic.

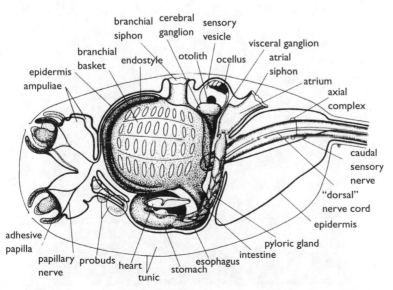

Fig. 2. Diagrammatic compound ascidian larva of *Distalpia occidentalis*. Details of the atrium, inner cuticular layer, and cells of the tunic are omitted. The length of the trunk is 1 mm and of the entire larva, including the caudal fin, 3.2 mm. (From Cloney, 1982)

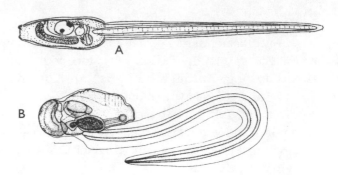

Because of the short pelagic duration and small range of dispersal of most tadpole larvae (Svane and Young 1989), the likelihood of encountering them in the plankton is low. Special care should be taken to not confuse them with the more abundant larvaceans, which if superficially examined outside their "houses" are somewhat similar in morphology to tadpole larvae (Fig. 3). The most obvious feature differentiating larvaceans from ascidian tadpole larvae is the position of the tail (S. Bassham, pers. comm.). In ascidian tadpole larvae, the tail protrudes from the posterior end of the trunk; in larvaceans, the tail is shifted to a position about halfway between the mouth and the posterior end of the trunk. Other features present in larvaceans but not tadpole larvae include a mouth at the anterior end of the trunk and gonad masses (in ripe larvaceans).

Identification of Local Taxa

Lacking morphological information on most larval ascidian, this chapter presents a compilation of useful diagnostic characteristics with pictures included when available. From this information identification to species level is not possible for most larvae, though in some cases one may identify the

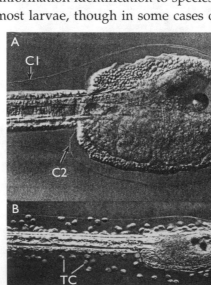

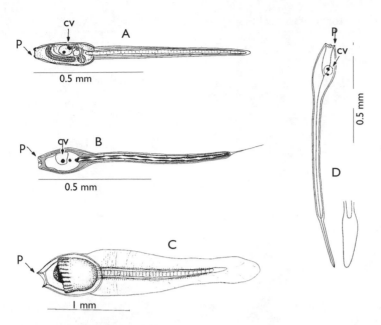

Fig. 5. Solitary ascidian larvae: (A) *Ciona intestinalis*. (B) *Boltenia villosa*. (C) *Metandrocarpa taylori*. (D) *Styela* sp. Species or genera found locally are in bold. Abbreviations: adhesive papillae (p), cerebral vesicle (cv). (A from Berrill, 1947, Fig. 1; B from Cloney, 1961, Plate 1-1; C from Abbott, 1955, Fig. 1; D from Grave, 1944, Fig. 1)

family or genus. At present, the only way to identify a larval type to species is to raise larvae to the adult stage. See Strathmann (1987) for information on culturing larvae.

Relatively few characteristics are available to key ascidian larvae without difficult dissections and tissue analysis. Easily observed morphological characteristics such as color, length, and shape of trunk are useful when differentiating between solitary and colonial ascidian larvae. To distinguish among the less differentiated solitary ascidian larvae usually requires closer examination through a compound microscope to examine the cuticular layers of the trunk tunic and the attached test cells. All ascidian larvae have two cuticular layers of tunic (Fig. 4), but the inner cuticular layer is absent in the tail of some species. The larvae of some species of solitary ascidian have prominent, firmly attached test cells on the outer cuticular

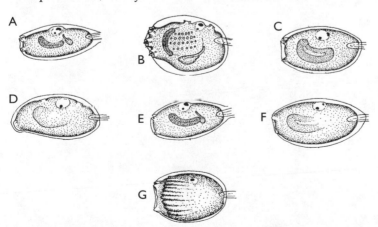

Fig. 6. Larvae of solitary ascidians (tails not shown). (A) *Ciona intestinalis*. (B) *Perophora* listeri. (C) *Pyura* microcosmus. (D) *Molgula* citrina. (E) *Ascidia* mentula. (F) *Styela* partita. (G) *Dendrodoa* grossularia. Species or genera found locally in bold. Figures not to scale. (From Millar, 1971, Fig. 5)

Table 1. Diagnostic information for solitary ascidian species

Species	Mean Length (mm)	Color	Tail with Inner Cuticular Layer[1]	Attached Test Cells[1]
Order Pleurogona				
Suborder Phlebobranchia				
Family Cionidae				
Ciona intestinalis	1.1	Transparent	+	+
Family Perophoridae				
Perophora annectens		Transparent	+	+
Family Corellidae				
Chelyosoma columbianum				
Chelyosoma productum	1.2	Transparent	+	+
Corella inflata	0.9	Transparent	+	+
Corella willmeriana	0.9	Transparent	+	+
Family Ascidiidae				
Ascidia callosa	1.2	Transparent	-	+
Ascidia paratropa	1.2	Transparent	-	+
Ascidia prunum	1.2	Transparent	-?	+
Ascidia ceretodes	1.2	Transparent	-?	+
Suborder Stolidobranchia				
Family Styelidae				
Cnemidocarpa finmarkiensis		Pale Rose, Orange	+	-
Dendrodoa abbotti		Transparent		
Metandrocarpa dura	2.5			
Metandrocarpa taylori		Vermillion		
Styela clavata			+	-
Styela coriacea	1.1	Orange	+	-
Styela gibbsii		Orange	+	-
Styela montereyensis		Orange		
Styela truncata				
Styela (barnharti) clava		Orange		
Family Pyuridae				
Boltenia villosa	0.8-1.2	Transparent	+	-
Halocynthia aurantium	1.5	Pale Yellow	+	-
Halocynthia igaboja	1.5-1.9	Pale Yellow	+	-
Pyura haustor	1.6	Transparent	+	-
Pyura mirabilis				
Family Molgulidae				
Molgula cooperi			-?	-
Molgula manhattensis	0.7		-?	-
Molgula oregonia			-?	-
Molgula pacifica			-?	-
Molgula pugetiensis			-?	-

1 Presence of character: + = yes; - = no.

layer of tunic which are visible under a compound microscope (Burighel and Cloney, 1997).

Solitary Ascidians

All solitary ascidian larvae are small (<1.5 mm length) and have relatively simple internal structures (Figs. 5, 6). All but the family Molgulidae have three simple attachment papillae with a coniform shape and a triangular configuration (see Fig. 5). The Molgulidae lack attachment papillae. *Ciona intestinalis* and all species from the family Ascidiidae have firmly attached test cells on the outer cuticular layer of tunic. The inner cuticular layer of the tunic is absent in the tail of the following species: *Corella inflata, Chelyosoma productum, Ascidia callosa, A. paratropa* (probably all ascidiids), and *Molgula occidentalis* (probably all molgulids) (R. A. Cloney, pers. comm.) (see Fig. 4). *Corella* and *Chelyosoma* species have trunks compressed in the sagittal plane. *Halocynthia* species have long trunks that taper down in

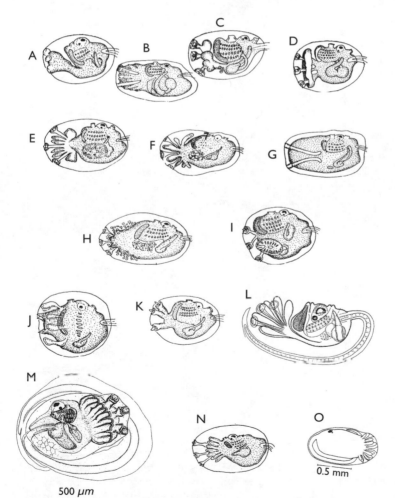

Fig. 7. Larvae of compound ascidians.
(A) *Clavelina lepadiformis.*
(B) *Pycnoclavella stanleyi.* (C) *Distaplia rosea.* (D) *Cystodytes dellechiajei.*
(E) *Eudistoma illotum.*
(F) *Synoicum georgianum.*
(G) *Euherdmania claviformis.*
(H) *Apidium nordamanni.*
(I) *Diplosoma listerianum.*
(J) *Botrylloides leachi.*
(K) *Sycozoa sigillinoides.*
(L) *Archidistoma aggregatum.*
(M) *Didemnum albidum.*
(N) *Didemnum helgolandicum.*
(O) *Ritterela rubra.*
Species or genera found locally are in bold. Figures A-L, N are not to scale. (A–K, N from Millar, 1971, Fig. 5; L from Berrill, 1948, Fig. 2; M from Marks 1996, Fig. 4A; O from Abbott and Trason, 1968, Fig. 1C)

500 μm

0.5 mm

diameter toward the papillae. All other species listed have trunks that are round to ovoid in cross section at the middle of the trunk. Additional diagnostic information on other solitary species is compiled in Table 1.

Table 2. Diagnostic information for compound ascidian larvae

Species	Length (mm)	Color	Adhesive Papillae Shape[1]	Adhesive Papillae Configuration	Tail Twisted Counter-Clockwise
Order Enterogona					
Suborder Aplousobranchia					
Family Clavelinidae					
Archidistoma molle					
Archidistoma psammion					
Eudistoma ritteri	3.0				
Eudistoma purpuropunctatum			Scy	Sagittal	
Clavelina huntsmani		Orange	Scy	Sagittal	
Clavelina sp.					
Cystodytes lobatus					
Distaplia occidentalis	3.2		Scy	Triangle	+
Distaplia smithi			Scy	Triangle	+
Pycnoclavella stanleyi			T-inv		
Family Polyclinidae					
Aplidium arenatum	2.5		Goblet	Sagittal	
Aplidium californicum	2.5		Goblet	Sagittal	
Aplidium glabrum	2.5		Goblet	Sagittal	
Aplidium propinquum	2.5		Goblet	Sagittal	
Aplidium solidum	2.5		Goblet	Sagittal	
Aplidium sp.	2.5		Goblet	Sagittal	
Euherdmania claviformis	3.0		T-inv		+
Ritterella aequalisiphonis					
Ritterella pulchra					
Ritterella rubra	0.8				
Synoicum parfustis					
Synoicum sp.					
Family Didemnidae					
Didemnum albidum	2.7				
Didemnum carnulentum					
Diplosoma listerianum	2.4		Scy	Sagittal	+
Trididemnum opacum					
Trididemnum strangulatum					
Suborder Stolidobranchia					
Family Styelidae					
Botrylloides sp.			Sim		
Botryllus sp.	1.5		Scy	Triangle	

[1]Scy = scyphate, the papillae appear to be set in cup-shaped invaginations in the trunk (e.g., Distaplia, see Fig. 2); T-inv = inverted (invaginated) tubular papillae that extend deeply into the body cavity of the trunk (e.g., Pycnoclavella, Fig. 7B); Goblet = goblet-shaped papillae with elongate tubular stalks (e.g., Aplidium, Fig. 7H).

Compound Ascidians

All larvae from colonial ascidians are greater than 1.5 mm in length and have relatively complex internal structures. All have three adhesive papillae except *Euherdmania claviformis*, which has two. Additional diagnostic characteristics of the larvae of local colonial ascidians are compiled in Table 2, and compound ascidian larvae are illustrated in Fig. 7.

References

Abbott, D. P. (1955). Larval structure and activity in the ascidian Metandrocarpa taylori. J. Morph. 97:569–94.

Abbott, D. P. and W. Trason (1968). *Ritterella rubra* and *Distaplia smithi*: Two new colonial ascidians from the west coast of North America. Bull. So. Calif. Acad. Sci. 67:143–53.

Berrill, N. J. (1947). The development and growth of Ciona. Jour. Mar. Biol. Assoc. 26:616–25.

——— (1948). Structure, tadpole and bud formation in the ascidian Archidistoma. Jour. Mar. Biol. Assoc. (UK) 27:380–88.

——— (1950). The Tunicata with an account of the British species. London: The Ray Society.

——— (1975). Chordata: tunicata. In: Reproduction of Marine Invertebrates, A. C. Giese and J. S. Pearse (eds.), pp 241–82. Academic Press, New York.

Burighel, P. and R. A. Cloney (1997). Urochordata: Ascidiacea. In: Microscopic Anatomy of Invertebrates, F. W. Harrison and F.-S. Chia (eds.), pp. 221–347. Wiley-Liss, Inc., New York.

Cloney, R. A. (1961). Observations on the mechanism of tail resorption in ascidians. Amer. Zool. 1:67–87.

——— (1978). Ascidian metamorphosis: review and analysis. In: Settlement and Metamorphosis of Marine Invertebrate Larvae, F.-S. Chia and M. Rice (eds.), pp. 255–82. Elsevier, New York.

——— (1982). Ascidian larvae and the events of metamorphosis. Amer. Zool. 22:817–826.

Grave, C. (1944). The larva of Styela (Cynthia) partita: structure, activities and duration of life. J. Morph. 75:173–91.

Katz, M. J. (1983). Comparative anatomy of the tunicate tadpole, *Ciona intestinalis*. Biol. Bull. 164:1-27.

Marks, J. A. (1996). Three sibling species of didemnid ascidians from northern Norway: *Didemnum albidum* (Verrill, 1871), *Didemnum polare* (Hartmeyer, 1903), and *Didemnum romssae* sp. nov. Can. J. Zool. 74:357–79.

Millar, R. H. (1971). The biology of Ascidians. Adv. Mar. Biol., Vol. 9:1-100.

Strathmann, M. F. (1987). Reproduction and Development of Marine Invertebrates of the Northern Pacific Coast. University of Washington Press, Seattle.

Svane, I. B. and C. M. Young (1989). The ecology and behaviour of ascidian larvae. Oceanogr. Mar. Biol. Annu. Rev. 27:45–90.

Index

Numbers in bold type refer to illustrations

Abalone, 109
Abarenicola claparedi oceanica, 49; *A. pacifica*, 49; *A. vagabunda oceanica*, 49
Abietinaria spp., 17
Acanella sp., 19
Acanthina, 89
Acanthinephyra curtirostris, 183; *A. hispidus*, 184
Acanthochitonidae, 153
Acanthochitonina, 152, 153
Acanthodoris, 90; *A. brunnea*, 120; *A. hudsoni*, 120; *A. nanaimoensis*, 120; *A. rhodoceras*, 120
Acanthogorgiidae Family, 19
Acarnus erithacus, 11
Acila castrensis, 132, **139**
Acmaea, 89
Acmaeidae, 89
Acorn worms, 293-95
Acotylea Suborder, 26
Acrocirridae Family, 68
Acrocirrus heterochaetus, 42
Acteocina, 89
Acteonidae Family, 103
Acteonidae, 89
Actiniaria Order, 15, 20-22
Actiniidae Family, 19
Actinocyclidae, 90
Actinostolidae Family, 19
Actinotroch larvae, 253
Actinotrocha A, 257; *Actinotrocha B*, 256; *Actinotrocha D*, 254, 257, **258**
Actinula, 16
Adalaria, 90; *Adalaria* sp., 120
Adocia, 9; *A. gellindra*, 11
Adula californiensis, 132; *A. diegensis*, 132; *A. falcata*, 132
Aedicira pacifica, 54
Aegires, 90, **98**; *A. albopunctatus*, 96, 120; *A. punctilucens*, 98
Aeolidia, 90; *A. papillosa*, 121
Aeolidiidae, 90
Aequorea victoria, 17
Aequoreidae Family, 17
Aetea sp., 262
Aeteidae Family, 262
Agalmidae Family, 17
Aglaja, 89, 104
Aglajid opisthobranch, 103
Aglajidae, 89
Aglaopheniidae Family, 17
Agriodesma saxicola, 135
Akentrogonida, 175
Akentrogonidae, 175
Alaskan Gaper, **148**
Alciopa reynaudi, 55
Alciopidae Family, 47, **55**
Alciopina tenuis, 55
Alcyonacea Order, 19
Alcyonaria Subclass, 19
Alcyoniddae Family, 19
Alcyonidiidae Family, 262

Alcyonidium gelatinosum, 262, 267; *A. mammilatum*, 262; *A. parasiticum*, 262; *A. pedunculatum*, 262; *A. duplex*, 262; *A. mytili*, 262, **264**; *A. polyoum*, 262
Alcyoniina Suborder, 19
Alcyonium spp., 19
Alderia, 90; *A. modesta*, 105, 119
Alderina brevispina, 262
Alderinidae Family, 262
Aldisa, 90; *A. cooperi*, 120; *A. sanguinea*, 120
Aldisidae, 90
Algaophenia spp., 17
Alia, 89
Allia ramosa, 54
Allocentrotus fragilis, 273, **279**
Allopora petrograpta, 17; *A. porphyra*, 17; *A. venusta*, 17; *A. verrilli*, 17
Alpheidae Family, 182, 196
Alteration of generations, 13
Alvania, 89
Amage anops, 74
Amblysyllis lineata var. alba, 66; *Amblysyllis* sp., 66
Ammochares, 55
Ammotrypane aulogaster, 53
Ampharete acutifrons, 74; *A. finmarchica*, 74; *A. goesi goesi*, 74; *A. arctica*, 74; *A. grubei*, 74; *A. gagarae*, 74
Ampharetidae Family, 74
Amphiblastula larvae, 6,7
Amphicteis gunneri floridus, 75; *A. mucronata*, 74; *A. scaphobranchiata*, 74
Amphictene auricoma, 75; *A. moorei*, **75**
Amphinomida Order, 48
Amphinomidae Family, 45, 48
Amphiodia occidentalis, 274; *A. urtica*, 274
Amphioplus macraspis, 274; *A. strongyloplax*, 274
Amphipholis pugetana, 274; *A. squamata*, 274
Amphisamytha bioculata, 74
Amphissa, 89
Amphitrite, 76; *A. cirrata*, 76; *A. scionides dux*, 76; *A. palmata*, 76
Amphoriscidae Family, 9
Anaata brepha, 11; *A. spongigartina*, 11
Anaitides groenlandica, 60, **61**; *A. hartmanae*, 60; *A. medipapillata*, 60; *A. mucosa*, 60, **62, 62**; *A. multiseriata*, 60; *A. williamsi*, 60, **62**
Anaspidea, 119
Anaspidean opisthobranch, 106
Anatoma, 89
Anchinoe, 9
Anchinoidae Family, 11
Ancistrolepis, 89
Ancistrosyllis aff. groenlandica, 63
Ancula, 90; *A. pacifica*, 120
Ancystrosyllis, 62, **63**

Angulosasscus tenuis, 158, 175
Anisodoris, 90; *A. lentiginosa*, 120; *A. nobilis*, 120
Anobothrus gracilis, 74
Anomiidae, 133
Anomura, 184: Anomuran megalopa, 184, 198; Anomuran zoea, **190**
Anthoarcuata graceae, 11
Anthopleura artemisia, 19; *A. elegantissima*, 13, 18-20; *A. xanthogrammica*, 19, 21
Anthoptilidae Family, 19
Anthoptilum grandiflorum, 19
Anthothela pacifica, 19
Anthothelidae Family, 14, 19
Anthozoa Class, 13, 18-22
Antinoe, 63
Antinoella, 63
Antipatharia Order, 19
Antipathes sp., 19
Antipathidae Family, 19
Antipathina Suborder, 19
Antiplanes, 89; *A. perversa*, **102**; *A. voyi*, **102**
Antomedusae, 17
Aphrodita japonica, 55; *A. longipalpa*, 55; *A. magellanica*, 55; *A. parva*, 55; *A. refulgida*, 55
Aphroditidae Family, 48, 55
Apidium nordamanni, 301
Apistobranchidae Family, 69
Apistobranchus ornatus, 69
Aplidium, 302; *A. arenatum*, 302; *A. californicum*, 302; *A. glabrum*, 302; *A. propinquum*, 302; *A. solidum*, 302; *Aplidium* sp., 302
Aplousobranchia Suborder, 302
Aplysia, 90; *A. californica*, 106, 119
Aplysiidae, 90
Aplysilla, 8; *A. ?glacialis*, 12
Aplysillidae Family, 12
Aplysiopsis enteromorphae, 106, 119
Apodida Order, 274
Apomatus geniculatus, 67; *A. timsi*, 67
Arabella iricolor, 51
Arabellidae Family, 51
Arachnidiidae Family, 262
Archaeogastropoda, 89
Archidistoma aggregatum, **301**; *A. molle*, 302; *A. psammion*, 302
Archidorididae, 90
Archidoris, 90; *A. montereyensis*, 120; *A. odhneri*, 120
Arcoida, 132
Arctonoe fragilis, 63; *A. pulchra*, 63, *A. vittata*, 63
Arenicola pusilla, **49**; *A. claparedi*, **49**; *A. marina*, 49
Arenicolidae Family, 48, 49
Argeia pugettensis, 178
Arginula, 89
Argis levior, 182; *A. alaskensis*, 182
Arhynchite pugetensis, 86
Aricia michaelseni, 54
Aricidea, 54; *A. wassi*, 54
Armandia bioculata, 53; *A. brevis*, **53**

Armina, 90; *A. californica*, 121
Arminidae, 90
Arndtanchora sp., 11
Artacama coniferi, 76
Artacamella hancocki, 76
Arthophryxus beringanus, 178
Articulata Class, 269-70
Asabellides lineata, 74; *A. sibrica*, 75
Asbestopluma occidentalis, 11
Ascidia callosa, 300, 301; *A. ceretodes*,
    300; *A. mentula*, **299**; *A. paratropa*,
    298, 300, 302; *A. prunum*, 300
Ascidiacea, 296-303
Ascidian tadpole larva, 296-303
Ascidians, 296-303: development,
    297-98; identification, 298-303;
    life cycle, **296**; reproduction, 287-
    98; solitary, 301-2; tadpole larvae
    296-303
Ascidiidae Family, 300, 301
Asclerocheilus beringianus, 53
Aspidochirotida Order, 273
Aspidosiphon sp., 81
Assiminea, 89
Assimineidae, 89
Astarte compacta, 133; *A. esquimalti*,
    133; *A. undata*, 133
Astartidae, 133
Asterina miniata, 273, **281**
Asteroid key, 284
Asteroidea Class, 281-84
Astraea, 89
Asychis disparidentata, 50
Atelecyclidae Family, 185
Athecanephria Order, 86, 87
Athecata Suborder, 17
Atlanta peroni, 113
Atlantia, 89
Atlantid heteropod, 113
Atlantidae, 89
Aulactinia incubans, 15, 19
Auricularia larva, 282, **285**
Autolytus varius, 66; *A. cornutus*, 66;
    A. prismaticus, 66; *Autolytus* sp.,
    66
Axiidae Family, 184, 186
Axinella sp., 11
Axinellida Order, 11
Axinellidae Family, 11
Axinopsida serricata, 133; *A. viridis*,
    133
Axiopsis spinulicauda, 184
Axiothella, 50: *A. rubrocincta*, 50
Axocielita originalis, 11

Balanoglossus clavigerus, **295**
Balanomorpha, 158, 165-72
Balanus balanus, 158, 162, **169**;
    B. crenatus, 158, 162, **167**;
    B. glandula, 158, 162, **166, 167, 168**;
    B. improvisus, 158, 162, **165, 166**;
    B. nubilus, 158, 162, **168, 169**;
    B. rostratus, 158
Balcis, 89
Balnophyllia elegans, 15, 20
Balthica Macoma, **145**
Balticina californica, 19;
    B. septentrionalis, 19

Bankia setacea, 135, 139, **140, 144**
Barleeia, 89
Barleeidae, 89
Barnacles, 157-77: development,
    157-59; identification, 159-64;
    local species, 158; morphology,
    157-59; parasitic, 158, 174-76
Barnea subtruncata, 135
Basket stars, 288
Bathybembix, 89
Batillaria, 89
Bentheogennema borealis, 182;
    B. burkenroadi, 182
Bergstroemia, 61
Beringius, 89
Berthella, 90: *B. californica*, 96, 119
Betaeus harrimani, 182; *B. setosus*,
    182
Bicellariella ciliata, **265**
Bicellariellidae Family, 262
Bicidium aequoreae, 19
Bicrisia edwardsiana, 261
Biemna rhadia, 11
Biemnidae Family, 11
Bimeria spp., 17
Bipinnariae larvae, 281
Bittium, 89, 109
Bivalves, 131-51: development, 131,
    136; identification, 137, **138**;
    periclymna larvae, 137, **139**;
    reproduction, 131, 136; veliger
    larvae, 139
Bocardiella hamata, 70
Boccardia, 70: *B. californica*, 70;
    B. columbiana, 70; *B. polybranchia*,
    70; *B. proboscidea*, 70; *B. uncata*, 70
Boccardiella hamata, 70, **72**
Bolitaena diaphana, 155, **156**
Boltenia villosa, **299**, 300
Bonelliidae Family, 86
Bonelloinea Order, 86
Bonneviella spp., 17
Bonneviellidae Family, 17
Bopyridae, 178
Bopyroides hippolytes, 178
Boreohydridae Family, 17
Botrylloides leachi, **301**; *Botrylloides*
    sp., 302
Botryllus sp., 302
Bougainvillia spp., 17
Bougainvilliidae Family, 17
Bowerbankia gracilis, 262, **265**, 267
Brachiolaria, **282**; brachiolaria
    larvae, **282**
Brachiopoda Phylum, 269-71
Brachyura, 185; *B. megalopae*, 197;
    B. zoea, 191
Branchiomaldane vincenti, 49;
    B. simplex, 49
Brarosccus callosus, 158, 175
Brisaster latifrons, 273, 275, **278**
Brittle stars, 287-89
Bryozoa, 260-68
Buccinidae, 89
Buccinium, 89
Bugula californica, 262; *B. cucllifera*,
    262; *B. flabellata*, 262, **265**, 266;
    B. pacifica, 262; *B. pugeti*, 262;
    B. plumosa, **265**; *B. simplex*, 266

Buskia nitens, 262
Buskiidae Family, 262
Byglides macrolepida, 63
Bythotiara huntsmani, 17

Caberea boryi, 262; *C. ellisi*, 262
Cadlina, 90: *C. flavomaculata*, 120;
    C. luteomarginata, 120; *C. modesta*,
    120
Cadulus aberrans, 154; *C. californicus*,
    154; *C. tolmiei*, 154
Caecid prosobranchs, 105
Caecidae, 89
Caenogastropoda, 89, 111
Calappidae Family, 185
Calastacus stilirostris, 184
Calcarea Class, 5, 7, 8, 9
Calcaronea, 9
Calcigorgia spiculifera, 19
Calinaticina, 89
Callianassa. See Neotrypaea
Callianassidae Family, 184, 196, 215
Callianopsis investigatoris, 184;
    C. goniophthalma, 184
Callioplanidae Family, 27
Calliostoma, 89, 108
Callistochiton crassicostatus, 153
Callistochitonidae, 153
Callogorgia kinoshitae, 19
Callopora armata, 262;
    C. circumclathrata, 262;
    C. corniculifera, 262; *C. horrida*,
    262; *C. lineata*, 262
Callyspongia, 9
Calocaris quinqueseriatus, 184
Calvopora occidentalis, 262
Calycella spp., 17
Calycophorae Suborder, 17
Calycopsidae Family, 17
Calyptraeidae, 89, 105, 109
Campanularia spp., 17
Campanulariidae Family, 17
Campanulinidae Family, 17
Cancellaria, 89
Cancellaridae, 112
Cancellariidae, 89
Cancer gracilis, 185, 198, 236;
    C. antennarius, 185, 198, 237;
    C. branneri (gibbosulus), 185;
    C. magister, 185, 198, 237;
    C. oregonensis, 185, 198, 237;
    C. productus, 185, 198, 238
Cancridae Family, 185, 193, 198,
    235-38
Cancridae zoeae, **235**
Capitella capitata, 49, 50
Capitellida Order, 49-50
Capitellidae Family, 48-50
Carcinus maenas, 186, 200, 246, **247**,
    248
Cardiidae, 133
Cardiomya californica, 135;
    C. oldroydi, 135; *C. pectinata*, 135;
    C. planetica, 135
Carditidae, 133
Caridea, 182, 197
Caridean zoea, **184**
Carinaria, 89

Carinariid heteropods, 113
Carinariidae, 89
Caryophyllia alaskensis, 19, 20;
    C. smithi, 18-20
Caryophylliidae Family, 19
Caryophylliina Order, 19
Catriona, 90: *C. columbiana*, 121;
    C. rickettsi, 121
Caulibugula californica, 262;
    C. ciliata, 262; *C. occidentalis*, 262
Caulleriella alata, 69; *C. gracilis*, 70;
    C. hamata, 69; *C. racilis*, 69
Cauloramphus brunea, 262;
    C. echinus, 262; *C. spiniferum*, 262
Cavolinia, 90, 116
Cavoliniid thecosomatous
    pteropods, **97**, 116, 117
Cavoliniidae, 90
Cellaria diffusa, 262; *C. mandibulata*,
    262
Cellariidae Family, 262
Cellepora pumicosa, 262; *Cellepora*
    sp., 262
Celleporidae Family, 262
Cephalaspidea, 119
Cephalaspidean optisthobranch,
    98, 104
Cephalopoda, 154-56
Ceramaster articus, 273;
    C. patagonicus, 273
Ceratonereis paucidentata, 59
Ceratostoma, 89
Cerberilla, 90
Ceriantharia Order, **15**, 19, 20
Cerianthidae Family, 18, 19
Ceriantipatharia Subclass, 19
Cerithiidae, 89, 109
Cerithiopsidae, 89, 110
Cerithiopsis, 89
Cestoda Class, 26
Cestoplana sp., 27
Cestoplanidae Family, 27
Chaetopleura apiculata, **152**;
    C. gemma, 153
Chaetopleuridae, 153
Chaetopteridae Family, 46, 69
Chaetopteridae, 69
Chaetopterus variopedatus, **69**
Chaetozone berkeleyorum, 70;
    C. setosa, 69
Chalinula, 9
Chama arcana, 133
Chamidae, 133
Chamylla, 90
Chapperiella condylata, 262; *C. patula*,
    262
Chapperiellidae Family, 262
Cheilonereis cyclurus, 59
Cheilopora praelonga, 263
Cheiloporinidae Family, 263
Chelonaplysilla polygraphis, 12
Chelophyes appendiculata, 17
Chelyosoma, 301: *C. columbianum*,
    300; *C. productum*, 300, 301
Chionoecetes bairdi, 185, 199, 242-**43**;
    C. angulatus, 185
Chitinopoma, 68; *C. groenlandica*, 68
Chitonia, 152, 153

Chitonina, 153
Chitons, 152-54
Chlamys behringiana, 132; *C. hastata*,
    132; *C. rubida*, 132
Chloeia entypa, 48; *C. pinnata*, 48
Chone gracilis, 67;
    C. infundibuliformis, 67, **68**;
    C. magna, 67; *C. minuta*, 67;
    C. mollis, 67; *C. ecaudata*, 67;
    C. teres, 67
Chorilia longipes, 185
Choristida Order, 11
Chromodorididae, 90
Chrysogorgiidae Family, 19
Chrysopetalidae Family, 46, 55
Chrysopetalidae, 55
Chrysopetalum, 55; *C. debile*, 55
Chthamalus dalli, 158, 162, 163, **172**;
    C. fissus, 172
Ciaperoecia californica, 261; *C. major*,
    261
Ciocalyptus penicillus, 11
Ciona intestinalis, **299**, 300, 301
Cionidae Family, 300
Circeis amoricana, 68; *C. spirillum*,
    68, **69**; *C. spirillum*, 68
Cirratulidae Family, 45, 69
Cirratus cirratus cingulatus, 70;
    C. cirratus cirratus, 70; *C. robustus*,
    70; *C. spectabilis*, 70
Cirriformia spirabranchia, 70
Cirripedia, 157-77
Cirrophorus lyra, 54
Cistenides brevicoma, 75;
    C. granulata, **75**
Cladocarpus spp., 17
Cladonema californicum, 17
Cladonematidae Family, 17
Clams, 131-51: Alaskan Gaper, **148**;
    Asiatic Freshwater, **144**; Butter,
    147; Little Gaper, **145**; Soft-shell,
    145-46
Clathriidae Family, 11
Clathrina, 8: *C. blanca*, 9; *C. coriacea*,
    9; *Clathrina* sp., 9
Clathrinidae Family, 9
Clathromangelia, 89
Clavelina huntsmani, 302;
    C. lepadiformis, **301**; *Clavelina* sp.,
    302
Clavelinidae Family, 302
Clavidae Family, 17
Clavoporidae Family, 262
Clavularia moresbii, 19; *Clavularia*
    spp., 19
Clavulariidae Family, 19
Clinocardium blandum, 133;
    C. californiense, 133; *C. cliatum*,
    133; *C. fucanum*, 133; *C. nuttalli*,
    133
Clio, 90, 116
Cliona 8, 90; *C. ?argus*, 11; *C. ?celata*,
    11; *C. limacina*, **19**, 120; *C. lobatta*,
    11; *C. ?warreni*, 11
Clionidae Family, 11
Clionidae, 90
Cliopsidae, 90
Cliopsis, 90

Clistosccus paguri, 158, 175
Clistrosaccidae, 175
Clymenella torquata, **50**
Clypeasteroida Order, 273
Clytia spp., 17
Cnemidocarpa finmarkiensis, 300
Cnidaria: development, 13-15;
    Phylum, 13-23; reproduction, 13-
    15
Cnidarian planulae, 6
Codonellina cribriformis, 263
Coelenterata, 13-24
Coelomate worms, 85-87
Coelosphaera, 9
Colletosia radiata, 262
Columbellidae, 89, 112
Colus, 89
Compsomyax subdiaphana, 134
Conchocele bisecta, 133
Conopeum reticulum, 262, **264**
Cooperella subdiaphana, 134
Cooperellidae, 134
Copepoda, **159**
Copidozoum protectum, 262;
    C. tenuirostre, 262
Corallimorpharia Order, 19
Corallimorphidae Family, 19
Corallimorphus sp., 19
Corals, 13, 14
Corambe, 90
Corambidae, 90
Corbicula fluminea, 134, 136, **140,
    143, 144**
Corbiculidae, 134
Cordagalma cordiformis, 17
Cordylophora caspia, 17
Corella, 301: *C. inflata*, **298**, 300, 301;
    C. willmeriana, 300
Corellidae Family, 300
Corolla, 90: *C. spectabilis*, 107
Coronate larvae, **260**, 265-67
Corymorpha sp., 17
Corymorphidae Family, 17
Corynactis californica, 19
Coryne sp., 17
Corynidae Family, 17
Cossura modica, 50
Cossurida Order, 50
Cossuridae Family, 50
Costazia costazia, 263; *C. robertsoniae*,
    263; *C. ventricosa*, 263
Cotylea Suborder, 26
Crab: Black-clawed, 248; Brackish-
    Water, 249; Butterfly, **221**-22;
    Dungeness, 237; Flat Porcelain,
    234; Flattop, 234; Graceful, 236;
    Graceful Decorator, **243**-44;
    Green, 246, **247**, 248; Hermit, **218,
    219**; Lined Shore, **241**-42; Mussel,
    244-**45**; Pacific Rock, 236; Pacific
    Sand, 219-**20**; Pubescent
    Porcelain, 234; Purple Shore, 240;
    Pygmy Rock, 237; Red Rock, 238;
    Rhinoceros, **222**; Shore, 246, **247,
    248**; Striped Shore, **241**-42;
    Tanner, **242**-43; Thick-Clawed
    Porcelain, 234; Turtle, **221**-22;
    Yellow Shore, 240

Crangon alba, 182; *C. alaskensis*, 182;
  C. franciscorum, 182; *C. handi*, 182;
  C. nigricauda, 182; *C. stylirostris*,
  182
Crangonidae, 182, 196
Crania, **271**: *C. californica*, 269, 270
Craniella villosa, 11; *C. spinosa*, 11
Craniopsis, 89
Crassicardia crassidens, 133
Crassostrea, 133: *C. gigas*, 133, **140,
  142, 144**; *C. virginica*, 133, **140,
  142, 144**
Crenella decussata, 132
Crepidula, 89: *C. adunca*, 105, 109
Crepipatella, 89
Creseis, 90: *C. virgula*, 116
Cresiella producta, **261**
Cribrilina annulata, 262; *C. corbicula*,
  262; *C. radiata*, 262
Cribrilinidae Family, 262
Cribrinopsis fernaldi, 15, 19;
  C. williamsi, 19
Crimora, 90: *C. coneja*, 120
Crinoid, 288, **289**
Crinoidea Class, 274, 289-90
Crisia eburnea, **261**; *C. elongata*, 261;
  C. occidentalis, 261; *C. operculata*,
  261; *C. pugeti*, 261; *C. serrulata*,
  261
Crisidia cornuta, 261
Crisiidae Family, 261
Crossaster borealis, 273
Crucigera irregularis, 67;
  C. zygophora, 67
Cryptobranchia, 89
Cryptochiton stelleri, **152**, 153
Cryptolaria spp., 17
Cryptolithodes sitchensis, 184;
  C. typicus, 184, **221**-22
Cryptomya californica, 134
Cryptoniscus larva, 178, **179**
Cryptosula pallasiana, 263
Ctenodrilida Order, 51
Ctenodrilidae Family, 51
Ctenodrilus serratus, 51
Ctmatiidae, 110
Cubozoa Class, 13
Cucumaria fallax, 274; *C. lubrica*, 274;
  C. miniata, 274; *C. piperata*, 274;
  C. pseudocurata, 274, 285;
  Cucumaria sp., **286**
Cultellidae, 134
Cumanotidae, 90
Cumanotus, 90: *C. fernaldi*, 121
Cumingia californica, 134
Cup corals, 14
Cuspidariidae, 135
Cuthona, 90: *C. abronia*, 121;
  C. albocrusta, 121; *C. cocoachroma*,
  121; *C. divae*, 121; *C. fulgens*, 121;
  C. lagunae, 121
Cuvierina, 90, 116
Cyathoceras quaylei, 19
Cyclichnidae, 89
Cyclocardia crebricostata, 133;
  C. ventricosa, 133
Cylichna, 89
Cylichnidae Family, 104

Cymakra, 89
Cymatiidae, 89, 107
Cymbulia peroni, 107
Cymbuliidae, 90
Cyphonautes larvae, **260**, 263-65
Cyprid: 159, 163; key to Cyprids,
  163-64
Cystodytes dellechiajei, **301**;
  C. lobatus, 302
Cytididae Family, 261

Dacrydium pacificum, 132
Dajidae, 178
Dakaria dawsoni, 263; *D. ordinata*,
  263; *D. pristina*, 263
Decapoda, 181-252: development,
  181, 186, **187**; keys megalopae,
  197-203; keys zoeae, 192-97; keys
  zoeae vs megalopae, 192; life
  cycle, 187; morphology, 187-88;
  reproduction, 181, 186, **187**
Decapodid, 187
Delectopecten randolphi, 132;
  D. vancouverensis, 132
Demonax media, 67, **68**
Demospongiae Class, 5, 8
Dendraster excentricus, 273, **278**
Dendrobeania curvirostrata, 262;
  D. laxa, 262; *D. lichenoides*, 262;
  D. longispinosa, 262; *D. murrayana*,
  262
Dendrobranchiata, 182
Dendroceratida Order, 8, 12
Dendrochirotida Order, 273
Dendrochiton flectens, 153;
  D. semiliratus, 153
Dendrodoa abbotti, 300;
  D. grossularia, 299
Dendrogaster sp., 158
Dendronotacean nudibranch, 107
Dendronotidae Family, 114
Dendronotidae, 90
Dendronotus, 90: *D. albopunctatus*,
  120; *D. diversicolor*, 120;
  D. frondosus, 120; *D. iris*, 120;
  D. subramosus, 120
Dendrophylliidae Family, 19
Dendrophylliina Suborder, 19
Dendropoma, 89
Dentaliida, 154
Dentaliidae, 154
Dentalium, 154, **155**: *D. agassizi*, 154;
  D. pretiosum, 154; *D. rectius*, 154
Dermasterias imbricata, 273
Desmophyes annectens, 17
Desmophyllum cristagalli, 19
Desmoxyidae Family, 11
Dexiospira, 68
D-hinge stage, 137, **138**
Diacria, 90, 116
Diaperoecia californica, 261;
  D. intermedia, 261; *D. johnstoni*,
  261; *D. obelium*, 261
Diaphana, 89, 104: *D. californica*, 119
Diaphanidae, 89
Diaphorodoris, 90: *D. lirulatocauda*,
  120
Diastoporidae Family, 261

Diaulula, 90: *D. sandiegensis*, 120
Dictyoceratida, **10**
Dictyociona asodes, 11
Didemnidae Family, 302
Didemnum albidum, **301**;
  D. carnulentum, 302
Dimophyes arctica, 17
Diodora, 89
Diogenidae Family, 184, 195, 201,
  202, 218
Diopatra cuprea, 52; *D. ornata*, 52
Diphasia spp., 17
Diphyidae Family, 17
Diplodonta impolita, 133; *D. orbellus*,
  133
Diplosoma listerianum, **301**
Dirona, 90: *D. albolineata*, 121;
  D. aurantia, 121; *D. picta*, 121
Dironidae, 90
Discocytis canadensis, 261
Discodorididae, 90
Discurria, 89
Disoma multisetosum, 74
Disporella fimbriata, 261; *D. hispida*,
  261; *D. separata*, 261
Dissoconch, 137
Distaplia rosea, **301**
Distaplia occidentalis, **297**, 302;
  D. smithi, 302
Distylia rugosa, 67
Dodecaceria choncharum, 70;
  D. fewkesi, **70**; *D. fistulicola*, 70
Doliolaria larvae, **285, 286, 289**
Doridacean nudibranch, **98**
Doridella, 90: *D. steinbergae*, 108, 120
Dorvillea, 51: *D. atlantica*, 51;
  D. moniloceras, 51;
  D. pseudorubrovittata, 51;
  D. rudolphi, 51
Dorvilledae Family, 45, 51
Doryporella alcicornis, 262
Doto, 90: *D. amyra*, 107, 108, 120;
  Doto form B, 120; *D. kya*, 120
Dotoidea, 90
Drilonereis falcata, 51; *D. filum*, 51
Dynamena spp., 17
Dysidea fragilis, 11
Dysideidae Family, 11

Echinodermata, 272-91: collecting,
  276-77; development, 272-76;
  local species, 273-74; morphology,
  278-90; observing, 276-77;
  reproduction, 272-76
Echinoid key, 280-81
Echinoida Order, 273
Echinoidea Class, 278-81
Echinoplutei, 278: larvae, 278
Echinopluteus, 278
Echinospira larvae, **97**
Echiura Phylum, 85
Echiuridae Family, 85-86
Echiuroinea Order, 85-86
Echiurus echiurus subsp. *Alaskanus*,
  86
Ectyomyxilla parasitica, 11
Edwardsia sipunculoides, 19
Edwardsiidae Family, 10

Ehlersia cornuta, 66
Eirenidae Family, 17
Electra crustulenta, 262, **264**, 265;
   E. pilosa, **264**
Electridae Family, 262
Ellisina levata, 262
Elysia, 90: *E. hedgpethi*, 119
Elysidae, 90
Emerita analoga, 184, 194, 219-**220**
Enipo, 63: *E. cirrata*, 63
Enterocoelous, 293-95
Enterogona Order, 302
Enteropneusta Class, 293
Enteroxenos, 89: *E. parastichopoli*, 117
Entoconchid prosobranch, 117
Entoconchidae, 89
Entodesma pictum, 135
Entoniscidae, 178
Ephesia, 65
Ephesiella, 65
Epiactis fernaldi, 15; *E. prolifera*, 15;
   E. ritteri, 15
Epicaridea, 178; larvae, 178
Epistomiidae Family, 262
Epitoniidae, 89, 110
Epitonium, 89
Epizoanthidae Family, 19
Epizoanthus scotinus, 15, 19, 20
Errinopora pourtalesii, 17
Esperiopsidae Family, 11
Eteone bistriata, 60; *E. californica*, 60;
   E. longa, 60, **61, 62**; *E. maculata*, 60;
   E. pacifica, 60; *E. spitsbergensis var.*
   pacifica, 60
Eualus avinus, 182; *E. barbatus*, 182;
   E. berkeleyorum, 182; *E. biunguis*,
   182; *E. fabricii*, 182; *E. lineatus*,
   182; *E. macrophthalmus*, 182;
   E. suckleyi, 183, **204**
Eubranchidae, 90, 114
Eubranchus, 90: *E. olivaceus*, 121;
   E. rustyus, 121
Euchone analis, 67; *E. barnardi*, 67;
   E. incolor, 67; *E. rosea*, 67; *E. sp.cf.*
   hancocki, 67; *E. trisegmentata*, 67
Euclymene, 50: *E. reticulata*, 67
Eudendriidae Family, 17
Eudendrium spp., 17
Eudistoma illotum, **301**;
   E. purpuropunctatum, 302;
   E. ritteri, 302
Eudistylia abbreviata, 67; *E. plumosa*,
   67; *E. polymorpha*, 67; *E. tenella*, 67;
   E. vancouveri, 67
Euherdmania claviformis, **301**, 302,
   303
Eulaeospira, 68
Eulalia aviculiseta, 61; *E. bilineata*,
   60; *E. levicornuta*, 60;
   E. nigrimaculata, 61;
   E. quadrioculata, 61; *E. sanguinea*,
   61, 62; *E. viridis*, 61, **62**
Eulima, 89
Eulimidae Family, 110
Eulimidae, 89
Eumastia sitiens, 11
Eumida, 61
Euniccidae Family, 45, 46, 51

Eunice kobiensis, 52; *E. segregata*, 52;
   E. valens, 52
Eunicida Order, 51
Eunicidae, 51
Eunoe barbata, 63; *E. nodosa*, 63;
   E. oerstedi, 63; *E. senta*, 63
Eupentacta pseudoquinquesemita, 274;
   E. quinquesemita, 274, **286**
Euphrosine bicirrata, 48; *E. hortensis*,
   48
Euphrosinidae Family, 48
Euphysa ruthae, 17; *Euphysa* spp., 17
Euphysidae Family, 17
Eupolymnia crescentis, 76;
   E. heterobranchia, 76
Eurylepta aurantiaca, 27; *E. leoparda*,
   27
Euryleptidae Family, 27
Eurystomella bilabiata, 263
Eurystomellidae Family, 263
Eusergestes similis, 182
Eusyllis assimilis, 66; *E. blomstrandi*,
   66
Euthecoscome pteropods, 116
Eutonina indicans, 17
Euzonus mucronata, 53; *E. williamsi*,
   53
Evasterias troschelii, 273
Exilioidea, 89
Exogone gemmifera, 66; *E. lourei*, 66;
   E. naidina, 66; *E. uniformis*, 66;
   E. verugera, 66

Fabia subquadrata, 186, 199, 244-**45**
Fabricia brunnea, 67; *F. dubia*, 67;
   F. oregonia, 67; *F. sabella*, 67
Fabriciola berkeleyi, 67; *F. pacifica*, 67;
   F. sabella, 67
Facelinidae, 90
False Angel Wing, **147**
Fartulum, 89
Fauveliopsida Order, 52
Fauveliopsidae Family, 52
Fauveliopsis armata, 63
Feather stars, 289
Fenestrulina malusii, 263
Filellum spp., 17
Filicrisia franciscana, 261;
   F. geniculata, 261
Fiona, 90: *F. pinnata*, 121
Fionidae Family, 114
Fionidae, 90
Fissurellidae, 89
Flabelliderma commensalis, 51
Flabelligera affinis, 53;
   F. infundibularis, 53
Flabelligerida Order, 52-53
Flabelligeridae Family, 52
Flabellina, 90: *F. fusca*, 121;
   F. trilineata, 121
Flabellinidae, 90
Flatworms, 26
Florometra serratissima, 274, **289**
Flustrellidra corniculata, 262;
   F. hispida, **264**
Flustrellidridae Family, 262
Foersteria spp., 17
Fondiporidae Family, 261

Forcepia ?japonica, 11
Forcipulatida Order, 273
Freemania litoricola, 27
Fulifascigera fasciculata, 261
Funiculina parkeri, 19
Funiculinidae Family, 19
Fusinidae, 89
Fusinus, 89
Fusitriton, 89: *F. oregonensis*, 110

Gadilida, 154
Gadilidae, 153
Galathealinum brachiosum, 87
Galatheidae Family, 184, 195
Galeommatidae, 133
Gari californica, 134
Garveia annulata, 17; *G. groenlandica*,
   17
Gastropoda: Class, 88-130;
   classification, 89; Families, 89-90;
   genera, 89-90; identification, 94-
   95; illustration sources, 123; key,
   96-117; larval form, 91;
   morphology, 88-91
Gastropteridae, 89
Gastropteron, 89, 104: *G. pacificum*,
   98, 103, 104, 119
Gattyana, 63
Geitodoris, 90: *G. heathi*, 120
Gemelliporella inflata, 263
Genetyllis, 61
Gennadas incertas, 182;
   G. propinquus, 182; *G. tinayrei*, 182
Geodia mesotriaena, 11
Geodiidae Family, 11
Geodinella robusta, 11
Gersemia rubiformis, 19
Glans carpenteri, 1333
Galucothoe, 187
Gleba sp., **107**
Glycera americana, 56; *G. capitata*, 56,
   57; *G. convoluta*, **56, 57**;
   G. gigantea, 56; *G. nana*, 56;
   G. oxycephala, 56; *G. robusta*, 56;
   G. tenuis, 56; *G. tesselata*, 56
Glyceridae Family, 45, 55, 56
Glycinde armigera, 57; *G. picta*, 56;
   G. polygnatha, 57
Glycymerididae, 132
Glycymeris corteziana, 132;
   G. subobsoleta, 132
Golfingia elongata, 81;
   G. margaritacea, 80; *G. minuta*, 81;
   G. misakiana, 81; *G. pugettensis*, 80,
   81; *G. vulgaris*, 81
Golfingiidae Family, 80
Goniada brunnea, 56
Goniadidae Family, 45, 56, 57
Goniodorididae, 90
Gonionemus vertens, 17
Gonothyraea spp., **16**, 17
Gorgonocephalus eucnemis, 274
Götte's larvae, 26, **27**
Grammaria spp., 17
Grantia comoxensis, 9; *G. compressa*,
   7, 9
Grantiidae Family, 9
Granulina, 89

Grapsidae Family, 185, 193, 200, **238-242**
Gymnolaematea, 261-63, **262**
Gymnomorph slugs, 89
Gymnomorpha, 89, 90
Gymnosomata, 120
Gymnosomatous pteropods, **97, 99,** 116
Gyptis, 57: *G. arenisola glabra*, 57

Hadromerida Order, 8, 11
Halcampa crypta, 19;
 H. decemtentaculata, 18, 19, 21
Halcampidae Family, 19
Halcampoides purpurea, 19
Halcampoididae Family, 19
Haleciidae Family, 17
Halecium spp., 17
Halichondria, 8, **10**, 11:
 H. bowerbanki, 11; *H. melanadocia*, 10; *H. moorei*, 8; *H. panicea*, 11
Halichondriida Order, 8, 10, 11
Halichondriidae Family, 11
Haliclona **8**, 9: *H.?ecbasis*, 11;
 H.?permollis, 11
Haliclonidae Family, 11
Halimedusidae Family, 17
Haliotidae, 89
Haliotids, 109
Haliotis, 109
Haliplanella lineata, 19
Haliplanellidae Family, 19
Halisarca, 8: *H. sacra*, 12
Halisarcidae Family, 12
Hallaxa, 90: *H. chani*, 120
Halmedusa typus, 17
Halochondria spp., 11
Haloclavidae Family, 19
Halocynthia, 301: *H. aurantium*, 300;
 H. igaboja, 300
Haloscoloplos, 53, 54
Halosydna brevisetosa, 63, **64**
Hamacanthidae Family, 11
Hamigera ?lundbecki, 11
Haminaea, 89, 104
Haminoeidae, 89
Hanleyidae, 153
Haplogaster grebnitzkii, 184
Haplosclerida Order, 9, **10**, 11
Haploscoloplos alaskensis, 54;
 H. pugettensis, 53
Haplosyllis spongicola, 66
Harmothoe extenuata, 63, **64, 65**;
 H. fragilis, 63; *H. hartmanae*, 63;
 H. imbricata, 63, **64, 65**;
 H. lunulata, 63, **64, 65**;
 H. multisetosa, 63; *H. pellucelytris*, 63; *H. rarispina*, 63; *H. tenebricosa*, 63; *H. triannulata*, 63
Harrimania, 293-95
Harrimaniidae Family, 293
Hataia parva, 17
Havelockia benti, 274
Heart urchin, 278
Hebella spp., 17
Helicoptilum rigidum, 19
Hemectyon hyle, 11
Hemiarthrus abdominalis, 178

Hemichordata Phylum, 293-95
Hemigrapsus nudus, 185, 200, 240;
 H. oregonensis, 185, 200, 241
Hemioniscidae, 178
Hemioniscus balani, 178
Hemipenaeus spinidorsalis, 183
Hemipodus borealis, **56, 57**
Hemithiris psittacea, 269
Henleya oldroydi, 153
Henricia spp., 273, **283**
Heptabrachia ctenophora, 87
Heptacarpus brevirostris, 183;
 H. camtschaticus, 183; *H. carinatus*, 183; *H. decorus*, 183; *H. flexus*, 183;
 H. herdmani, 183; *H. kincaidi*, 183;
 H. littoralis, 183; *H. moseri*, 183;
 H. paludicola, 183; *H. pictus*, 183;
 H. pugettensis, 183; *H. sitchensis*, 183; *H. stimpsoni*, 183; *H. taylori*, 183; *H. tenuissimus*, 183
Hermadion truncata, 63
Hermaea, 90; *H. vancouverensis*, 119
Hermaeid sacoglossan opisthobranch, 106
Hermaeidae, 90
Hermissenda, 90; *H. crassicornis*, 121
Hesionidae Family, 48, 57
Hesperonoe adventor, 63;
 H. complanata, 63
Hesperophyllum, 61
Heterobranchia Subclass, 89
Heteromastus filobranchus, 49;
 H. filiformis, 50
Heteropoda, 89, 105
Heteropora alaskensis, 261; *H. magna*, 261; *H. pacifica*, 261
Heteroporidae Family, 261
Heterostrophic shell growth, 102, 103, 104
Hexactinellida Class, 5
Hexadella sp., 12
Hiatella arctica, 135, **140, 142, 145**
Hiatellidae, 135
Higginsia sp., 11
Hincksina alba, 262; *H. pallida*, 262
Hincksinidae Family, 262
Hinnites gigantea, 132
Hippasteria spinosa, 273
Hippidae Family, 184, 194, 219
Hippodiplosia insculpta, 263, **265,** 266; *H. reticulato-punctata*, 263
Hippolyte clarki, 183
Hippolytidae Family, 182, 197, 204
Hippomonavella longirostrata, 263
Hipponicidae, 89
Hipponix, 89
Hippoporella nitescens, 263
Hippoporinidae Family, 263
Hippothoa divaricata, 262; *H. hyalina*, 263
Hippothoidae Family, 262
Hobsonia florida, 75
Holaxonia Suborder, 19
Holophryxus alaskensis, 178
Holoplocamia, 9
Holoporella brunnea, 263
Holothuroid key, 287
Holothuroidea Class, 273, 285-87

Homalopoma, 89
Homoscleromorphida Order, 11
Hopkinsia, 90: *H. rosacea* 120
Hormathiidae Family, 19
Humilaria kennerlyi, 134
Hybocodon, 16: *H. prolifer*, 17
Hyboscolex pacificus, 53
Hydractinia, 17: *H. aggregata*, 17;
 H. laevispina, 17; *Hydractinia* sp., 17
Hydractiniidae Family, 17
Hydrallmania spp., 17
Hydrocorals, 13, 17
Hydrodendron spp., 17
Hydroida Order, 13, 17
Hydroids, 13
Hydromedusae, 16
Hydrozoa Class, 13, 17
Hymedesanisochela rayae, 11
Hymedesmiidae Family, 11
Hymedesmio spp., 11
Hymenamphiastra cyancrypta, 11
Hymenanchora sp., 11
Hymendecyon lyoni, 11
Hymeniacidon, 10: *H.?perleve*, 11;
 H. sinapium, 11; *Hymeniacidon* sp., 11; *H. ungodon*, 11
Hymenodora acanthitelsonis, 183;
 H. frontalis, 183; *H. glacialis*, 183

Idanthyrsus armatus, 75;
 I. ornamentatus, 75
Immergentiidae Family, 262
Immergetia sp., 262
Inarticulata Class, 270-71
Ione cornuta, 178
Iothia, 89
Ischnochiton abyssicola, 153;
 I. interstinctus, 153; *I. trifidus*, 153
Ischnochitonidae, 153
Iselica, 89
Isididae Family, 19
Isocirrus longiceps, 50
Isopoda, 178-89

Janolus, 90: *J. fuscus*, 121
Janthina, 89
Janthinidae, 89, 110
Janura rugata, 70
Jellyfish, 13
Jones amaknakensis, 11

Kaburakia excelsa, 27
Katharina tunicata, 153
Kellia suborbicularis, 133, 136, **142, 143, 145**
Kelliidae, 133
Kentrogonida, 175
Keyhole limpets, 89
Kophobelemnidae Family, 19
Kophobelemnon affine, 19;
 K. biflorum, 19; *K. hispidum*, 19
Kurtzia, 89
Kurtziella, 89

Lacerna fistulata, 263
Lacuna, 89
Lacunid larvae, 111

Lacunidae, 89, 111
Laeospira, 68
Laevidomus, 171
Lafoea spp., 17
Lafoeidae Family, 17
Lagenipora punctulata, 263;
    L. socialis, 263
Lagisca, 63
Laila, 90: *L. cockerelli*, 120
Lamellaria, 89
Lamellibranchs, 131-51
Lamellisabella coronata, 87; *L. zachsi*,
    87
Langerhansia, 66
Lanice chonchilega, 76
Laodiceidae Family, 17
Laonice cirrata, 70, **72**
Laqueus californianus, 269
Larvaceans, 298
Lasaea subviridis, 133
Lasaeidae, 133
Latrunculia sp., 11
Latrunculiidae Family, 11
Laxosuberites sp., 11
Leanira calcis, 65
Lebbeus schrencki, 183; *L.
    washingtonianus*, 183; *L.
    groenlandicus*, 183, 204, **205**
Lebrunia, 15
Leitoscoloplos elongatus, 53, **54**;
    L. panamensis, 54
Lensia baryi, 17; *L. conoidea*, 17
Lepadomorpha, 158, 173-74
Lepas anatifera, 158, **173**;
    L. fascicularis, 158; *L. hilli*, 158;
    L. pacifica, 158; *Lepas* spp., 164
Lepeta, 89
Lepetidae, 89
Lepidasthenia longicirrata, 63
Lepidochitona cinereus, **152**;
    L. dentiens, 153; *L. fernaldi*, 153;
    L. hartwegii, 153
Lepidochitonidae, 153
Lepidonotus caekorus, 63;
    L. squamatus, 63, **64, 65**
Lepidopleurina, 153
Lepidopleurus asellus, **152**
Lepidozona cooperid, 153;
    L. mertensii, 153; *L. retiporosa*, 153;
    L. scabricostata, 153; *L. willetti*, 153
Lepraliella bispina, 263
Leptasterias hexactis, 273
Leptochiton nexus, 153; *L.rugatus*,
    153
Leptochitonidae, 153
Leptomedusae, 17
Leptoplana vesiculata, 27
Leptoplanidae Family, 27
Leptosynapta clarki, 274; *L. roxtona*,
    274; *L. transgressor*, 274
Leucandra heathi, 9; *L. pyriformis*, 9;
    L. taylori, 9; *L. ?levis*, 9
Leucckartiara spp., 17
Leucilla nuttingi, 9
Leucopsila stylifera, 9
Leucosolenia eleanor, 9; *L. nautilia*, 9;
    Leucosolenia spp., 9; *L. variabilis*, **8**
Leucosoleniida Order, 9

Leucosoleniidae Family, 9
Lichenopora novae-zelandiae, 261;
    L. verrucaria, 261
Lichenoporidae Family, 261
Limacina, 90: *L. helicina*, 103, 104,
    120
Limacinidae, 90
Limatula subauriculata, 132
Limidae, 132
Limnomedusae Suborder, 17
Limoida, 132
Limpets, 89
Lingula, 270, **271**
Liponema brevicornis, 19
Liponematidae Family, 19
Liriopsidae, 178
Liriopsis pygmaea, 178
Lirularia, 89
Lissocrangon stylirostris, 182
Lissodendoryx, 9; *L. firma*, 11;
    Lissodendoryx sp., 11
Lithodes couesi, 184
Lithodidae Family, 184, 195, 221-22
Lithophaga plumula, 132
Lithopoma, 89
Littorina, 89
Littorinidae, 89, 111
Loimia medusa, 76
Lopadorhynchidae Family, 45, 46,
    58
Lopadorhyncus uncinatus, **58**;
    L. varius, 58
Lophelia californica, 19
Lopholithodes foraminatus, 184;
    L. mandtii, 184
Lophon chelifer, 11; *L. piceus*, 11
Lophopanopeus bellus, 186, 199;
    L. bellus bellus, 186, 248; *L. bellus
    diegensis*, 186
Lottia, 89
Lottidae, 89
Loxothylacus panopaei, 158, 175
Lucina tenuisculpta, 133
Lucinidae, 133
Lucinoma annulata, 133
Luidia foliata, 273, **281**
Lumbrineridae Family, 45, 46, 52
Lumbrineris aff. *abyssicola*, 52;
    L. bifurcata, 52; *L. bricirrita*, 52;
    L. californiensis, 52; *L. cruzensis*,
    52; *L. japonica*, 52; *L. latreilli*, 52;
    L. zonata, 52
Lyocyma fluctuosa, 134
Lyonsia californica, 135
Lyonsiidae, 135
Lyrula hippocrepis, 262
Lysippe annectens, 75

Macoma calottensis, 134;
    M. eliminata, 134; *M. nasuta*, 134;
    M. secta, 134; *M. yoldiformis*, 134;
    M. balthica, 134, **140, 145**
Mactra californica, 133
Mactridae, 133
Madreporia, 19
Magelona sp., **70**
Magelonidae Family, 45, 70
Majidae Family, 185, 194, 198, 199,
    200, 242-44

Maldane robusta, 50; *M. sarsi*, 50
Maldanella harai, 50
Maldanidae Family, 50
Malmgrenia, 63: *M. nigralba*, 63
Margarites, 89; *M. pupillus*, 103
Marginellidae, 89
Marphysa stylobranchiata, 51
Marseniid prosobranchs, **97, 115**
Marseniidae, 89
Marsenina, 89
Marseniopsis, 89
Mediaster aequalis, 273, **283**
Mediomastuf californiensis, 50
Megacrenella columbiana, 133
Megalomma splendida, 67
Megalopa, 187, **188**
Melampidae, 90
Melanochlamys, 104: *M. diomedea*,
    103, 119
Melibe, 90: *M. leonina*, 107, 120
Melibe leonina, 107
Melinna cristata, 75; *M. denticulata*,
    75
Membranipora membranacea, 262,
    264
Membraniporidae Family, 262
Merriamum oxeota, 11
Mesochaetopterus taylori, 69
Mesocrangon munitella, 182
Mesogastropoda, 89
Metacrangon acclivis, 182;
    M. munita, 182; *M. spinosissima*,
    182; *M. variabilis*, 182
Metandrocarpa dura, 300; *M. taylori*,
    299, 300
Metridiidae Family, 19
Metridium senile, **15**, 19; *Metridium*
    spp., 18, 19, 21
Micranellum, 89
Microciona, 9; *M. coccinea*, **8**;
    M. microjoanna, 11; *M. primitiva*,
    11; *M. prolifera*, 11
Microglyphis, 89
Micronereis bodegae, 59;
    M. nanaimoensis, 59, **60**;
    M. variegata, 59
Microniscus, 178
Microporella californica, 262;
    M. ciliata, 262; *M. coriacea*, 262;
    M. setiformia, 262; *M. umbonata*,
    262; *M. vibraculifera*, 262
Microporellidae Family, 263
Microporidae Family, 262
Microporina borealis, 262
Micrura wilsoni, 30
Mimulus foliatus, 185
Minuspio cirrifera, 70
Miontodiscus prolongatus, 133
Mitella polymerus, 158
Mitrella, 89
Mitrocoma spp., 17
Mitrocomella spp., 17
Mitrocomidae Family, 17
Modiolus modiolus, 132, **140, 142,
    145**; *M. rectus*, 132
Molgula citrina, **299**; *M. cooperi*, 300;
    M. manhattensis, 300; *M. oregonia*,
    300; *M. pugetiensis*, 300;

M. occidentalis, 301; *M. pacifica*, 300
Molgulidae Family, 300, 301
Molpadia intermedia, 274
Molpadiida Order, 274
Monobrachium parasiticum, 17
Monogtenea Class, 27
Montaculidae, 132
Mooreonuphis stigmatis, 52
Mopalia ciliata, 153; *M. cirrata*, 153; *M. cithara*, 153; *M. egretta*, 153; *M. hindsii*, 153; *M. imporcata*, 153; *M. laevior*, 153; *M. lignosa*, **152**, 153; *M. muscosa*, 153; *M. phorminx*, 153; *M. porifera*, 153; *M. sinuata*, 153; *M. spectabilis*, 153; *M. swanii*, 153
Mopaliidae, 153
Mucronella ventricosa, 263
Muggiaea atlantica, 17
Müller's larva, 26, **27**
Munida quadrispina, 184
Munidion parvum, 178
Munidopsis quadrata, 184
Muricidae, 89, 112
Mursia gaudichaudi, 185
Musculista senhousia, 132
Musculus discors, 132; *M. niger*, 132; *M. taylori*, 132
Mussel: Blue, **146**; Northern Horse, **145**
Mya arenaria, 134, **140**, **142**, 145, **146**; *M. truncata*, 134
Mycale, **8**, 9: *M. adhaerens*, 11; *M. bamfieldense*, 11; *M. bellabellensis*, 11; *M. hispida*, 11; *M. macilenta*, 8; *M.?toporoki*, 11; *M. macginitiei*, 11; *M. richardsoni*, 8
Mycalecarmia lobata, 11
Mycalidae Family, 11
Myidae, 134
Myoida, 134
Myoisophagos, 32: *M. sanguineus*, 30, 32
Myriochele oculata, **55**; *M. heeri*, 55
Myriozoidae Family, 263
Myriozoum coarctatum, 263; *M. subgracile*, 263; *M. tenue*, 263
Mysella tumida, 133
Mysotella, 90
Mytilidae, 132
Mytilimeria nuttalli, 135
Mytiloida, 132
Mytilus californianus, 132, **146**; *M. trossulus (edulis)*, 132, **140**, **142**, **146**
Myxicola aesthetica, 67; *M. infundibulum*, 67
Myxilla behringensis, 11; *M. lacunosa*, 11
Myxillidae Family, 11

Naineris berkeleyorum, 54; *N. cara*, 17; *N. dendritica*, **54**; *N. laevigata*, 54; *N. uncinata*, 54
Nassariidae, 89, 111, 112
Nassarius, 89
Natantia, 181-83

Natica, 89
Naticidae, 89, 113
Nauplius, **157**, **159**, 160: key, 162-63; key stages, 161-62; larva, 186; stages, **161**
Neanthes, 58
Nellobia eusoma, 86
Nematostella vectensis, 19
Nemertea, 28-38: development, 28-30; identification, 29; reproduction, 28-29
Nemertopsis gracilis, 30
Nemocardium centrifilosum, 133
Neoamphitrite robusta, 76
Neocrangon abyssorum, 182; *N. communis*, 182; *N. resima*, 182
Neoesperiopsis digita, 11; *N. infunibula*, 11; *N. rigida*, 11; *N. vancouvernensis*, 11
Neogastropoda, 89
Neoleanira areolata, 65
Neoleprea spiralis, 76
Neoloricata, 153
Neopygospio laminifera, 71
Neotrypaea californiensis, 184, **215**-16; *N. gigas*, 184
Neoturris spp., 17
Nephtheidae Family, 19
Nephtyidae Family, 45, 58
Nephtys assignis, 58; *N. caeca*, **58**; *N. caecoides*, 58; *N. californiensis*, 58; *N. cornuta cornuta*, 58; *N. cornuta franciscana*, 58; *N. discors*, 58; *N. ferruginea*, 58; *N. longosetosa*, 58; *N. paradoxa*, 58; *N. punctata*, 58; *N. rickettsi*, 58; *N. schmitti*, 58
Neptunea, 89
Neptunidae, 89
Nereidae Family, 46, 58, 59
Nereis brandti, 59; *N. callaona*, 59; *N. diversicolor*, 59; *N. eakini*, 59; *N. grubei*, 59; *N. japonica*, 59; *N. lighti*, 59; *N. limnicola*, 59, **60**; *N. mediator*, 59; *N. natans*, 59; *N. neoneanthes*, 59; *N. pelagica*, 59, **60**; *N. procera*, 59; *N. vexillosa*, 59; *N. virens*, 59, **60**; *N. zonata*, 59
Nerine, 71
Netastoma rostrata, 135
Neverita, 89
Nicomache lumbricalis, 50; *N. personata*, 50
Nicon moniloceras, 59
Ninoe gemmea, 52
Nitidiscala, 89; *N. tincta*, 110
Nolella stipata, 262
North Atlantic Lepton, **145**
Notaspidea, 119
Notaspidean heterobranchs, 104
Notaspidean opisthobranchs, 96, 98
Nothria, 52: *N. abyssalis*, 52; *N. geophiliformis*, 52; *N. iridescens*, 52; *N. occidentalis*, 52
Notocirrus californiensis, 51
Notodorididae, 90
Notomastus balanoglossi, 50; *N. giganteus*, 50; *N. lineatus*, 50; *N. magnus*, 50; *N. tenuis*, 50

Notophyllum imbricatum, 60; *N. tectum*, 60
Notoplana atomata, 27; *N. celeris*, 27; *N. inquieta*, 27; *N. inquilina*, 27; *N. longastyletta*, 27; *N. natans*, 27; *N. rupicola*, 27; *N. rupicola rupicola*, 27; *N. sanjuania*, 27
Notoproctus pacificus, 50
Notopygos labiatus, 48
Notostomus japonicus, 183
Nucella, 89
Nucellidae, 89
Nucula delphinodonta, **139**; *N. proxima*, **139**; *N. tenuis*, 132
Nuculana cellutita, 132; *N. hamata*, 132; *N. minuta*, 132
Nuculanidae, 132
Nuculidae, 132
Nuculoida, 132, 136
Nudibranchia: Aeolidacea, 114, 121; Arminacea, 121; Dendronotacea, 96, 120; Doridacea, 120
Nuttallina californica, 153
Nynantheae Suborder, 19

Obelia bidentata, 17; *O. dichotoma*, 17; *O. geniculata*, 17
Ocenebra, 89
Ococorallia, 19
Octopoda, 154-56
Octopodidae, 154
Octopus dofleini, 154, 155, **156**; *O. leioderma*, 154, 155; *O. rubescens*, 154, 155, **156**
Odontogena borealis, 133
Odontosyllis parva, 66; *O. phosphorea*, 66; *O. phosphorea var. nanaimoensis*, 66
Odostomia, 89
Oedignathus inermis, 184
Oenopota, 89; *O. levidensis*, 112
Oerstedia dorsalis, 30, **31**
Oikopleura sp., 298
Oligobrachia dogieli, 86; *O. webbi*, 86
Olindiasidae Family, 17
Olivella, 89
Olivellidae, 112
Olividae, 89
Onchidella, 90
Onchidiidae, 90
Onchidorididae, 90
Onchidoris, 90: *O. bilamellata*, 120; *O. muricata*, 120
Oncoscolex, 53
Oncousoecia ovoidea, 261
Oncousoeciidae Family, 261
Onoba, 89
Onuphidae Family, 46, 52
Onuphis, 52: *O. longibranchiata*, 52
Opalia, 89
Ophelia borealis, 53; *O. limacina*, 53
Ophelida Order, 53
Ophelidae Family, 46, 54
Ophelina acuminata, 54
Ophiodermella, 89
Ophiodromus pugettensis, 57
Ophiopholis aculeata, 274, **289**

Ophiopluteus, 288, **289**: larvae 288, **289**
Ophiopteris papillosa, 274
Ophiura leptoctenia, 274; *O. lütkeni*, 274; *O. sarsi*, 274
Ophiurida Order, 274
Ophiuroidea Class, 274, 287-89
Ophiuroids, 274, 287-89
Ophlitaspongia, 8, 9: *O. pennata*, 11; *O. seriata*, 8
Opisthobranchia, 89: Anaspidea, 90; Cephalaspidea, 89; Gymnosomata, 90; Notaspidea, 90; Nudibranchia, Aeolidacea, 90; Nudibranchia, Arminacea, 90; Nudibranchia, Dendronotacea, 90; Nudibranchia, Doridacea, 90; Sacoglossa, 90; Thecosomata, Euthecosomata, 90; Thecosomata, Pseudothecosomata, 90
Opisthobranchs comparative data, 118-22
Oplophoridae Family, 183
Orbinia, 53
Orbiniida Order, 53-54
Orbiniidae Family, 46, 53
Oregonia bifurca, 185; *O. gracilis*, 186, 198, **243**-244
Orina sp., 11
Oriopsis gracilis, 67
Orthasterias koehleri, 273
Orthopagurus minimus, 184; *O. schmitti*, 203
Orthopyxis spp., 17
Oscarella, 8
Osteoida Order, 132
Ostracods, **159**
Ostrea conchaphilia, 133; *O. lurida*, 133, **140, 143, 147**
Ostreidae, 133
Ostreoida, 132
Ovatella, 90
Owenia fusiformis, **55**
Oweniida Order, 54
Oweniidae Family, 46, 54
Oysters, 135: Eastern, **144**; Giant Pacific, 144; Native Pacific, **147**

Pachastrellidae Family, 11
Pachycerianthus fimbriatus, 15, 19, 20
Pachychalina spp., 11
Pachycheles, 201: *P. pubescens*, 185, 202, 234; *P. rudis*, 185, 202, 235
Pachygrapsus crassipes, 185, 194, 201, **241**-42
Paguridae Family, 184, 195, 201, 202, 203, 223-32: key megalopae, 231-32; key zoeae I, 223-**24**; key zoeae II, 225; key zoeae III, 225, 227; key zoeae IV, **226-231**; key zoeal stages, 223
Pagurus aleuticus, 184; *P. armatus*, 184; *P. beringanus*, 185, 203; *P. capillatus*, 185; *P. caurinus*, 185; *P. confragosus*, 185; *P. cornutus*, 185; *P. dalli*, 185; *P. granosimanus*, 185, 202; *P. hemphilli*, 185; *P. hirsutiusculus*, 185, 203;

P. ochotensis, 185, 204; *P. quaylei*, 185; *P. samuelis*, 185; *P. setosus*, 185; *Pagurus* sp. C, 224, 227, 231, 232; *Pagurus* sp. I, 224, 227, 231; *Pagurus* sp. J, 229, 232; *P. tanneri*, 185; *P. turgidus*, 184, 202, **218**, 219; *P. ulreyi*, 184
Palaeonemertea Order, 28-31
Paleanotus bellis, 55, **56**; *P. chrysolepis*, 55; *P. occidentale*, 55
Palio zosterae, 120
Pandalidae Family, 183, 197, 205
Pandalopsis ampala, 183; *P. dispar*, 183, 205, **206**
Pandalus borealis, 183, **207**; *P. danae*, 183, **208**; *P. eous (borealis)*, **207**; *P. gurneyi*, 183; *P. jordani*, 183, **209**, 210; *P. platyceros*, 183, **211**, 212; *P. stenolepis*, 183, **214**, 215; *P. tridens*, 183;
Pandeidae Family, 17
Pandora bilirata, 135; *P. filosa*, 135; *P. glacialis*, 135; *P. punctata*, 135; *P. wardiana*, 135
Pandoridae, 135
Panomya chrysis, 135
Panope abrupta, 135
Paracaudina chilensis, 274, **286**
Paracornulum, 9
Paracrangon echinata, 182
Paractinostola faeculenta, 18, 19
Paracyathus stearnsi, 19
Paradexiospira violacea, 68; *P. vitrea*, 68
Paragorgia pacifica, 19
Paragrogiidae Family, 19
Paralomis multispina, 184; *P. verrilli*, 184
Paranaitis, 61
Paranemertes peregrina, 30
Paraonidae Family, 54
Paraonis, 54; *P. ivanovi* 54
Parapaguridae Family, 185, 195, 201
Parapagurus pilosimanus, 185
Parapasiphae sulcatifrons, 183
Paraprionospio pinnata, 70, **72**
Parasabella, 67; *P. maculata*, 67
Parasitic barnacles, 158, 174-76
Parasmittina collifera, 263
Paraspidosiphon fischeri, 81
Parastennella sp., 19
Parastichopus californicus, 273, **285**; *P. leukothele*, 273; *P. parvimensis*, 273; *Parastichopus* sp., **285**, 286
Parenchymella larvae, 6, 7
Paresperella psila, 11
Parvamussium alaskensis, 132
Pasiphaea pacifica, 183; *P. tarda*, 183
Pasiphaeidae Family, 183
Patellogastropoda Order, 89, 96, **97**, 114
Patinopecten caurinus, 132
Peachia quinquecapitata, 18, 19
Peanut worms, 80-84
Pectinaria, 75: *P. belgica*, 75; *P. californiensis*, 75
Pectinaridae Family, 44, 75

Pectinidae, 132
Pedivelige, 137, **138**
Pelagobia longicirrata, **58**
Pelagodiscus atlanticus, **271**
Pelagosphera: larvae, 80-81; stage, 80-81
Pelecypods, 131-51
Pelonia corrugata, 297
Peltogaster, **175**: *P. boschmae*, 158, 175; *P. gracilis*, 158, 175; *P. paguri*, 158, 175
Peltogastridae, 175
Penaeidae Family, 182, 196
Penares cortius, 11
Penetrantia sp., 262
Penetrantiidae Family, 262
Penitella conradi, 135; *P. gabbii*, 135; *P. penita*, 135; *P. turnerae*, 135
Pennatula phosphorea, 19
Pennatulacea Order, 15, 19
Pennatulidae Family, 19
Pentamera lissoplaca, 274; *P. populifera*, 274; *P. pseudocalciegera*, 274; *P. trachyplaca*, 274
Pereonites, 179
Periclymma larva, 137, **139**
Perigonimus spp., 17
Perinereis monterea, 58
Periwinkles, 89
Perophora annectens, 300; *P. listeri*, **299**
Perophoridae Family, 300
Petalidium subspinosum, 182
Petricola carditoides, 134; *P. pholadiformis*, 134, **140, 147**
Petricolidae, 135
Petrolisthes, 201: *P. cinctipes*, 185, 202, 234; *P. eriomerus*, 185, 201, 235
Petrosiida Order, 11
Petrosiidae Family, 11
Phakettia ?beringensis, 11
Phascolion cryptus, 81; *P. strombi*, 81
Phascolopsis gouldi, 81
Phascolosoma agassizii, 80, 81, **82**; *P. antillarum*, 81; *P. perlucens*, 81; *P. varians*, 81
Phascolosomatidae Family, 80
Pherusa inflata, 53; *P. neopapillata*, 53; *P. papillata*, 53; *P. plumosa*, 53
Phidolopora labiata, 263
Philine, 89, 104: *P. auriformis*, 119
Philinidae, 89
Philobrya setosa, 132
Philobryidae, 132
Phlebobranchia Suborder, 300
Pholadidae, 134
Pholadomyoida, 135
Pholoe caeca, 65, **66**; *P. minuta*, 65; *P. tuberculata*, 65
Pholoides aspera, 60
Pholoididae Family, 60
Phorbas, 9, **10**
Phoronida, 253-59: early pelagic stages, **258**; identification, 253; local taxa, 257; reproduction and development, 253

Phoronis architecta, 254, 255, **256**;
P. *ijimai*, 254; *P. muelleri*, 256;
P. *pallida*, 254, **255**; *P.
vancouverensis*, **254**
Phoronopsis harmeri, 256; *P. viridis*,
254, 256, **257**
Phylactellidae Family, 263
Phyllaplysia, 90: *P. taylori*, 119
Phyllochaetopterus prolifica, 69
Phyllodoce, 60: *P. castanea*, 61;
P. *polynoides*, 61
Phyllodocella, 59
Phyllodocidae Family, 48, 60-62
Phyllodocidae Order, 55-66
Phyllodurus abdominalis, 178
Phyllolithodes papillosus, 184
Phylloplana viridis, 27
Phylo felix, 54
Physonectae Suborder, 17
Physophora hydrostatica, 17
Physophoridae Family, 17
Pilargiidae Family, 44, 62
Pilargus berkeleyi, 63
Pileolaria langerhansi, 68;
P. *quadrangularis*, 68; *P. potswaldi*,
68, **69**
Pilidium larvae, 32-36: history of
classification, 34-35; new
observations, 35-36
Pinnixa eburna, 186; *P. faba*, 186;
P. *littoralis*, 186; *P. occidentalis*, 186;
P. *schmitti*, 186; *Pinnixa* spp., 186,
199; *P. tubicola*, 186
Pinnotheres pugettensis, 186;
P. *taylori*, 186, **246**
Pinnotheridae Family, 186, 193, 194,
197-199, 244-246
Pionosyllis gigantea, 66
Pisaster brevispinus, 273; *P.
giganteus*, 273; *P. ochraceus*, 273,
282
Pista brevibranchiata, 76; *P. cristata*,
76; *P. elongata*, 76; *P. fasciata*, 76;
P. *fimbriata*, 76; *P. moorei*, 76;
P. *pacifica*, 76
Placida, 90: *P. dendritica*, 119
Placiphorella rufa, 153; *P. verlata*, 153
Plagioecia patina, 261
Plakina, 8: *P. ?brachylopha*, 11;
Plakina sp., 11; *P.?trilopha*, 11
Plakinidae Family, 11
Planes cyaneus, 185; *P. marinus*, 185
Planula, 14, 16, **21**
Platyasterida Order, 273
Platyhelminthes Phylum, 26
Platynereis, 59
Platyodon cancellatus, 134
Platynereis bicanaliculata, 59, **60**;
P. *dumerili var. agassizi*, 59
Plectodon scaber, 135
Pleocyemata, 184
Pleraplysilla sp., 12
Pleurobranchaea californica, 96, 119
Pleurobranchia, 90
Pleurobranchidae, 90
Pleurogona Order, 300
Plexauridae Family, 19
Plicifusus, 89

Plocamia karykina, 11
Plocamiidae Family, 11
Plocamilla illgi, 11; *P. lambei*, 11
Plotohelmis. 55
Plumularia spp., 17
Plumulariidae Family, 17
Pluteus larvae, 275
Pneumoderma, 90
Pneumodermatidae, 90
Pneumodermopsis, 90
Podarke, 57
Podarkeopsis brevipalpa, **57**
Pododesmus cepio, 133
Podotuberculum hoffmanni, 11
Poecillastra rickettsi, 11
Poecilosclerid larvae, 11
Poecilosclerida Order, 9, **10, 11**
Pogonophora Phylum, 86-87
Polinices, 89: *P. lewisii*, 113
Pollicipes polymerus, 158, 162, 162,
174
Polybrachia canadensis, 87
Polybrachiidae Family, 87
Polycera, 90: *P. atra*, 120
Polyceratidae, 90
Polychaeta, 41-79: development,
41-42; morphology, 41-44;
reproduction, 41-42
Polycirrus caliendrum, 76;
P. *californicus*, 76
Polyclades, 26
Polycladida Order, 26
Polyclinidae Family, 302
Polydora, 71: *P. alloporis*, 71;
P. *armata*, 71; *P. brachycephala*, 71;
P. *caeca var. mangna*, 71; *P. cardalia*,
71; *P. ciliata var. spongicola*, 71;
P. *commensalis*, 71, **72**; *P. giardi*, 71;
P. *kempi japonica*, 71; *P. lighti*, 71,
72; *P. ligni*, 71, **72**; *P. plena*, 71;
P. *pygidialis*, 71; *P. socialis*, 71, **72**;
P. *spongicola*, 71; *P. websteri*, 71, **72**
Polymastia, 8: *P. pacifica*, 11;
P. *pachymastia*, 11
Polymastiidae Family, 11
Polynoe canadensis, 63; *P. gracilis*, 63
Polynoidae Family, 48, 63-65
Polyplacophora, 152-54
Polytrochous larvae, **99**
Porcellanidae: Family, 185, 194, 201,
232-34; key to zoeae, **232-233**
Porella columbiana, 263; *P. concinna*,
263; *P. porifera*, 263
Porifera: larvae, 6; reproduction, 5
Portunidae Family, 186, 194, 200,
246-48
Portunion conformis. 178
Postlarva, 187
Potamididae, 89
Potamilla californica, 67; *P.
intermedia*, 67; *P. neglecta*, 67;
P. *occelata*, 67
Praxillella affinis var. pacifica, 50;
P. *gracilis*, 50
Praya dubia, 17; *P. reticulata*, 17
Prayidae Family, 17
Prianos problematicus, 11
Primnoa willeyi, 19

Primnoidae Family, 19
Prionospio, 70: *P. lighti*, 71;
P. *malmgreni*, 71; *P. ornata*, 70;
P. *steenstrupi*, 71, **72**
Proboscidactyla flavicirrata, 17
Proboscidactylidae Family, 17
Proboscina incrassata, 261
Proboscis, 31, **32**
Procephalothrix filiformis, **31**;
P. *simulus*, **31**; *P. spiralis*, 30
Propagules, 17
Prophryxus alascensis, 178
Prosobranchia Subclass, 89
Prosuberites sp., 11
Protocapitella, 49
Protodorvillea gracilis, 51;
P. *kefersteini*, 51; *P. recuperata*, 51
Protoglossidae Family, 293
Protohydra leuckarti, 17
Protohydridae Family, 17
Protolaeospira eximia, 68
Protoptilidae Family, 19
Prototimia staminea, 134; *P. tenerrina*,
134
Psammobiidae, 134
Psammopemma sp., 12
Psephidia lordi, 134; *P. ovalis*, 134
Pseudaxinella ?rosacea, 11
Pseudione galacanthae, 178; *P. giardi*,
178
Pseudoceros canadensis, 27
Pseudocerotidae Family, 27
Pseudochama exogyra, 133
Pseudochitinopoma occidentalis, 68
Pseudomelatoma, 89
Pseudopolydora kempi, 71, **72**
Pseudopotamilla, 67: *P. reniformis*, 67
Pseudopythina compressa, 133;
P. *rugifera*, 133
Pseudosabellides, 74
Pseudostylochus burchami, 27;
P. *ostreophagus*, 27
Pseudosuberites spp., 11
Pseudothecosome pteropod, 107
Psolidium bulatum, 273
Psolus chitonoides, 273, **286**;
P. *squamatus*, 273
Pteraster tesselatus, 273, **283**
Pterotrachea, 89; *P. coronata*, 105
Pterotracheidae, 89; pterotracheid
heteropod, 105
Ptilosarcus gurneyi, 18, 19
Ptychoderidae Family, 293, 295
Ptychogena spp., 17
Puellina setosa, 262
Pugettia gracilis, 186, 200;
P. *producta*, 186, 201; *P. richii*, 186;
Pugettia sp., 198
Pulmonata, 89, 90
Pulsellidae, 154
Pulsellum salishorum, 154
Puncturella, 89
Pycnoclavella stanleyi, **301**, 302
Pycnopodia helianthoides, 273
Pygospio elegans, 71, **72**
Pyramidellacea, 89
Pyramidellid: heterobranchs, 103,
104; snails, 89

Pyramidellidacea Superorder, 89
Pyramidellidae, 89
Pyura haustor, 300; *P. microcosmus*, **299**; *P. mirabilis*, 300
Pyuridae Family, 300

Radiceps sp., 19
Raspailiidae Family, 11
Rathkea octopunctata, 17
Rathkeidae Family, 17
Reginella furcata, 262; *R. nitida*, 262
Reniera, 9; *R. mollis*, 11
Renilla sp., 15
Reptantia, 181, 184-86
Reteporidae Family, 263
Retusidae, 89, 104
Rhamphidonta retifera, 133
Rhamphostomella cellata, 263; *R. costata*, 263; *R. curvirostrata*, 263
Rhinolithodes wossnessenskii, 184, **222**
Rhizocaulus verticillatus, 17
Rhizocephala, 158, 174-76
Rhizogeton sp., 17
Rhodine bitorquata, 50
Rhynchozoon tumulosum, 263
Rhynocragon alata, 182
Rhysia sp., 17
Rhysiddae Family, 17
Rictaxis, 89; *R. punctocaelatus*, 119
Rissoidae, 89, 111
Rithropanopeus harrisii, 186, 201, 249
Ritterella aequalisiphonis, 302; *R. pulchra*, 302; *R. rubra*, **301**, 302
Romanchella, 68
Rostanga, 90; *R. pulchra*, 107, 120
Rostangidae, 90
Rostaria larva, **48**
Runcina, 89; *R. macfarlandi*, 119
Runcinidae, 89

Sabella, 67: *S. aulaconota*, 67; *S. crassicornis*, 67
Sabellaria alveola, 75, **76**; *S. cementarium*, 75; *S. gracilis*, 75
Sabellariidae Family, 45, 46, 75
Sabellida Order, 67-68
Sabellidae Family, 47, 67-68
Saccoglossus bromophenolosus, 293; *S. horsti*, **294**; *Saccoglossus* spp., 293
Sacculina carcini, **175**
Sacculinidae, 175
Sacoglossa, 119: Sacoglossan opisthobranchs, 106
Sagitella kowalevskii, 66
Salmacina dysteri var. tribranchiata, 68; *S. tribranchiata*, **68**
Sand dollars, 278-81
Sarcodictyon sp., 19
Sarsia japonica, 17; *Sarsia* spp., 17
Sarsonuphis elegans, **52**; *S. lepta*, 52
Saxicave, **145**
Saxidomus giganteus, 134, **140, 143, 147**
Scalibregma inflatum, 53
Scalibregmidae Family, 53
Scalpellum columbianum, 158

Scaphopoda, 154
Schistocomus hiltoni, 75
Schistomeringos longicornis, **51**
Schizobranchia insignis, 67
Schizomavella auriculata, 263
Schizoplax brandtii, 153
Schizoporella cornuta, 263; *S. linearis*, 263; *S. unicornis*, 263
Schizoporellidae, 263
Scintillona bellerophon, 133
Scionella japonica, 76
Scissurellidae, 89
Scleractinia Order, **15**, 19, 20
Scleraxonia Suborder, 19
Scleroplax granulata, 186
Scleroptilidae Family, 19
Scleroptilum sp., 19
Scolelepis acuta, 71; *S. foliosa*, 71; *S. squamata*, 71
Scoloplos, 53, 54: *S. acmeceps*, **54**; *S. armiger*, **54**
Scrobiculariidae, 134
Scruparia sp., **265**, 267
Scrupocellaria californica, 262; *S. reptans*, **265**; *S. varians*, 262
Scrupocellariidae Family, 262
Scypha compacta, 9; *S. mundula*, 9; *S. protecta*, 9; *Scypha* spp. 9
Scyphozoa Class, 13
Scyra acutifrons, 186
Sea anemones, 13
Sea cucumbers, 285-87
Sea pens, 14
Sea stars, 281-84
Sea urchins, 278-81
Searlesia, 89
Semele rubropicta, 134
Semibalanus balanoides, 158, 162, 163, 170, **171**; *S. cariosus*, 158, 162, 163, **167**, 169, **179**
Sepioidea, 155
Septibranchida, 135
Sergestidae Family, 182, 196
Sergia tenuiremis, 182
Serpula columbiana, 68; *S. vermicularis*, 68
Serpulidae Family, 47, 67-68
Serripes groenlandicus, 133
Sertularella spp., 17
Sertularia spp., 17
Sertulariidae Family, 17
Sessiliflorae Suborder, 19
Shell coiling, **101**
Shipworm: Common, **147**; Feathery, **139**
Shortscale Eualid, **204**
Shrimp: Bay Ghost, 215, **216**; Blue Mud, 216, **217**; Dock, **208**; Pacific Ocean, **209**-10; Rough Patch, **212**-13; Sidestriped, 205, **206**; Spot, **211**-12
Siboglinidae Family, 87
Siboglinum fedotovi, 87; *S. fiordicum*, **86**; *S. pusillum*, 87
Siderastraea sp., 15
Sigalionidae Family, 48, 65
Sigambra tentaculata, 63
Sigmadocia, 9: *S. edaphus*, 11; *Sigmadocia* spp., 11

Siliqua lucida, 134; *S. patula*, 134; *S. sloati*, 134
Sinistrella media, 68
Siphonaria, 90
Siphonariidae, 90
Siphonophora Class, 13
Siphonosoma cumanense, 81; *Siphonosoma* sp., 81
Sipuncula, 80-84: development, 80-81; identification, 81-82; reproduction, 80-81
Sipunculus nudus, 81; *S. polymyotus*, 81; *Sipunculus* sp., 81
Slipper shells, 105, 108
Smittina cordata, 263; *S. landsborovi*, 263
Smittinidae Family, 263
Snails: Bubble, 89; Lunged, 89; Moon, 113; Pyramidellid, 89; Turban, 89; Tusk, 154; Vermetid, 88
Solariella, 89
Solaster dawsoni, 273; *S. stimpsoni*, 273
Solecurtidae, 134
Solemya reidi, 132, 136, **139**; *S. velum*, **139**
Solemyidae, 132
Solemyoida, 132, 136
Solen sicarius, 134
Solenidae, 134
Solenosmilia variabilis, 19
Solidobalanus engbergi, 158; *S. hesperius*, 158, 162, **171**
Spat, 136
Spatangoida Order, 273
Spengelidae Family, 293
Sphaerodoridae Family, 45, 65
Sphaerodoridium, 65
Sphaerodoropsis biserialis, 65; *S. minuta*, 65; *S. sphaerulifer*, 65
Sphaerodorum, 65: *S. papillifer*, 65
Sphaeronectes gracilis, 17
Sphaeronectidae Family, 17
Sphaerosyllis californiensis, 66; *S. hystrix*, 66; *S. pirifera*, 66
Sphenia ovoidea, 134
Spicules, 6
Spinulosida Order, 273, **283**
Spiny Lebbeid, 204
Spio filicornis, 71, **72**
Spiochaetopterus costarum, **69**
Spionida Order, 68-73
Spionidae Family, 46, 70-73
Spiophanes berkeleyorum, 71; *S. bombyx*, 71, **72**; *S. cirrata*, 71; *S. foliosa*, 71, **72**; *S. squamata*, 71, **72**
Spirontocaris arcuata, 183; *S. holmesi*, 183; *S. lamellicornis*, 183; *S. prionota*, 183; *S. sica*, 183; *S. spina*, 183; *S. synderi*, 183; *S. truncata*, 183
Spirophorida Order, 11
Spirorbidae Family, 47, 68
Spirorbis, 68: *S. ambilateralis*, 68; *S. bifurcatus*, 68; *S. moerchi*, 68; *S. racemosus*, 68; *S. rugatus*, 68; *S. semidentatus*, 68; *S. variabilis*, 68

Spirularina Suborder, 19
Spisula falcata, 133
Spongia, **10**
Spongionella sp., 11
Stauronereis, 51: S. articulata, 51
Staurophora spp., 17
Stegopoma spp., 17
Stelletta clarella, 11
Stellettidae Family, 11
Stelodoryx alaskensis, 11
Stelotrochota hartmani, 11
Stenolaemates, 261
Stenoplax fallaz, 153; S. heathiana, 153
Stephanauge annularis, 19
Stephanosella vitrea, 263
Stereogastrula larvae, 6
Sternaspida Order, 74
Sternaspidae Family, 74
Sternaspis affinis, 74; S. fossor, 74; S. scutata, 74
Sthenelais berkeleyi, 65; S. fusca, 65; S. verruculosa, 65
Stiliger, 90: S. fuscovittatus, 119
Stiligeridae, 90
Stolidobranchia Suborder, 300, 302
Stolonifera Suborder, 19
Stomachetosella cruenta, 263; S. limbata, 263; S. sinuosa, 263
Stomachetosellidae Family, 263
Stomatopora granulata, 261
Stomphia coccinea, 19; S. didemon, 19; Stomphia sp., 19
Stone coral, 15
Straight-hinge stage, **138**
Streblospio benedicti, 71
Strongylocentrotus droebachiensis, 273; S. franciscanus, 273, **278, 279**; S. purpuratus, 51, 273, **278, 279**; Strongylocentrotus sp., **278**
Styela (barnharti) clava, 300; S. clavata, 300; S. coriacea, 300; S. gibbsii, 300; S. montereyensis, 300; S. partita, **299**; Styela sp., **299**; S. truncata, 300
Styelidae Family, 300, 302
Stylasterias forreri, 273
Stylasteridae Family, 17
Stylasterina Order, 13, 17
Stylatula elongata, 19
Styleroides, 53
Stylinos sp., 11
Styliola, 90: S. subula, 117
Stylissa stipitata, 11
Stylochidae Family, 27
Stylochoplana chloranota, 27
Stylochus, 26; S. atentaculatus, 27; S. tripartitus, 27
Stylopus arndti, 11
Stylostomum album, 27; S. sanjuania, 27
Suberites montiniger, 11; S. ?suberea, 11; S. simplex, 11; Suberites sp., 11
Suberitidae Family, 11
Subselliflorae Suborder, 19
Swiftia kofoidi, 19; S. simplex, 19; S. spauldingi, 19; S. torreyi, 19
Sycandra ?utriculus, 9

Sycettida Order, 9
Sycettidae Family, 9
Syhmelmis aff. klatti, 63
Syllidae Family, 47, 66
Syllis, 66: S. adamantea, 66; S. alternata, 66; S. armillaris, 66; S. elongata, 66; S. heterochaeta, 66; S. hyalina, 66; S. oerstedi, 66; S. pulchra, 66; S. spenceri, 66; S. variegata, 66
Sylon hippolytes, 158, 175
Sylonidae, 175
Symplectoscyphus spp., 17
Synnotum aegyptiacum, 262
Synoicum georgianum, **301**; S. parfustis, 302; Synoicum sp., 302
Syringella amphispicula, 11
Systellaspis braueri, 183; S. cristata, 183

Tachyrynchus, 89
Tadpole larvae, 298-303
Tagelus californianus, 134
Tapes philippinarum, 134
Taranis, 89
Tauberia gracilis, 54
Tealia filina. See Urticina
Tedania, 9: T. fragilis, 11; T. gurjanovae, 11
Tedaniidae Family, 11
Tedanione obscurata, 11
Tegella aquilirostris, 262; T. armifera, 262; T. robertsonae, 262
Tegula, 89
Telepsavus, 69
Tellina bodegensis, 134; T. carpenteri, 134; T. modesta, 134; T. nuculeoides, 134
Tellinidae, 134
Telmessus cheiragonus, 185
Tenellia, 90: T. adspersa, 114, 121
Tenthrenodes sp., 9
Terebellidae Family, 75-76
Terebellidae Order, 74-76
Terebellides stroemi, 76
Terebratalia transersa, 269
Terebratalina, **270**
Teredinidae, 135
Teredo navalis, 135, **140, 147**
Tergipedidae, 90, 114
Terrebratulina unguicula, 269
Tethidae Family, 11
Tethya, 8: T. californiana, 11
Tethyidae, 90
Tetillidae Family, 11
Tetrastemma candidum, 30; T. nigrifrons, 30, T. phyllospadicola, 30
Teuthoidea, 155
Thalassinidae, 184, 197
Thalysias laevigata, 11
Tharyx, 60: T. multifilis, 70; T. multifilis var. parvus, 70; T. parvus, 70
Thecanephria Order, 87
Thecata Suborder, 17
Thecocarpus spp., 17
Thecosomata, 120

Thelepus cincinnatus, 76; T. crispus, 76; T. hamatus, 76; T. setosus, 76
Themiste alutacea, 81; T. dyscrita, 80; T. lageniformis, 81; T. pyroides, 80, 81
Thompsonia sp., 175, 176
Thoracica, 165-74
Thoracophelia, 53
Thracia beringi, 134; T. challisiana, 134; T. curta, 134; T. trapezoides, 134
Thraciidae, 134
Thuiaria spp., 17
Thyasira barbarensis, 133; T. cygnus, 133; T. gouldii, 133
Thyasiridae, 133
Thyonicola, 89
Thysanocardia nigra, 80
Tiarannidae Family, 17
Tiaropsidium spp., 17
Tiaropsis spp., 17
Tochuina, 90
Tomopteridae Family, 45, 66
Tomopteris cavalli, 66; T. elegans, 66; T. pacifica, 66; T. renata, 66; T. septentrionalis, 66
Topsentia disparilis, 11; T. insignis, 153; T. lineata, 153
Tornaria larva, 294, **295**
Toxidocia spp., 11
Transenella confusa, 134; T. tantilla, 134
Travisia brevis, 53; T. carnea, 53; T. pupa, 53
Travisiopsis lobifera, 66
Trematoda Class, 26
Tresus capax, 134, **140, 148**; T. nuttalli, 134
Tricellaria gracilis, 262; T. occidentalis, 262, 266; T. praescuta, 262; T. ternata, 262
Trichobranchidae Family, 76
Trichobranchus glacialis, 76
Trichotropididae, 89
Trichotropis, 89
Trichydra pudica, 17
Trichydridae Family, 17
Tricolia, 89
Trididemnum opacum, 302; T. strangulatum, 302
Tridonta alaskensis, 133
Trimusculidae, 90
Trimusculus, 90
Triopha, 90: T. catalinae, 120; T. maculata, 120
Triticella pedicellata, 262
Triticellidae Family, 262
Tritonia, 90: T. diomedea, 120; T. festiva, 120
Tritoniidae, 90
Trochidae, 89
Trochochaeta franciscanum, 74; T. multisetosa, 74
Trochochatidae Family, **74**
Trochophore larvae, 93, **94**
Trophonia, 52
Trophonopsis, 89
Trypanosyllis gemmipara, 66; T. ingens, 66

Trypetesa sp., 158
Trypostega claviculata, 263
Tubilipora pulchra, 261
Tubulanus capistratus, 30;
   T. polymorphus, 30; *T. sexlineatus*,
   30; *Tubulanus* sp., **31**
Tubularia, 16: *T. crocea*, 17;
   T. harrimani, 17; *T. indivisa*, 17;
   T. marina, 17
Tubulariidae Family, 17
Tubulipora flabellaris, 261; *T. pacifica*,
   261; *T. phalangea*, **261**; *T. tuba*, 261
Tubuliporidae Family, 261
Tunicates, 298-303
Turbellaria Class, 26
Turbinidae, 89
Turbonilla, 89
Turrid prosobranchs, 102
Turridae, 89, 112
Turritellidae, 89, 109
Turritellopsis, 89
Turtonia minuta, 133
Turtoniidae, 133
Typhloscolecidae Family, 66
Typosyllis, 66

Ulosa, 10
Umbellula lindahli, 19
Umbellulidae Family, 19
Umbo stage, **138**
Umbonula arctica, 263
Umbonulidae Family, 263
Ungulinidae, 133
Upogebia pugettensis, 184, 196, **216-**
   17
Upogebiidae Family, 184, 196, 216
Urechidae Family, 86
Urechis caupo, **85**, 86
Urochordata, 298-303
Urticia crassicornis, 19, 22;
   U. lofotensis, 19; *U. piscivora*, 19;
   Urticia sp., 19
Urticina columbiana, 19; *U. coriacea*,
   19; *U. crassicornis*, 19; *U. filina*, **15**,
   19

Valvatida Order, 273
Veliconcha, **138**: Veliconcha stage,
   138
Veliger, 137, 139: Veliger larvae, 91,
   92, 93
Velutina, 89: *V. plicatilis*, 115;
   V. velutina, 115
Velutinid prosobranchs, **97**
Velutinidae, 89
Veneridae, 134
Veneroida, 133
Vermetidae, 89
Vermetus, 89
Verongiida Order, 12
Verongiidae Family, 12
Vesiculariidae Family, 262
Vetigastropoda, 89, 96, 108
Virgularia spp., 19
Virgulariidae Family, 19
Volvulella, 89

Waltonia, **270**
Watersipora arcuata, **265**
Weberella ?verrucosa, 11
Whelks, 89
Williamia, 90
Worms: Acorn, 293-95; Coelomate,
   85-87; Common Ship-, **147**;
   Feathery Ship-, **139**; Flat-, 26;
   Peanut, 80-84; Ribbon, 30

Xanthidae Family, 186, 193, 199,
   201, 248-49: key to zoeae, 248
Xenopneusta Order, 86
Xestospongia trindanea, 11; *X. vanilla*,
   11; *X. washingtona*, 135
Xylophagaidae, 135

Yellow Leg Pandalid, **214**-15
Yoldia amygdalea, 132; *Y. limatula*,
   132, **139**; *Y. myalis*, 132;
   Y. scissurata, 132; *Y. thraciaeformis*,
   132
Yoldiidae, 132

Zephyrinidae, 90
Zirfaea pilsbryii, 135
Zoantharia Subclass, 19
Zoanthidea Order, 15
Zoanthinaria Order, 19, 20
Zoea, 187-88
Zygherpe hyaloderma, 11
Zygonemertes virescens, 30
Zygophylax spp., 17